METHODS AND PROTOCOLS IN FOOD SCIENCE

For further volumes:
http://www.springer.com/series/16556

Methods and Protocols in Food Science series is devoted to the publication of research protocols and methodologies in all fields of food science.

Volumes and chapters will be organized by field and presented in such way that the readers will be able to reproduce the experiments in a step-by-step style. Each protocol will be characterized by a brief introductory section, followed by a short aims section, in which the precise purpose of the protocol will be clarified.

Multidimensional Characterization of Dietary Lipids

Edited by

Christelle Lopez, Claude Genot and Alain Riaublanc

INRAE, UR BIA, Nantes, France

Humana Press

Editors
Christelle Lopez
INRAE
UR BIA, Nantes, France

Claude Genot
INRAE
UR BIA, Nantes, France

Alain Riaublanc
INRAE
UR BIA, Nantes, France

ISSN 2662-950X ISSN 2662-9518 (electronic)
Methods and Protocols in Food Science
ISBN 978-1-0716-3757-9 ISBN 978-1-0716-3758-6 (eBook)
https://doi.org/10.1007/978-1-0716-3758-6

This Humana imprint is published by the registered company Springer Science+Business Media, LLC, part of Springer Nature.
The registered company address is: 1 New York Plaza, New York, NY 10004, U.S.A.

Paper in this product is recyclable.

Preface to the Series

Methods and Protocols in Food Science series is devoted to the publication of research protocols and methodologies in all fields of food science. The series is unique as it includes protocols developed, validated, and used by food and related scientists, as well as theoretical basis are provided for each protocol. Aspects related to improvements in the protocols, adaptations, and further developments in the protocols may also be approached.

Methods and Protocols in Food Science series aims to bring the most recent developments in research protocols in the field as well as very well-established methods. As such the series targets undergraduate, graduate, and researchers in the field of food science and correlated areas. The protocols documented in the series will be highly useful for scientific inquiries in the field of food sciences, presented in such a way that the readers will be able to reproduce the experiments in a step-by-step style.

Each protocol will be characterized by a brief introductory section, followed by a short aims section, in which the precise purpose of the protocol is clarified. Then, an in-depth list of materials and reagents required for employing the protocol is presented, followed by a comprehensive and step-by-step procedures on how to perform that experiment. The next section brings the dos and don'ts when carrying out the protocol, followed by the main pitfalls faced and how to troubleshoot them. Finally, template results will be presented and their meaning/conclusions addressed.

The *Methods and Protocols in Food Science* series will fill an important gap, addressing a common complain of food scientists, regarding the difficulties in repeating experiments detailed in scientific papers. With this, the series has a potential to become a reference material in food science laboratories of research centers and universities throughout the world.

Campinas, Brazil ***Anderson S. Sant'Ana***

Preface

Dietary lipids are important chemical and structural components of foods, involved in their techno-functional and organoleptic properties as well as in fundamental nutritional and biological functions in humans. These lipids recovered in foods, that are also called oil or fat depending on their composition and physical state, include lipids that are present in various raw food materials, ingredients, or processed food products from animal sources (meat, milk and dairy products, fish, eggs) and vegetable sources (cereal products, fiber-rich vegetables, oleaginous, algae, etc.).

Lipids represent a wide range of molecules (e.g., triacylglycerols, fatty acids, diacylglycerols, monoacylglycerols, phospholipids, sphingolipids, glycolipids, and components from the unsaponifiable fraction including the lipophilic pigments carotenoids, and the antioxidant molecules such as tocopherols and the sterols) that can be present in various respective amounts in foodstuffs and in the whole human diet. These dietary lipids can evolve during storage or processing for example due to their hydrolysis and oxidative alteration. The quantitative evaluation and identification of the secondary lipids produced after hydrolysis of the primary components (e.g., free fatty acids and the partial glycerides monoacylglycerols and diacylglycerols formed after hydrolysis of triacylglycerols), and of the lipid derivates (oxidized fatty acids, conjugated dienes, secondary lipid oxidation products, hydroperoxides, aldehydes, isoprostanoids) formed in oxidizing oils or foods require specific methods.

From the lipid molecules (molecular scale) to the supramolecular lipid assemblies (nano to microscopic scales) dispersed in complex food systems and in interaction with other components (e.g., proteins, polysaccharides), lipids require to be analyzed with a multi-scale approach to understand their functions in foods. In foods, lipids can be organized under various structures, as lipid membranes formed by phospholipids, dispersed in oil-in-water emulsions that can be natural (e.g., oil bodies in seeds and nuts, fat globules in milk, lipoproteic assemblies in egg yolk) or processed (e.g., mayonnaise, homogenized milk, processed cheeses). They can also be organized in water-in-oil emulsions (e.g., butter and margarine). Food processing can change the structural organization of lipids by altering natural structures or during the preparation of emulsions or the transformation of raw materials into processed food products (e.g., from milk to cheeses and butter).

The other aspect of primary importance concerning lipids in foods is their physical state. Due to their chemical composition, mainly the level of high melting point saturated fatty acids, dietary lipids can be liquid at room temperature (e.g., rapeseed and sunflower oils, algae, and fish oils rich in $\omega 3$ fatty acids), solid or partially solid (e.g., cocoa butter in chocolate, milk fat in dairy products). The crystallization properties of dietary lipids affect the functional and rheological products of food products.

The multidimensional description of the chemical composition, structure, and physical properties of dietary lipids require the use of specific methods and research protocols, alone or in combination. The volume *Multidimensional Characterization of Dietary Lipids* is the first book in the area of food science that provides readers a comprehensive knowledge and a detailed step-by-step description of experimental protocols dedicated to the analysis of dietary lipids with the originality to consider lipids from the molecules to their macroscale organization and physical state. Undoubtedly, this book represents a valuable, pedagogic, and helpful source of information for a wide audience including undergraduate and graduate students, biochemical engineers and technologists, researchers, academics, and industry personnel that are interested in the characterization of lipids in foods.

This volume contains 26 chapters that are split into 5 parts: Part I, Extraction and Fractionation of Dietary Lipids (Chaps. 1–4); Part II, Analysis of Dietary Lipid Composition (Chaps. 5–11); Part III, Evaluation of Dietary Lipid Oxidation (Chaps. 12–19); Part IV, Preparation and Characterization of Dietary Emulsions (Chaps. 20–22); Part V, Characterization of Physical Properties of Dietary Lipids and Lipid Structures (Chaps. 23–26).

Written in the format of the Methods and Protocols in Food Science (MeFS) series, the chapters include an introduction to the respective topic, list necessary materials and reagents, detail well-established and validated methods for readily reproducible laboratory protocols, and contain notes on how to avoid or solve typical problems. In each part of this volume, the chapters guide readers through comprehensive information provided by experienced food and lipid technicians, engineers, and scientists.

The expert contributions of a wide diversity of authors from several countries are greatly acknowledged with gratitude. Dr. Christelle Lopez warmly thanks the MeFS series editor Prof. Anderson Sant'Ana for the kind invitation to edit this book. Special acknowledgments are also extended to Springer Nature publishing team, especially Monica Suchy, for their support during the course of this book project.

INRAE, UR BIA, Nantes, France ***Christelle Lopez***

Contents

Contributors

VALÉRIE BEAUMAL • *INRAE, UR BIA, Nantes, France*
THAIS CRISTINA BENATTI GALLO • *INRAE, UR BIA, Nantes, France; São Paulo State University (Unesp), Institute of Biosciences, Humanities and Exact Sciences, São José do Rio Preto, Department of Food Engineering and Technology, São José do Rio Preto, São Paulo, Brazil*
CLAIRE C. BERTON-CARABIN • *INRAE, UR BIA, Nantes, France; Wageningen University & Research, Laboratory of Food Process Engineering, Wageningen, The Netherlands*
VINCENT BOERKAMP • *Laboratory of Food Chemistry, Wageningen University & Research, Wageningen, The Netherlands*
CLAIRE BOURLIEU-LACANAL • *INRAE, UMR IATE, Montpellier, France*
VALÉRIE BULTEL-PONCÉ • *Pôle Recherche Chimie Balard, IBMM, UMR 5247, Equipe SLB, Montpellier, France*
ERWANN DURAND • *CIRAD, UMR Qualisud, Montpellier, France; Qualisud, Univ Montpellier, Avignon Université, CIRAD, Institut Agro, IRD, Université de la Réunion, Montpellier, France*
THIERRY DURAND • *Pôle Recherche Chimie Balard, IBMM, UMR 5247, Equipe SLB, Montpellier, France*
ANIS ELLOUMI • *Pôle Recherche Chimie Balard, IBMM, UMR 5247, Equipe SLB, Montpellier, France*
SHIVA EMAMI • *Department of Food Science and Technology, University of California, Davis, CA, USA*
MARIE-HÉLÈNE FAMELART • *INRAE, UMR STLO, Institut Agro, Rennes, France*
BIANCA FERRAZ TEIXEIRA • *Department of Food Science and Technology, University of California, Davis, CA, USA; ESALQ Food, College of Agriculture "Luiz de Queiroz", University of São Paulo, Piracicaba, SP, Brazil*
JEAN-MARIE GALANO • *Pôle Recherche Chimie Balard, IBMM, UMR 5247, Equipe SLB, Montpellier, France*
CLAUDE GENOT • *INRAE, UR BIA, Nantes, France*
BÉATRICE GLEIZE • *INRAE, Avignon Université, UMR SQPOV, F-84000 Avignon, France*
PASCALE GOUPY • *INRAE, Avignon Université, UMR SQPOV, F-84000 Avignon, France*
VALÉRIE GROS • *Pôle Recherche Chimie Balard, IBMM, UMR 5247, Equipe SLB, Montpellier, France*
ALEXANDRE GUY • *Pôle Recherche Chimie Balard, IBMM, UMR 5247, Equipe SLB, Montpellier, France*
MARIE HENNEBELLE • *Laboratory of Food Chemistry, Wageningen University & Research, Wageningen, The Netherlands*
EMMA B. A. HINDERINK • *TiFN, Wageningen, the Netherlands; Laboratory of Food Process Engineering, Wageningen, the Netherlands*
CHARLOTTE JACOBSEN • *National Food Institute, Technical University of Denmark, Kgs. Lyngby, Denmark*
ALICE KERMARREC • *INRAE, UR BIA, Nantes, France*
BRIGITTE LAILLET • *INRAE, UR BIA, Nantes, France*

Amandyne Linares-Maurizi • *Pôle Recherche Chimie Balard, IBMM, UMR 5247, Equipe SLB, Montpellier, France*

Christelle Lopez • *INRAE, UR BIA, Nantes, France*

Sébastien Marze • *INRAE, UR BIA, Nantes, France*

Donny Merkx • *Unilever Food Innovation Centre, Wageningen, The Netherlands; Laboratory of Food Chemistry, Wageningen University & Research, Wageningen, The Netherlands; Laboratory of Biophysics, Wageningen University & Research, Wageningen, The Netherlands*

Anne Meynier • *INRAE, UR BIA, Nantes, France*

Françoise Michel Salaun • *Diana Pet Food, ZA du Gohélis, ELVEN, France*

Camille Oger • *Pôle Recherche Chimie Balard, IBMM, UMR 5247, Equipe SLB, Montpellier, France*

Yurika Otoki • *Laboratory of Food Function Analysis, Graduate School of Agricultural Science, Tohoku University, Sendai, Miyagi, Japan*

Coralie Pasquier • *ICSM, CEA, CNRS, ENSCM, Univ Montpellier, Marcoule, Bagnols sur Ceìze, France*

Guillaume Reversat • *Pôle Recherche Chimie Balard, IBMM, UMR 5247, Equipe SLB, Montpellier, France*

Alain Riaublanc • *INRAE, UR BIA, Nantes, France*

Lucie Ribourg-Birault • *INRAE, UR BIA, Nantes, France*

Leonard M. C. Sagis • *Laboratory of Physics and Physical Chemistry of Foods, Wageningen, the Netherlands*

Karin Schroën • *Laboratory of Food Process Engineering, Wageningen, the Netherlands*

Qing Shen • *Department of Food Science and Technology, University of California, Davis, CA, USA; College of Biosystems Engineering and Food Science, Zhejiang University, Hangzhou, China*

Ha Phuong Ta • *INRAE, UR BIA, Nantes, France*

Ameer Y. Taha • *Department of Food Science and Technology, University of California, Davis, CA, USA*

John van Duynhoven • *Unilever Food Innovation Centre, Wageningen, The Netherlands; Laboratory of Biophysics, Wageningen University & Research, Wageningen, The Netherlands*

Michelle Viau • *INRAE, UR BIA, Nantes, France*

Claire Vigor • *Pôle Recherche Chimie Balard, IBMM, UMR 5247, Equipe SLB, Montpellier, France*

Pierre Villeneuve • *CIRAD, UMR Qualisud, Montpellier, France; Qualisud, Univ Montpellier, Avignon Université, CIRAD, Institut Agro, IRD, Université de la Réunion, Montpellier, France*

Zhichao Zhang • *Department of Food Science and Technology, University of California, Davis, CA, USA*

Bingqing Zhou • *Pôle Recherche Chimie Balard, IBMM, UMR 5247, Equipe SLB, Montpellier, France*

Part I

Extraction and Fractionation of Lipids

Chapter 1

Cold Extraction of Total Lipids from Biological Materials and Food Matrices with Mixtures of Polar and Apolar Solvents

Alice Kermarrec, Anne Meynier, Claude Genot, and Michelle Viau

Abstract

Lipid extraction from its original matrix is a crucial step to determine the lipid content of a food product or to get, before further analysis, a representative lipid extract not biased by selective extraction. Extraction should also avoid the deterioration of the original lipids and ensure the elimination of nonlipid materials. Pretreatment of the samples and precautions required to disintegrate the matrix and improve lipid accessibility to the solvents, deactivate lipolytic and oxidative enzymes, and avoid oxidation during all steps of the procedures are first underlined. Then protocols to extract at ambient temperature total lipids from complex food matrices with mixtures of nonpolar and polar (alcohols) organic solvents are detailed. Three of them, based on the phase diagram of the chloroform/methanol/water ternary mixture, were directly adapted from the methods of Folch et al. (J Biol Chem 226:497–509, 1957) and Bligh and Dyer (Can J Biochem Physiol 37:911–917, 1959). The last one takes the advantage of the hexane/isopropanol mixture (Hara and Radin, Anal Biochem 90:420–426, 1978). The criteria for choosing one or the other method for a given matrix are presented. The parameters to adapt and the crucial steps and/or conditions for extraction are also highlighted.

Key words Pretreatment, Lipid extraction, Total lipids, Organic solvents, Ternary mixture, Liquid–liquid phase separation

1 Introduction

The aim of lipid extraction is to quantify the lipids present in a given material and/or to prepare a lipid extract ready for further analysis. Indeed, the knowledge of the amounts of lipids in a food product (lipid content) is a main parameter for its nutritional evaluation (calculation of energy intake), for ensuring its functional and organoleptic quality, and also to evaluate its commercial value. Lipid extraction from the matrices of animal or vegetable origins is also the first step before lipid analysis. Ideally, the extraction procedure should minimize selective extraction or losses of specific

Christelle Lopez et al. (eds.), *Multidimensional Characterization of Dietary Lipids*, Methods and Protocols in Food Science, https://doi.org/10.1007/978-1-0716-3758-6_1,

compounds and avoid any lipid degradation, namely, lipolysis and oxidation, before and during the extraction. It should also ensure the removal of any nonlipid compounds from the extract. Although lipids are defined by their solubility in nonpolar solvents such as n-hexane, diethyl ether, or petroleum ether, they are often embedded in a matrix and counteract interactions with other food components, i.e., starch–lipid complexes and protein–lipid interactions. They also are part of cell membranes and tissue structures. Thus, extraction by nonpolar solvents, by pressing or by water/oil partition to get crude oils from many sources, gives access to only neutral lipids accompanied by a small fraction of polar lipids (so-called "free lipids").

Many lipid extraction methods can be found in the literature because biological tissues and food products present a great diversity in terms of structure, texture, water, protein and lipid contents, and lipid composition. For example, refer to the Web site dedicated to lipids http://cyberlipid.gerli.com/techniques-of-analysis/extraction-handling-of-extracts/ but also to reviews and book chapters such as [4–8].

It is essential to minimize degradations and changes within the samples before, during, and after the extraction procedure. For example, enzymatic degradation could be limited by the inhibition of the targeted enzymes by heat, specific inhibitor, or the addition of alcohol. To avoid lipid oxidation, it is recommended to work under limited oxygen pressure or under a nitrogen atmosphere, to add a metal chelator or antioxidant such as butylated hydroxytoluene (BHT) to the extraction solvent.

Briefly, all lipid extraction methods include the following steps, some being proceeded simultaneously depending on the chosen method: The temperature should also be controlled throughout the sampling, extraction, and storage processes. Some food matrices contained light-sensitive components that were required to avoid light exposition and consequently required dark-brown glassware.

Briefly, all lipid extraction methods include the following steps, some being proceeded simultaneously depending on the chosen method:

- Sampling
- Sample pretreatment (previous digestion of the surrounding material may be necessary in specific cases) and (as short as possible) storage
- Grinding and homogenization with the solvent
- Separation of the organic phase from the non-extractable matter
- Purification of the organic phase: washing and elimination of nonlipid substances

- Concentration of the lipid extract (solvent evaporation)
- Storage of the lipid extract until further analysis.

Further analytical characterization of the lipids implies the use of ultrapure solvents and clean glassware and their protection against oxidation during extraction and further storage of the extracts.

The reader interested in the preparation of the lipid extract for metabolomic or lipidomic analyses from animal or human biopsies can also refer to the following ref. [9]. Considering the improvement of the analytical methods dedicated to lipid characterization, relatively low amounts of extracted lipids (<100 mg) are now requested for a full characterization of a lipid extract.

The methods detailed here to recover total lipids from food matrices are based on cold extraction by mixtures of nonpolar and polar solvents in proportions allowing the breaking down of lipids to nonlipid interactions to ensure the quantitative recovery of all lipid classes.

The Folch method is the most widely quoted extraction procedure in the world [1], accounting for over 60,000 citations in WOS in 2021. Originally, the method was set up to extract lipids from animal tissues. It is based on the use of chloroform/methanol (2/1; v/v) as the initial solvent system, followed by filtration and the separation of the bottom apolar organic phase dissolving the lipids from the upper alcoholic water phase allowed by the ternary mixture chloroform/methanol/water (8/4/3; v/v/v). The initial water content of the sample should thus be known to respect these proportions. Finally, the lipid extract is washed 2 times with 0.2 volumes of water to eliminate nonlipid compounds or by 0.73% (w/v) NaCl solution (or KCl 0.88%) and ensure a good recovery of acidic lipids [10].

The method developed by Bligh and Dyer [2] is also based on the chloroform/methanol/water ternary diagram; the water carried by the sample is also taken into account (see also cyberlipids Web site http://cyberlipid.gerli.com/techniques-of-analysis/extraction-handling-of-extracts/,). The extraction of lipids follows three steps: (i) the sample is mixed with the solvents to obtain the monophasic mixture chloroform/methanol/water (2/1/0.8; v/v/v) that simultaneously breaks down the lipid–protein interactions and extracts the lipids; (ii) chloroform is added to the mixture to obtain chloroform/methanol/water (2/2/0.8; v/v/v), which is just at the demixion line between monophasic and biphasic systems; and (iii) finally, water is added to reach chloroform/methanol/water (2/2/1.8; v/v/v), allowing a clear phase separation between the upper methanol/water phase containing some soluble nonlipid substances, the in-between semi-solid slurry containing most aggregated proteins and carbohydrates, and the bottom chloroform phase containing the extracted lipids. This bottom phase is tricked out or pipetted depending on the glassware to recover the lipids. The method however gives significantly lower estimates of lipid content in samples containing more than 2 wt% lipid, and this underestimation increases with increasing lipid content of the samples [11]. This is why a second extraction of the upper phase and the slurry with chloroform is recommended to achieve a better recovery of total lipids (>99% instead of about 95% for single-step extraction). This method is suitable for the lipid extraction of incubation media, tissue homogenates, or cell suspensions.

Hara and Radin [3] described an alternative extraction procedure adapted to nervous tissues. The tissue sample is homogenized with 18 volumes of a mixture of hexane/2-propanol (3/2; v/v), the suspension is then filtered, and the filtrate is rinsed with 3 × 2 vol of the same solvent. Let "quiet" until phase separation. The upper hexane phases containing the extracted lipids are almost free from nonlipid compounds [12]. The method has the advantage of facilitating lipid recovery, as it is easier to withdraw the upper hexane phase than the bottom chloroform phase, in particular in the case of extraction carried out on small samples. It was thus further adapted to other materials, including selected foods such as egg yolk products and in vitro digestats [13–15].

The purpose of lipid extraction (only quantification of lipid content, further analysis of fatty acid composition, or detailed analysis of the lipid components) will condition the chosen method and the sampling aliquot size. Depending on the food matrices, physical aspect, and water content, some minor changes to the original Folch et al. or the Bligh and Dyer procedures would be required to reach the appropriate proportion of $CHCl_3/CH_3OH/H_2O$ during the extraction and to extract a sufficient amount of lipid for further analysis.

The four presented protocols detailed below aim at extracting total lipids from specific food matrices.

The first protocol adapted from the Folch method (protocol 1) for dry samples such as dehydrated animal product, concentrates, or isolates of legume proteins is presented below. Then two protocols are based on the chloroform/methanol/water diagram as described by Bligh and Dyer. The first one has been adapted for samples containing high amounts of water (>70%) and low lipid content (<10%) (protocol 2) and the second for precious samples containing very low amounts of lipids (protocol 3). Finally, a protocol adapted from Hara & Radin, using hexane/isopropanol as the extracting solvent for aqueous media such as emulsions, is described (protocol 4).

2 How to Get a Nondegraded Lipid Extract?

2.1 Sample Pretreatment and Storage Before Extraction

After sample collection, lipid extraction should be done as soon as possible to minimize changes during storage. If this is not possible, the samples should be kept at as low a temperature as possible and stored in the absence of oxygen and light. In samples from different sources, such as brain, heart, fish tissues and offal, insects, algae, seeds, flours, plant tissues, and extracts, enzymes can still be active and degrade lipids. Plant phospholipases are inactivated by macerating the sample in isopropanol or by plunging the tissues in boiling water for about 1 min [16]. If the samples have to be stored before extraction, they should be frozen and stored in sealed glass containers under nitrogen at ideally −80 °C. Indeed, at −20 °C and reduced water activity, lipolytic enzymes remain active and continue to hydrolyze lipids.

Freeze-drying is recommended for materials of high water content because it simplifies lipid extraction [16]. The dry sample is more easily grounded and dispersed in the solvent mixture than the wet sample and a smaller volume of solvents is required [16]. Lyophilization was also shown to facilitate the recovery of the polar lipids in phospholipid-stabilized emulsions [17]. As the process is carried out at a high vacuum and low temperature, lipids are not degraded during dehydration. It is, however, advised to control the water content of the freeze-dried samples and to add back some water to the extraction solvents to ensure the complete lipid extraction and facilitate the decantation process.

2.2 Storage of Lipid Extracts

Unsaturated lipids that are left dry in air and in the daylight autoxidize. To minimize autoxidation reactions, lipid extract should not be stored in a dry state. It should be dissolved in a small amount of organic solvent. Natural or synthetic antioxidants such as tocopherols or butylated hydroxytoluene (BHT) can also be added to the solvent. BHT concentrations in a solvent at about 0.02% (p/v) or tocopherol/lipid ratio around 100–300 ppm are recommended. Notice that an antioxidant at excessive concentration can also act as a pro-oxidant [5].

The extracts should be put in air-tight glass containers in an atmosphere of nitrogen or argon and kept at −20 °C or below. Plastic containers have to be avoided because plasticizers are easily leached out by organic solvents [5].

3 Protocols for Extraction of Lipids by Mixtures of Solvents

3.1 Mandatory Equipment (Table 1)

Table 1
Mandatory equipment

Equipment	Protocol number
Analytical precision balance, sensitivity: 0.01 mg	1/2/4
Magnetic stir plate and bar magnets	1
Vacuum system	1/2
Vortex mixer	3/4
Rotary evaporator	1/2/3/4
Nitrogen generator or nitrogen cylinders (Standard quality)	1/2/3/4
Blender	2
Temperature-controlled chamber at −20 °C	1/2/3/4
Temperature-controlled chamber at 4 °C	1
100 mL amber glass Pyrex bottles	1
10 or 25 mL Pyrex glass tube	3/4
Glass funnel filter No.1 with sintered glass disc	1/2
Whatman paper No. 1	1/2
100, 250 mL graduated measuring cylinder	1/2
250 mL Separating funnels	1
50, 100 or 250 mL flat-bottomed flasks	1/2/3/4
Glass conical funnels	1
Glass wool previously hand-washed with dichloromethane and dried under air	1
Aqueous pipette (5 mL) and disposable tips (5 mL)	1
Positive displacement pipettes (0.25, 1, and 10 mL) and disposable capillary piston tips (0.25, 1, and 10 mL)	3/4
Glass Pasteur pipettes	2/3/4
10 mL storage vials	1/2/3/4
Centrifuge	3/4

3.2 Chemical Products (Table 2)

Prepare solutions using ultrapure water (conductivity = 18 MΩ cm^{-1} at 25 °C) and/or HPLC-grade reagents. Diligently follow safety instructions and waste disposal regulations when disposing of waste materials.

3.3 Protocol 1: Adapted from Folch et al. (1957) [1]

Chemicals:

Chloroform, methanol, NaCl (0.73%w/v in water), and anhydrous sodium sulfate (dried overnight at 100 °C). See Table 1 for specifications.

Table 2
Chemical compounds or solvents used in the various extraction protocols

Chemical	Chemical formula	CAS number	Molecular weight (g/mol)	Quality	Safety instructions	Protocol number
Chloroform	$CHCl_3$	67-66-3	119.38	HPLC grade		1/2/3
Hexane	C_6H_{14}	110-54-3	86.18	HPLC grade		3
Isopropanol	C_3H_8O	67-63-0	60.1	ACS reagent, ≥99.5%		3
Methanol	CH_4O	67-56-1	32.04	HPLC grade		1/2
Sodium chloride	NaCl	7647-14-5	58.44	ACS reagent		1/3
Sodium sulfate	Na_2SO_4	7757-82-6	142.04	Anhydrous, ≥99.0%		1

Procedure:

In this example, we made an extraction from flour containing 15% lipids. Accordingly, from 3 g of flour, 0.45 g of lipid extract was expected. Prior solubilization in water was needed in order to reconstitute the initial water content of the sample (70%):

1. Weigh 3 g of flour (*see* **Note 1**) into a 100 mL amber glass Pyrex bottle (adjust the amount of sample to obtain a sufficient amount of lipids; *see* **Notes 2** and **3**).
2. Add 7 mL of water (to be calculated in order to reach the initial water content of the sample; *see* **Note 4**).
3. After dispersion, the whole mixture is agitated overnight on a magnetic stir plate (4 °C in the dark).
4. Prepare the volume of chloroform/methanol (2:1; v/v) and 0.73% (w/v) NaCl solution you will need for all your samples.
5. Add 70 mL of chloroform/methanol (2:1; v/v).
6. Leave the mixture under magnetic stirring for 30 min.
7. Pour the mixture onto a glass funnel filter No. 1 covered with Whatman paper No. 1.
8. Filter the mixture under vacuum.
9. Re-suspend the remaining solid residue in 70 mL of chloroform/methanol (2:1; v/v).

10. Leave the mixture under magnetic stirring for 30 min.
11. Repeat **steps 7–10** and then once more 7 and **8**.
12. Combine the three filtrates in a 250 mL measuring cylinder.
13. Measure the total volume of filtrate.
14. Transfer the filtrate into a 250 mL separating funnel.
15. Add one-fourth of the total volume of the combined filtrates of 0.73% (w/v) NaCl solution (*see* **Note 5**).
16. Shake 5 times the separating funnel, degas it, and repeat this operation 3 times.
17. Place the separating funnel in a cold room at 4 °C overnight, and place a tared 250 mL flat-bottomed flask to avoid liquid loss in case of leaking.
18. Place prewashed glass wool at the entrance of a glass conical funnel (diam. 80 mm) and recover the wool with 4–5 spatulas of anhydrous sodium sulfate.
19. Draw off the lower organic phase from the separating funnel on the conical funnel and collect the organic phase in the tared flat-bottomed flask.
20. Rinse the aqueous phase remaining in the separating funnel with 50 mL of chloroform. Shake 5 times the separating funnels and then degas it. Repeat this operation 3 times. Leave to settle for 30 min.
21. Again, draw off the organic phase in the conical funnel and recover the organic phase in the glass flask.
22. Repeat **steps 20** and **21** (*see* **Notes 6** and 7).
23. Evaporate the solvent under a vacuum with a rotary evaporator set to 45 °C (*see* **Note 8**).
24. Dry the lipid extract under a stream of nitrogen and weigh the flask in order to calculate the weight of the lipid extract by difference.
25. Store the lipid extract in a small volume of chloroform or hexane and in an atmosphere of nitrogen in a sealed glass container kept at −20 °C (Fig. 1).

3.4 Protocol 2: Adapted from Bligh & Dyer (1959) for Samples Containing Above 70% Water and Less Than 10% Lipids [2]

As for the previous protocol, it is necessary to know the water content of the sample.

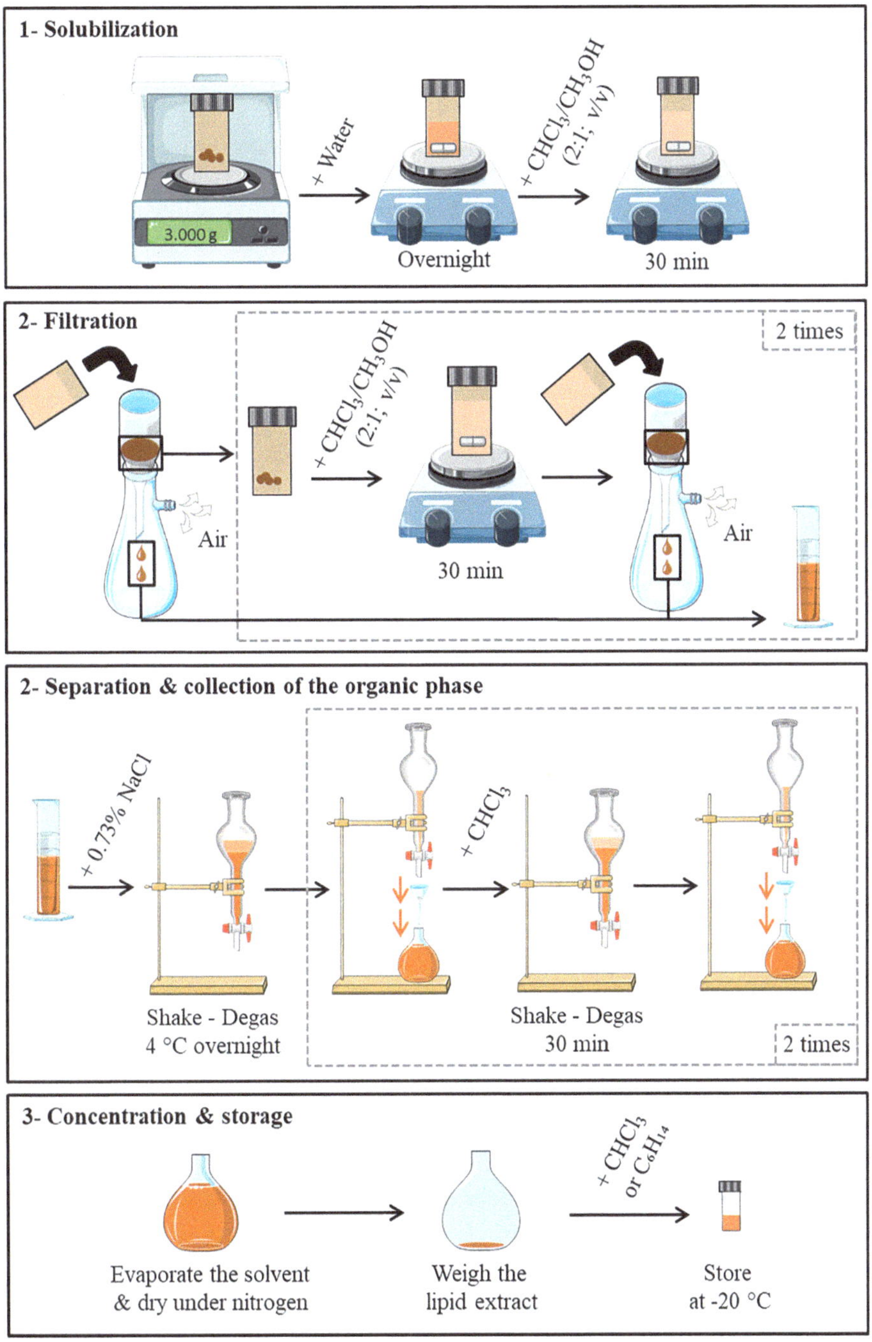

Fig. 1 Schematic diagram of protocol 1

Reagents:
Chloroform, methanol, and water.

Procedure:

1. Prepare the volume of chloroform/methanol (2:1; v/v) you will need for all your samples.
2. Homogenize 10 g of the sample (*see* **Notes 2** and **9**) in a blender for 2 min with 30 mL chloroform/methanol (1/2; v/v) (to a final volume 3 times the volume of the sample: 1 g in 3 mL of extraction mixture).
3. Add 10 mL of chloroform.
4. Homogenize for 30 s.
5. Add 10 mL of water (*see* **Note 10**).
6. Homogenize for 30 s.
7. Filter the mixture through a glass funnel filter No. 1 equipped with a folded Whatman paper No. 1.
8. Re-suspend the remaining solid residue in the same volume of chloroform as before.
9. Homogenize for 3 min and filter once more.
10. Combine the two filtrates in a 100 mL measuring cylinder (*see* **Notes 5** and **11**).
11. Allow to settle a few minutes for complete separation and clarification.
12. Measure the volume of the chloroform phase (lower phase).
13. Remove the upper phase and the interfacial material by aspiration (*see* **Note 12**).
14. The extracted lipids are contained in the chloroform layer.
15. Evaporate, dry, and store the lipid extract like in the previous protocol (*see* **Note 8**).

3.5 Protocol 3: Modified "Bligh & Dyer" Method for Precious Samples Containing Very Low Amounts of Lipids

Reagents:
Chloroform, methanol, and water.

Procedure:
In this example, we made an extraction from a sample containing a very low amount of lipids (≈10 mg) and diluted in a buffer.

1. Take the number of 10 mL Pyrex glass tubes needed depending on the number of samples (*see* **Note 1**).
2. In a 10 mL Pyrex glass tube containing your sample (*see* **Notes 2** and **10**), add 1.25 mL chloroform and 2.5 mL methanol for every milliliter of water in the sample in order to reach 3:2:0.8 chloroform/methanol/water (v/v/v).
3. Vortex the mixture for 60 s.

4. Add 1.25 mL chloroform and 1.25 mL water for every milliliter of water initially present in the sample in order to reach 2:2:1.8 chloroform/methanol/water (v/v/v).
5. Vortex the mixture for 60 s.
6. Centrifuge the sample for 5 min at 2000 g.
7. Collect the lowest phase in a 50 mL flat-bottomed flask with a Pasteur pipette.
8. Wash the upper phase with chloroform (twice that of the initial volume of water).
9. Repeat **steps 5–8** and then once more **5** and 7.
10. Evaporate the solvent under a vacuum with a rotary evaporator set to 45 °C (*see* **Note 8**).
11. Dry the lipid extract under a stream of nitrogen and weigh the flask in order to calculate by difference the weight of the lipid extract.
12. Store the lipid extract in a small volume of chloroform or hexane and in an atmosphere of nitrogen in a sealed glass container kept at −20 °C (Fig. 2).

3.6 Protocol 4: Adapted from Hara & Radin (1978) for Liquid Media Such as Emulsions and Digestion Media

In our laboratory, we have applied this method to egg yolks, emulsions up to 30% W/W lipids, and samples recovered during the in vitro digestion of the lipid-containing material.

Reagents:
Hexane, isopropanol, and NaCl solution (150 mM in water).

Procedure:

1. Prepare the volume of hexane/isopropanol mixture (3:2; v/v) and 150 mM NaCl solution you will need for all your samples.
2. Take the number of 100 mL flat-bottomed flasks needed depending on the number of samples (*see* **Note 1**).
3. Number and tare each 100 mL flat-bottomed flask.
4. Place your emulsion (≈4 g; adjust the amount of the sample to obtain a sufficient amount of lipids) in a 25 mL Pyrex glass tube (*see* **Note 2**).
5. Add 10 mL of hexane/isopropanol mixture.
6. Add 0.2 mL of 150 mM NaCl.
7. Vortex all samples for 60 s.
8. Centrifuge the samples for 5 min at 2000 g.
9. Collect the hexane phases using Pasteur pipettes and collect them in the flat-bottomed flasks.
10. Add 6 mL of hexane over the water–isopropanol phase.

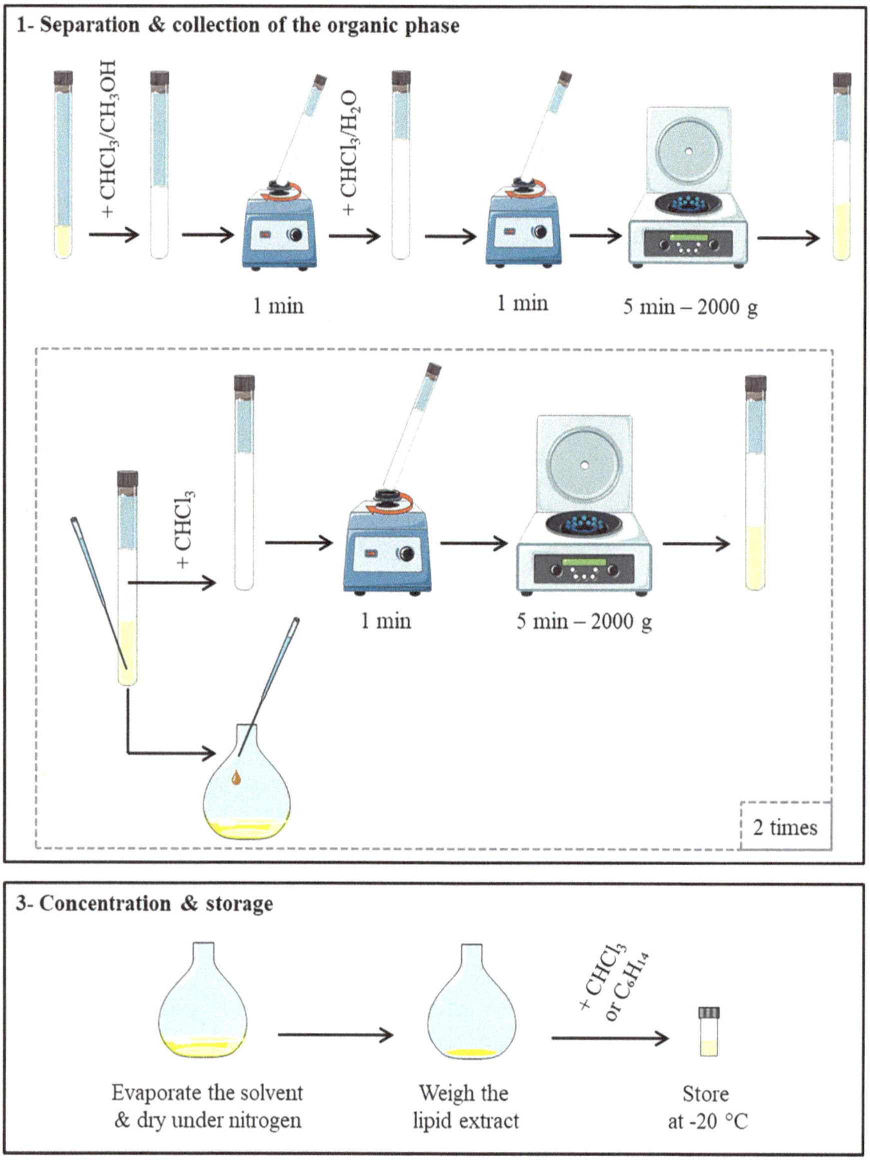

Fig. 2 Schematic diagram of protocol 3

11. Repeat **steps 7–10** and then once more 7 and **9**.
12. Evaporate the solvent under a vacuum with a rotary evaporator set to 45 °C (*see* **Note 8**).
13. Dry the lipid extract under a stream of nitrogen and weigh the flask in order to calculate by difference the weight of the lipid extract.
14. Store the lipid extract in a small volume of chloroform or hexane and in an atmosphere of nitrogen in a sealed glass container kept at −20 °C (Fig. 3).

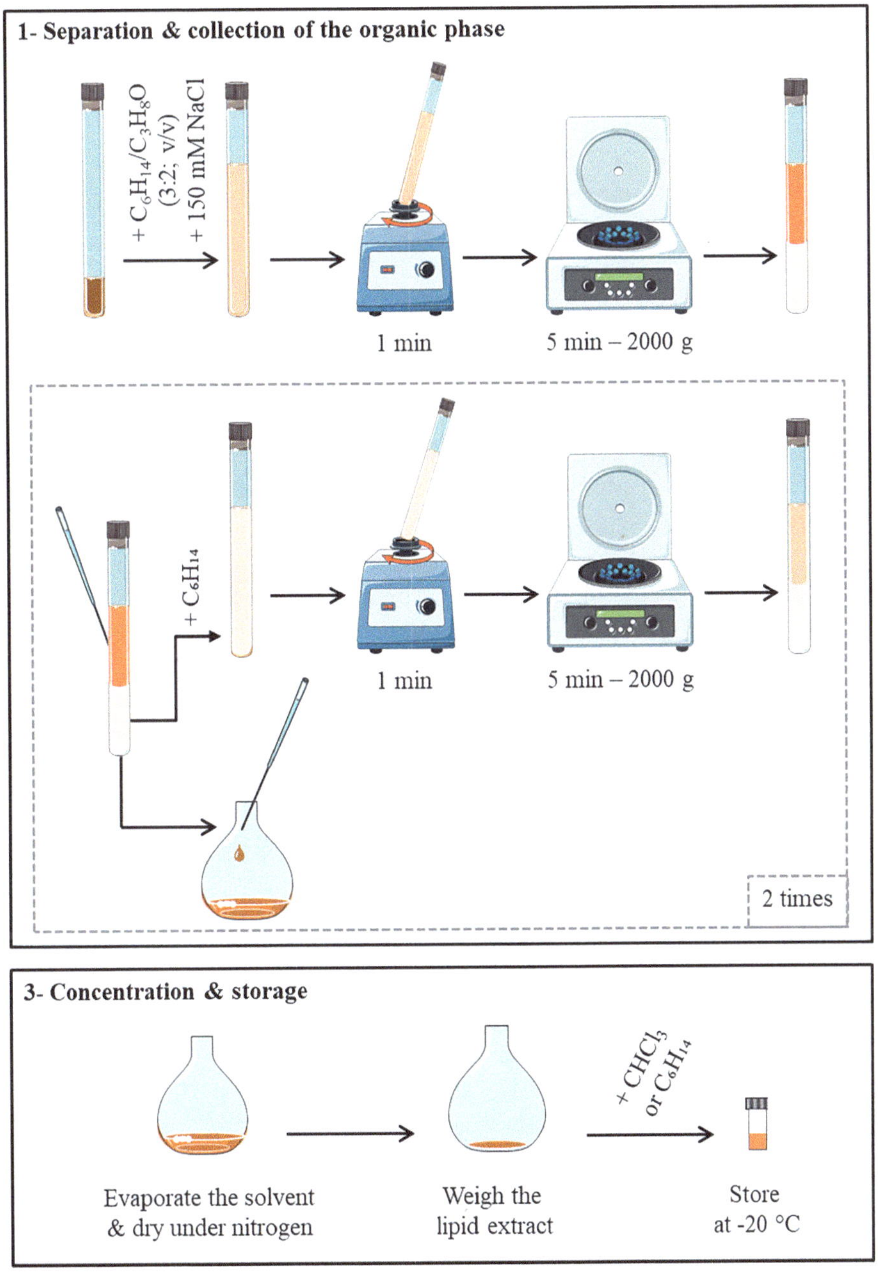

Fig. 3 Schematic diagram of protocol 4

4 Notes

1. An operator can perform 6–8 extractions at the same time.
2. When weighing the sample, take into account the final quantity of lipids desired. It is necessary that the quantity extracted can be weighed (>10 mg, ideally>50–100 mg) and sufficient for the overall characterization of the extract.

3. Depending on the amount of lipids present in the sample and the amount of lipids desired at the end of the extraction, the method can be adapted to much smaller samples by adapting the glassware.
4. It is important to take into account the water content of the sample to calculate the solvent proportions.
5. The amount of acidic lipids in the upper phase decreases with the use of a salt solution rather than water.
6. To save time, it is possible to suppress **step 22** to make only washing of the aqueous phase, but it increases the risk of losing some lipids.
7. A second wash of the lowest phase can be done to remove the remaining nonlipid contaminants with a mixture with a similar composition to that of the upper phase.
8. The evaporation of the solvent should be done at a temperature not greater than 50 °C. The lipid extract must remain at the minimum time possible at the water-bath temperature. Remove the remaining solvent by blowing a stream of nitrogen to preserve lipids from autoxidation, especially very unsaturated lipids. To minimize the autoxidation in samples containing high LC PUFA or carotenoids, it is also recommended to use an amber flat-bottomed flask.
9. If the sample contains a large amount of lipid or if its supply is limited, the size of the sample can be reduced, and thus, solvent quantities can be scaled down.
10. The water present in the sample must be taken into account in order to respect the proportions 3:2:0.8 and 2:2:1.8 (chloroform/methanol/water; v/v/v).
11. One-fourth of the total volume of the combined filtrates, of salt solution (0.05% $CaCl_2$, 0.04% $MgCl_2$, 0.73% NaCl, or 0.88% KCl), can be added after **step 10** to improve the phase separation.
12. To ensure a complete removal of the interfacial material, a small volume of the chloroform phase is also removed.

References

1. Folch J, Lees M, Sloane Stanley GH (1957) A simple method for the isolation and purification of total lipids from animal tissues. J Biol Chem 226:497–509. https://doi.org/10.1016/s0021-9258(18)64849-5
2. Bligh EG, Dyer WJ (1959) Can J Biochem Physiol 37:911–917
3. Hara A, Radin NS (1978) Lipid extraction of tissues with a low-toxicity solvent. Anal Biochem 90:420–426. https://doi.org/10.1016/0003-2697(78)90046-5
4. Furse S, Egmond MR, Killian JA (2015) Isolation of lipids from biological samples. Mol Membr Biol 32:55–64. https://doi.org/10.3109/09687688.2015.1050468
5. Christie WW, Han X (2010) Lipid analysis isolation, separation, identification and lipidomic

analysis, 4th edn. Woodhead Publishing/The Oily Press, Cambridge

6. Christie WW (1982) Lipid analysis: isolation, separation, identification, and structural analysis of lipids, 2nd edn. Pergamon Press, New York
7. Shahidi F (2005) Extraction and measurement of total lipids. Handb Food Anal Chem 1–2: 425–435. https://doi.org/10.1002/0471709085.ch12
8. Hamilton RJ, Hamilton S (1992) Lipid analysis: a practical approach. IRL Press at Oxford University Press, Oxford/New York
9. Chen S, Hoene M, Li J et al (2013) Simultaneous extraction of metabolome and lipidome with methyl tert-butyl ether from a single small tissue sample for ultra-high performance liquid chromatography/mass spectrometry. J Chromatogr A 1298:9–16. https://doi.org/10.1016/j.chroma.2013.05.019
10. Christie WW (1993) Advances in lipid methodology-Two. Oily Press, Dundee
11. Iverson SJ, Lang SLC, Cooper MH (2001) Comparison of the Bligh and Dyer and Folch methods for total lipid determination in a broad range of marine tissue. Lipids 36:1283–1287. https://doi.org/10.1007/s11745-001-0843-0
12. Eder K, Reichlmayr-Lais AM, Kirchgeβner M (1993) Studies on the extraction of phospholipids from erythrocyte membranes in the rat. Clin Chim Acta 219:93–104. https://doi.org/10.1016/0009-8981(93)90200-N
13. Meynier A, Leborgne C, Viau M et al (2014) n-3 fatty acid enriched eggs and production of egg yolk powders: an increased risk of lipid oxidation? Food Chem 153:94–100. https://doi.org/10.1016/j.foodchem.2013.12.028
14. Corstens MN, Berton-Carabin CC, Schroën K et al (2018) Emulsion encapsulation in calcium-alginate beads delays lipolysis during dynamic in vitro digestion. J Funct Foods 46: 394–402. https://doi.org/10.1016/j.jff.2018.05.011
15. Pizones Ruiz-Henestrosa VM, Ribourg L, Kermarrec A et al (2022) Emulsifiers modulate the extent of gastric lipolysis during the dynamic in vitro digestion of submicron chia oil/water emulsions with limited impact on the final extent of intestinal lipolysis. Food Hydrocoll 124:107336. https://doi.org/10.1016/j.foodhyd.2021.107336
16. Perkins EG (1993) Analyses of fats, oils and derivatives. American Oils Chemist's Society, Champaign
17. Kabri TH (2013) Conception raisonnée d'émulsions submicroniques vectrices d'acides gras polyinsaturés n-3. Impact de la structure moléculaire et supramoléculaire des lipides sur l'oxydation.pH-D thesis. LUNAM (Nantes)

Chapter 2

Separation of Lipid Classes by Solid Phase Extraction for Subsequent Analysis

Michelle Viau, Alice Kermarrec, and Anne Meynier

Abstract

Selective recovery of the desired compounds in a mixture can be obtained by combining an appropriate solid adsorbent and an eluting solvent to isolate the components of interest. This process is designated as solid-phase extraction (SPE). Three protocols using SPE for the separation of lipid classes are described. A simple procedure (protocol 1) allows the separation of main lipid classes, namely, nonpolar and polar lipids, free fatty acids, glycolipids, and phospholipids. Protocol 2 allows the separation of neutral lipid classes (cholesterol, cholesterol esters, monoglycerides, diglycerides, triglycerides, and free fatty acids) and polar lipids. The last one is dedicated to the isolation and purification of phospholipids in edible oil for further analysis (protocol 3).

Key words Solid-phase extraction, Total lipids, Oils, Lipid classes, Neutral lipids, Polar lipids, Phospholipids, Isolation, Free fatty acids, Glycolipids, TAG, MAG, DAG, Cholesterol and esters

1 Introduction

The fractionation of lipids by liquid–solid chromatography or solid phase extraction (SPE) has been widely applied in lipid research [1–4]. SPE is considered one of the most powerful techniques currently available for rapid and selective sample preparation or enrichment. In an SPE process, the lipid components of interest are retained by the adsorbent packed in a cartridge. The versatility of SPE allows it to be used for a large number of purposes as sample purification or compound isolation. Presently, a wide variety of supports or adsorbents is available for lipid separation.

As SPE is based on the same principle as liquid chromatography, it uses strong but reversible interactions between the analyte and the stationary phase in the cartridges. For instance, nonpolar interactions occur between the C-H functions of the analyte and of the adsorbent (Van der Waals forces). The most common support for nonpolar interactions is C18 cartridges (C8 and C2 are also

Christelle Lopez et al. (eds.), *Multidimensional Characterization of Dietary Lipids*, Methods and Protocols in Food Science,
https://doi.org/10.1007/978-1-0716-3758-6_2,

Table 1
Properties of the SPE support or adsorbents

Separation mode	Phase	Properties
Normal phase	Silica [–Si–OH]	As the support is polar, the compounds are successively eluted according to their polarity (neutral compounds first, then more and more polar compounds) by eluents of increasing polarity
Normal or reverse phase	NH_2 (aminopropyl) [–Si–$(CH_2)_3NH_2$] CN (cyanopropyl) [–Si–(CH_3) $(CH_2)_3CN$] Diol (2,3-dihydroxypropoxypropyl) [–Si–$(CH_2)_3OCH_2CH$ $(OH)CH_2OH$]	Alternative to silica, moderately polar. The moisture has less effect on the capacity for lipid adsorption by the adsorbent than on silica
Reverse phase	C18 (octadecyl) [–Si–C_{18} H_{37}]	Strong hydrophobicity
	C8 (octyl) [–Si–C_8 H_{17}]	Moderate hydrophobicity
	C2 (ethyl) [–Si–C_2 H_5] Phenyl [–Si–Ph]	Low hydrophobic
Ion exchange	Quaternary amine [–Si–$(CH_2)_3N^+(CH_3)_3$]	Strong anion exchanger used to extract negatively charged compounds
	Propylbenzenesulfonate [–Si–$(CH_2)_3C_6H_4$–SO_3^-] Propylsulfonate [–Si–$(CH_2)_3SO_3^-$]	Strong cation exchanger used to extract positively charged basic compounds

available). Polar interactions include hydrogen bonding, dipole-dipole forces, etc. These interactions are typical of cyano (CN), amino (NH_2), diol, or silica (Si) supports. Finally, ion exchange support is also available for specific separation involving charged classes (Table 1).

Three detailed analytical protocols with different supports are proposed below. The analyst must choose the most appropriate one for a given extract and the specific aim of the research.

- *Protocol 1*: It describes the simple separation of nonpolar and polar fractions from lipid extracts or oils on a cartridge packed with silica gel. In the lipid area, this is the most widely used solid-phase extraction method.
- *Protocol 2*: It presents the isolation of seven lipid classes among which neutral lipids: cholesterol, cholesteryl esters, triglycerides, diglycerides, monoglycerides, free fatty acids, and phospholipids

fractions as a whole with a cartridge packed with the aminopropyl-bonded phase.

- *Protocol 3*: The low concentration of phospholipids in vegetable oils requires a concentrating step before HPLC analysis. This can be achieved by SPE diol cartridges.

2 Solid-Phase Extraction Chromatography

2.1 Mandatory Equipment

- Positive displacement pipettes and disposable capillary piston tips (1000 μL, 100 μL).
- Glass Pasteur pipettes.
- SPE vacuum manifold with individual control valves and stainless-steel solvent guide needles.
- Solvent drying cabinet with nitrogen supply.
- 10 mL screw-capped (Teflon-lined) tubes in case of SPE vacuum manifold or SPE cartridges.
  - *Si Cartridge (Protocol 1)*

    Sep-Pak Silica, 500 mg, 6 cc (Waters, USA) or equivalent
  - *NH_2 Cartridge (Protocol 2)*

    Chromabond NH_2, 500 mg, 6 cc (MACHEREY-NAGEL SAS, France) or equivalent
  - *Diol Cartridge (Protocol 3)*

    Discovery® DSC-Diol SPE Tube, 1 g Sorbent, 6 cc (Supelco, USA) or equivalent.

2.2 Chemical Products

The chemical products necessary for the lipid's separation following different protocols are given in Table 2.

2.3 Fractionation of Main Lipid Classes Using Si Cartridges (Protocol 1)

Juaneda and Rocquelin published one of the most emblematic works (549 citations in WOS August 2021) on the use of SPE for neutral lipid (NL) and polar lipid (PL) separations [5]. Their separation procedure has been applied successfully for many years by a considerable number of authors. The procedure consists of the use of silica cartridges and solvents of increasing polarity. The solid phase is polar and consequently adsorbs strongly the more polar lipid. Thus, the lipids are eluted by increasing the polar solvents. The neutral lipids are first eluted with chloroform, but in case of the high amount of free fatty acid suspected, chloroform:acetic acid (100:1, v/v) is recommended, glycolipids (GL) are then recovered with acetone, and finally phospholipids (PL) with methanol. The elution scheme is summarized in Fig. 1:

1. Label and attach the required number of SPE Si 500 mg cartridges to the SPE vacuum manifold, making sure any unused

Table 2
Chemical compounds or solvents used in the various SPE protocols

Chemical	Chemical formula	CAS number	Quality	Safety instructions	Protocol number
Acetone	C_3H_6O	67-64-1	For analysis		1
Ammonia solution 25%	NH_4OH	1336-21-6	For analysis		3
Chloroform	$CHCl_3$	67-66-3	HPLC grade		1-2-3
Diethyl ether	$C_4H_{10}O$	60-29-7	For analysis		2
Ethyl acetate	$C_4H_8O_2$	141-78-6	HPLC grade		2
Glacial acetic acid	$C_2H_4O_2$	64-19-7	≥99.0%		2
Hexane	C_6H_{14}	110-54-3	HPLC grade		2-3
Isopropanol	C_3H_8O	67-63-0	ACS reagent, ≥99.5%		2
Methanol	CH_4O	67-56-1	HPLC grade		1-2-3
Methylene chloride	CH_2Cl_2	75-09-2	HPLC grade		2

ports are closed and there is a waste bucket underneath the cartridges for waste solvent collection (*see* **Note 1**).

2. Activate the cartridges with 6 mL of methanol, followed by 6 mL of chloroform. Since each cartridge holds approximately 6 mL and accuracy is not important at this step, we find it easier to use wash bottles filled with each SPE solvent and fill up the SPE cartridges rather than measuring out each solution for each cartridge.
3. Close the tap under each cartridge as soon as the last of the chloroform drop has flowed through.
4. Remove the waste bucket from the vacuum manifold, replacing it with clean, round-bottomed glass tubes under each SPE cartridge in a rack.

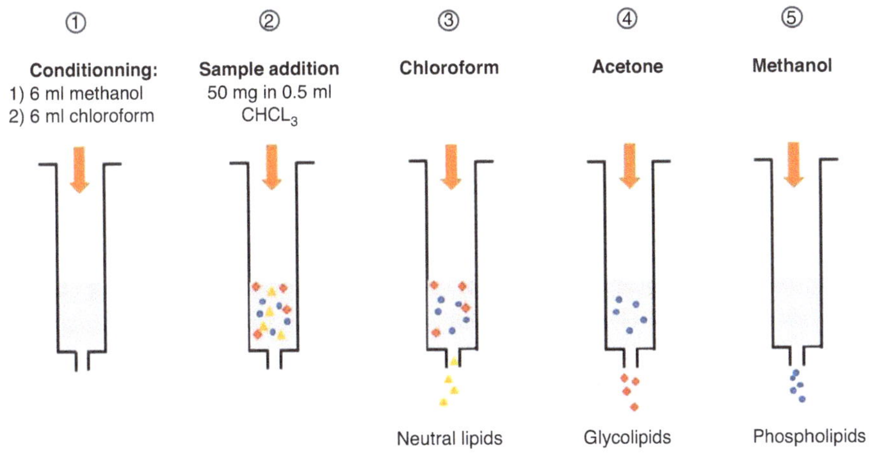

Fig. 1 Elution scheme for the fractionation of lipid extracts using Si cartridges

5. Prepare sample loading solution by solubilizing in a vial 50 mg of lipids in 0.5 mL of chloroform.
6. Transfer the samples onto the conditioned cartridges with glass Pasteur pipettes, and open the tap under each cartridge as it is filled. Wash the vial with another 0.5 mL and pour onto the column.
7. Let the sample flow through the SPE cartridges in a dropwise manner with the vacuum switched off (if needed, a low vacuum can be used for highly viscous samples). *Take care not to allow any air* to pass through the cartridge sorbent and close the tap under each cartridge as soon as the last of the sample has flowed through.
8. Elute the neutral lipids with chloroform (6 mL), *see* **Notes 2** and **3**.
9. Remove the tubes from the vacuum manifold and replace them with another series of tubes.
10. Elute the glycolipids (GL) with acetone (6 mL), *see* **Note 3**.
11. Remove the tubes from the vacuum manifold and replace them with another series of tubes.
12. Finally, elute the phospholipids (PL) with 6 mL of methanol, *see* **Note 3**.
13. The collected fractions are dried under a stream of nitrogen and finally weighed. The fractions are dissolved in chloroform and stored at −20 °C until further analysis.

The resulting crude fractions obtained can be further analyzed by HPTLC or HPLC or used for GC analysis of fatty chains.

Notes

1. If you want to deposit more lipids (>50 mg), use a 1000 mg Si Cartridge. For quantities greater than 100 mg, there is a risk of cartridge saturation, and a complete separation of fractions is not guaranteed.
2. If a high amount of free fatty acid is suspected, we recommend using chloroform:acetic acid (100:1, v/v) for a better desorption of free fatty acid (allows protonated form).
3. The volume of solvent required for a good isolation must sometimes be adjusted since Si-cartridges differ from one supplier to another and even among batches from the same source. The reason is the water content that greatly influences cartridge efficiency.

2.4 Fractionation of Neutral Lipid Classes with NH_2 Cartridges (Protocol 2)

This protocol is dedicated to the fractionation of a lipid extract into several neutral lipid fractions: free fatty acids, cholesterol esters, triglycerides, diglycerides, monoglycerides, and cholesterol plus one fraction where all phospholipids are gathered. The protocol uses three NH_2 cartridges to prepare in 1–1.5 h up to seven lipid fractions from a total lipid extract. This methodology is more complex than the one described in protocol 1, as can be seen in the diagram in Fig. 2 [6]:

1. Label and attach the required number of SPE NH_2 cartridges to the SPE vacuum manifold, making sure any unused ports are closed and there is a waste bucket underneath the cartridges for waste solvent collection.

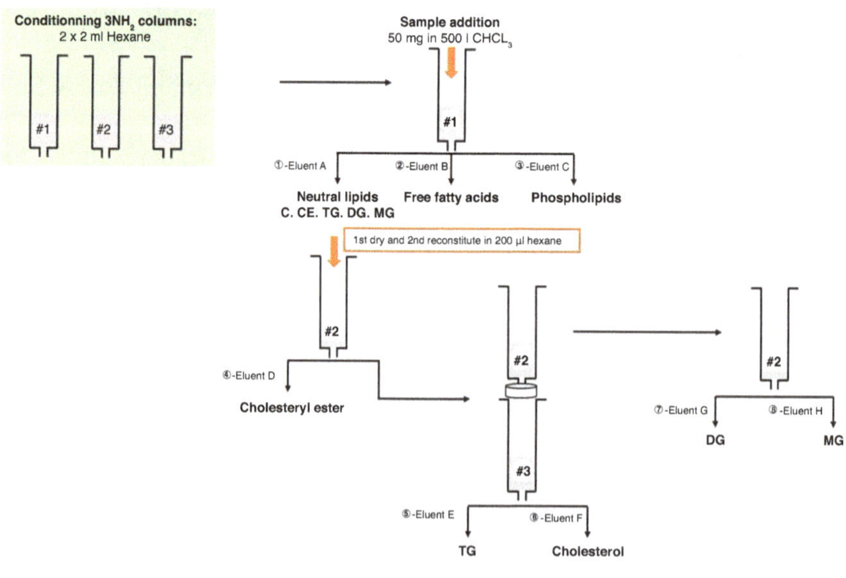

Fig. 2 Elution scheme for the fractionation of lipid extracts using aminopropyl-bonded phase (NH_2) cartridges

Table 3
Summary of solvent and collected fraction in protocol 2

Column number	Eluant name	Eluant composition	Column volume (mL)	Eluate number	Eluted lipids
1	A	Chloroform/2-propanol (2/1; v/v)	4	1	Neutral lipids (C, CE, TG, DG, MG)
	B	Diethyl ether/acetic acid (98/2; v/v)	4	2	Free fatty acids
	C	Methanol	4	3	Phospholipids
2	D	Hexane	4	4	Cholesteryl esters (CE)
2 + 3	E	Hexane/methylene chloride/ diethyl ether (89/10/1; v/v)	6	5	Triglycerides (TG)
	F	Hexane/ethyl acetate (95/5; v/v)	12	6	Cholesterol (C)
2	G	Hexane/ethyl acetate (85/15; v/v)	4	7	Diglycerides (DG)
	H	Chloroform/methanol (2/1; v/v)	4	8	Monoglycerides (MG)

2. Activate the cartridges with two × 2 mL of hexane.
3. Close the tap under each cartridge as soon as the last drop of hexane has flowed through.
4. Remove the waste bucket from the vacuum manifold, replacing it with clean, round-bottomed glass tubes under each SPE cartridge in a rack.
5. Prepare sample solution by solubilizing 50 mg of lipids in 0.5 mL of chloroform.
6. Prepare eluent A, B, E, F, G, and H (Table 3).

NH_2 Sorbent #1

7. Transfer the samples onto the conditioned cartridges (NH_2 Sorbent #1) with glass Pasteur pipettes, and open the tap under each cartridge as it is filled.
8. Let the sample flow through the SPE cartridges in a dropwise manner with the vacuum switched off (if needed, a low vacuum can be used for highly viscous samples). *Take care not to allow any air* to pass through the cartridge sorbent and close the tap under each cartridge as soon as the last of the sample has flowed through.
9. *Elute #1 (neutral lipids, eluent A)*. Elute the neutral lipids with 4 mL chloroform/2-propanol (2:1, v/v).

10. Remove the tubes from the vacuum manifold and replace them with another series of tubes.
11. Dry eluate#1 under nitrogen and set aside for further fractionation.
12. *Elute #2 (free fatty acids, eluent B).* Elute the free fatty acids with 4 mL of diethyl ether/acetic acid (98:2, v/v). Collect the free fatty acid fraction.
13. Remove the tubes from the vacuum manifold and replace them with another series of tubes.
14. *Elute #3 (phospholipids, eluent C).* Elute the phospholipids (PL) with 4 mL methanol and collect the phospholipid fraction.

NH_2 Sorbent #2

15. Solubilize the dried eluate from Elution #1 (neutral lipids) in 200 μL of hexane.
16. Place new round-bottomed glass tubes under each SPE cartridge.
17. Transfer the samples onto the conditioned cartridges (NH_2 Sorbent #2) using glass Pasteur pipettes, open the tap under each cartridge, and aspirate reconstituted eluate through the NH_2 sorbent #2.
18. *Elute #4 (cholesteryl esters, eluent D).* Elute the cholesteryl esters (CE) with 4 mL of hexane and collect the cholesteryl ester fraction.

NH_2 Sorbent #3

19. Since during the elution of the triglycerides cholesterol may be eluted too, a third sorbent is added. Attach NH_2 conditioned sorbent #3 below NH_2 sorbent #2 with an adaptor.
20. Place new tubes under each joined cartridge (sorbent #2 + sorbent #3).
21. *Elute #5 (triglycerides, eluent E).* Elute the triglycerides (TG) with 6 mL of diethyl ether/methylene chloride/hexane (1:10:89, v/v/v), *see* **Note 1**. Aspirate through both columns to collect the triglyceride fraction.
22. Remove the tubes from the vacuum manifold and replace them with another series of tubes.
23. *Elute #6 (cholesterol, eluent F).* Elute the cholesterol with 12 mL of hexane/ethyl acetate (95:5, v/v). Aspirate through both columns to collect the cholesterol fraction.

24. Disconnect the stacked columns, keeping only the top column (NH_2 #sorbent 2) for further elution.
25. Remove the tubes from the vacuum manifold and replace them with another series of tubes.
26. *Elute #7 (diglycerides, eluent G)*. Elute the diglycerides (DG) with 4 mL of hexane/ethyl acetate (85:15, v/v) and collect the diglyceride fraction.
27. Remove the tubes from the vacuum manifold and replace them with another series of tubes.
28. *Elute #8 (monoglycerides, eluent H)*. Elute monoglycerides (MG) with 4 mL of chloroform/methanol (2:1, v/v) and collect the monoglyceride fraction.
29. The collected fractions are dried under a stream of nitrogen (*see* **Note 2**). The fractions are finally dissolved in chloroform and stored at −20 °C until further analysis.

Notes

1. For samples rich in TG there is a risk of incomplete desorption, so we recommend eluate with more eluent E.
2. The solvents from each fraction may also be evaporated using a rotary evaporator at a temperature no greater than 50 °C.

2.5 Isolation of Phospholipids from Edible Oils (Protocol 3)

The low phospholipid concentrations (0.1–1.2%) present in crude vegetable oils require a method for concentrating them before HPLC or HPTLC analysis can take place. This can be achieved by diol SPE [7]. The elution scheme is summarized in Fig. 3.

1. Label and attach the required number of SPE diol cartridges to the SPE vacuum manifold, making sure any unused ports are closed and there is a waste bucket underneath the cartridges for waste solvent collection.
2. Activate the cartridges with 2 mL of methanol, followed by 2 mL of chloroform, and finally by 4 mL of hexane. Close the tap under each cartridge as soon as the last hexane has flowed through.
3. Prepare sample solution by solubilizing 150 mg of oil in 1 mL of chloroform.
4. Prepare PL eluent by mixing 0.5 mL of ammonia solution 25 vol% and 99.5 mL of methanol.
5. Transfer the samples onto the conditioned cartridges with glass Pasteur pipettes, opening the tap under each cartridge when it is filled.

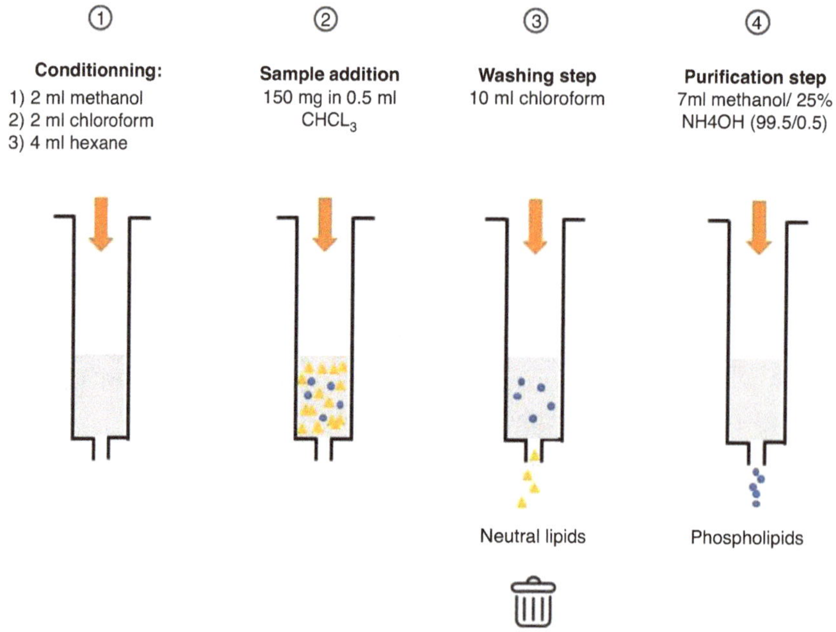

Fig. 3 Elution scheme for PL purification using diol cartridge (protocol 3)

6. Let the sample flow through the SPE cartridges in a dropwise manner with the vacuum switched off (if needed, a low vacuum can be used for highly viscous samples). *Take care not to allow any air* to pass through the cartridge sorbent and close the tap under each cartridge as soon as the last of the sample has flowed through.
7. Wash with 10 mL of chloroform (*see* **Note 1**). This elute is rejected.
8. Remove the waste bucket from the vacuum manifold, and replace it with clean, round-bottomed glass tubes under each SPE cartridge in a rack.
9. Elute the phospholipids with 7 mL of methanol/25% ammonia solution (99.5: 0.5/v/v) (eluant PL).
10. The collected fraction is dried under a stream of nitrogen, resuspended in 500 μL of chloroform, and stored at −20 °C until further analysis.

Notes

1. A large percentage of the triglycerides was released from the sorbent by passing 5 mL chloroform through. This volume needs to be adjusted and increased to eluate more triglycerides.

References

1. Mangold HD (1984) No Title. In: CRC handbook of chromatography. Lipids. CRC Press, Boca Raton, p 387
2. Christie WW (1982) Lipid analysis: isolation, separation, identification, and structural analysis of lipids, 2nd edn. Pergamon Press, New York
3. Ruiz-Gutiérrez V, Pérez-Camino MC (2000) Update on solid-phase extraction for the analysis of lipid classes and related compounds. J Chromatogr A 885:321–341. https://doi.org/10.1016/S0021-9673(00)00181-3
4. Christie WW, Han X (2010) Lipid analysis isolation, separation, identification and lipidomic analysis, 4th edn. Woodhead Publishing/The Oily Press, Cambridge
5. Juaneda P, Rocquelin G (1985) Rapid and convenient separation of pentachlorophenol from human fat using silica Sep-Pak™ cartridges. Lipids 20:40–41. https://doi.org/10.1016/S0021-9673(00)90537-5
6. Kaluzny MA, Duncan LA, Merritt MV, Epps DE (1985) Rapid separation of lipid classes in high yield and purity using bonded phase columns. J Lipid Res 26:135–140. https://doi.org/10.1016/s0022-2275(20)34412-6
7. Carelli AA, Brevedan MIV, Crapiste CH (1997) Quantitative determination of phospholipids in sunflower oil. J Am Oil Chem Soc 74:511–514. https://doi.org/10.1007/s11746-997-0173-2

Chapter 3

Purification of Lipids by Cold-Acetone Precipitation

Claude Genot

Abstract

Cold-acetone precipitation is a very simple way to isolate polar lipids from a total lipid extract. This chapter details a simple procedure to prepare phospholipids or polar (phospholipids + glycolipids) fractions at the laboratory scale from total lipid extracts or from commercial lecithins. The method includes the precipitation of the polar lipids at 4 °C followed by their recovery by centrifugation or filtration. The recovered fractions usually contain less than 1% neutral lipids.

Key words Acetone, Lecithin, Phospholipids, Polar lipids, Purification

1 Introduction

The cold-acetone purification method is a very simple way to isolate polar lipids from a total lipid extract [1]. The method is also useful for deoiling of commercial fluid lecithins that currently contain 30–40% neutral lipids [2–5]. The method is based on the insolubility in cold acetone of most phospholipids and glycolipids versus the solubility in the solvent of glycerides and neutral lipids. For instance, the solubility of soybean phospholipids in acetone at 5 °C is less than 0.03 g/L [6]. Acetone insolublity is also a criterion for evaluating lecithin quality according to the standard method [7].

The acetone-soluble compounds generally comprise neutral lipids: glycerides, sterols, sterol esters, free fatty acids, most carotenoids, lipid-soluble vitamins, cholesterol, selected glycolipids (monogalactosyl diacylglycerides, sterol glycosides), and only traces of phospholipids [8]. However, phospholipids with acyl chains less than C:10 were reported to be soluble in acetone [3]. A class of phosphatidyl-choline (PC) rich in EPA and DHA was also extracted from Antarctic krill oil by acetone at 25 °C and recovered in the 4 °C acetone-soluble fraction [9].

Christelle Lopez et al. (eds.), *Multidimensional Characterization of Dietary Lipids*, Methods and Protocols in Food Science, https://doi.org/10.1007/978-1-0716-3758-6_3,

The purity of the polar lipids, mainly phospholipids, recovered in the cold-acetone-insoluble fraction is usually greater than 95%. It increases with the increase in the acetone amount (acetone/lipid ratio), extraction time, and number. It can reach over 99%. Depending on the initial composition of the lipids, the fraction may contain glycolipids: di- or polyglycosyl diglycerides, cerebrosides, and sulfatides [10], sterol glycoside, galactodiacylglycerol, and carbohydrates (sucrose, raffinose, stachyose) [11], traces of cholesterol and pigments such as oxycarotenoids (lutein, zeaxanthin) [8]. It may also contain waxes [12–14].

Generally, acetone fractionation is performed at chilling temperatures (2–5 °C) as described below [4, 5], but in the literature, the actual temperature of the solvent was often not specified. Lower or higher temperatures were also reported (*see* **Note 3**).

Very importantly acetone can react with amino-phospholipids chemically to form acetone abducts [15–17]. These PL–acetone derivatives, among them mesityloxide, give the PL extract an unpleasant odor or aftertaste [18]. Therefore, a careful drying process is required to eliminate the residual acetone and prevent the formation of these undesired compounds.

Moreover, the lipid fractions should be protected against oxidation during the whole purification process and their further storage by avoiding exposure to light and oxygen as far as possible, flushing the solvents and containers with N_2, and storing the fractions as a solution.

The small-scale method described below is valuable for preparing grams of purified phospholipids with good recovery from animal lipids (lipids from muscle foods, fish and fish roes, krill, egg yolk [8, 18–20], milk [21], and from microbial lipids [10, 22, 23]. It also allows to prepare powdery phospholipids from soybean, rapeseed, peanut, and corn lecithins [6, 24]. In our laboratory, we have accordingly easily purified phospholipid fractions from meat/muscle of different animals (turkey, pork, etc.), egg yolk, buttermilk, and DHA-rich lecithin [25] with good yields and recovery of the different PL classes as determined by HPLC (see Chapter XX). The method can be easily scaled up to obtain powdered lecithin containing only 2–5% remaining oil at a higher scale, including the industrial level [6, 21, 22].

2 Materials

2.1 Chemicals (Table 1)

Table 1
Chemical required for purifying polar lipids with acetone

Chemical	Chemical formula	CAS number	Quality	Safety instructions
Chloroform	$CHCl_3$	67-66-3	HPLC grade	
Acetone[a]	C_3H_6O	67-64-1	For analysis	

[a]Stored at +4 °C

2.2 Mandatory Equipment

1. Temperature-controlled laboratory (+4 °C) and cold storage room (+4 °C) or ice bath.
2. Analytical precision balance, sensitivity: 0.01 mg.
3. (method 1) Glass centrifuge tubes (15 or 50 mL).
4. (method 1) Refrigerated centrifuge.
5. (method 2) Büchner funnel and paper filter.
6. (method 2) Magnetic stirrer and bar magnet.
7. Pipettes and disposable capillary piston tips or graduated glass pipettes and rubber bulb.
8. Graduated cylinder.
9. Volumetric glass flasks: 10 mL.
10. Vortex mixer.
11. Vacuum rotary evaporator and vacuum system.
12. Nitrogen generator or nitrogen cylinder.
13. Round-bottom flasks with ground-glass stopper, 250 mL.

3 Method

Method 1 with centrifugation:

1. Weigh 15 mL/50 mL centrifuge glass tubes.
2. Introduce in the tubes aliquots of the solution of lipid extracts containing about 500–1500 mg lipids.
3. Evaporate the solvent under nitrogen (*see* **Note 1**).
4. Add 2–5 mL acetone and dissolve thoroughly the lipids; a warm water bath may be used to facilitate solubilization.
5. Add 10–30 mL cold acetone (20–30 vol) (*see* **Note 2**).
6. Mix for 1 min on a vortex mixer.
7. Store at +4 °C for 1 h (*see* **Note 3**).

8. Centrifuge at 1000 × *g* for 10 min.
9. Remove the acetone supernatant containing the neutral lipids.
10. Repeat **steps 5–9** at least twofold and mix the acetone extracts.
11. Evaporate residual acetone in the tubes with a stream of nitrogen.
12. Weigh the tubes.
13. Dissolve the precipitated polar fraction at a known concentration in chloroform or alternate solvent (for instance, chloroform/methanol).
14. Recover the neutral fraction in the acetone extract by evaporation of the solvent with a rotary evaporator and dissolve it at a known concentration in chloroform.

Method 2 using filtration:

1. Weigh a 250 mL round-bottom flask.
2. Introduce an aliquot of the solution of total lipid extracts of known lipid concentration containing 2–5 g lipids in the flask.
3. Evaporate the solvent with a rotary evaporator and then under nitrogen (*see* **Note 1**).
4. (optional) Weigh the flask to control the amount of total lipids.
5. Add 5 mL acetone and dissolve thoroughly the lipids; a warm water bath may be used to facilitate solubilization.
6. Add 20–30 volumes of cold acetone [20].
7. Add a magnetic rod.
8. Close the flask with a glass stopper.
9. Mix for 1 min, then agitate with a magnetic barrel for 1 h at 4 °C (*see* **Note 3**).
10. Decant the mixture.
11. Remove carefully the acetone supernatant containing the neutral lipids with a pipette.
12. Repeat **steps 6–11** at least twofold and collect the acetone supernatants.

    Option instead of 10–12:

    10b. Remove the rod.

    11b. Filter the mixture on a Büchner filter equipped with a round paper filter and rinse abundantly the filter with cold acetone.

    12b. Recover the precipitated polar lipids in a weighed flask.
13. Evaporate the remaining solvent with a stream of nitrogen.
14. Weigh the flask to calculate the amount of precipitated polar lipids.

15. Dissolve the polar lipids at a known concentration in chloroform of dichloromethane and keep it at −20 °C under N_2.
16. Recover the neutral fraction dissolved in the acetone extract by the evaporation of the solvent with a rotary evaporator and dissolve it at a known concentration in chloroform.

4 Notes

1. A small amount of solvent can be let [10].
2. Kates [10] also mentioned the addition of 0.1 mL 10% $MgCl_2.6H_2O$ in methanol for 5 mL acetone.
3. Standard precipitation temperature is around 4 °C, but higher or lower temperatures were mentioned in the literature to isolate fractions from specific materials: PC from soybean lecithin [26], phospholipids from salmon head lipids [27], and PL fraction from egg yolk without cholesterol [19] were accordingly isolated by precipitation at −20 °C. Higher amounts of residual neutral lipids were however noticed at −20 °C than at 4 °C due to neutral lipid crystallization or to the viscosity of the mixture [28]. Marine or algae oils enriched in long-chain omega 3 polyunsaturated fatty acids (PUFA) were obtained by acetone precipitation at 7 °C [12]. Total polar lipids were recovered from milk samples at room temperature [29, 30]. A high-purity (97%) PC was also isolated from krill oil at 25 °C [9], while an acetone-soluble fraction enriched in EPA and DHA was recovered at 30 °C from free fatty acid salts prepared from a fish oil [31].

References

1. Christie W cyberlipids [Internet]. Available from: http://cyberlipid.gerli.com/techniques-of-analysis/fractionation-complex-extracts/solvent-fractionation-2/
2. Stopper HR, Saffer BA, Bauer ST (1953) Factors influencing the determination of the acetone insolubles of commercial lecithins. J Am Oil Chem Soc 30(10):408–410
3. Bueschelberger H-G (2004) Lecithins. In: (Oxford) B, editor. Emulsifiers in food technology. Wiley
4. El-Shattory Y, Aly SM (2002) Soybean lecithin: acetone insoluble residue fractionation and their volatile components. Grasas Aceites 53(3):319–323
5. Vandana V, Karuna MSL, Vijayalakshmi P, Prasad RBN (2001) A simple method to enrich phospholipid content in commercial soybean lecithin. J Am Oil Chem Soc 78(5):555–556
6. Liu D, Ma F (2011) Soybean phospholipids. Recent trends for enhancing the diversity and quality of soybean products. IntechOpen
7. AOCS official method Ja 4-46 (1994) Acetone-Insoluble matter. Official Methods and Recommended Practices of the American Oil Chemist's Society (4th edn). AOCS Press, Champaign
8. Bernardo J, Videira RA, Valentao P, Veiga F, Andrade PB (2019) Extraction of phospholipid-rich fractions from egg yolk and development of liposomes entrapping a dietary polyphenol with neuroactive potential. Food Chem Toxicol 133
9. Zhao J, Wei SS, Liu F, Liu DC (2014) Separation and characterization of acetone-soluble phosphatidylcholine from Antarctic krill (Euphausia superba) oil. Eur Food Res Technol 238(6):1023–1028

10. Kates M (1972) In: Work TS, Work E (eds) Techniques of lipidology: Isolation, analysis and identification of lipids. North-Holland Publishing Company, Amsterdam
11. van Nieuwenhuyzen W, Tomas MC (2008) Update on vegetable lecithin and phospholipid technologies. Eur J Lipid Sci Technol 110(5): 472–486
12. Callejon MJJ, Medina AR, Sanchez MDM, Cerdan LE, Moreno PAG, Lopez EN et al (2020) Obtaining highly pure EPA-rich lipids from dry and wet Nannochloropsis gaditana microalgal biomass using ethanol, hexane and acetone. Algal Research-Biomass Biofuels and Bioproducts, p 45
13. Huynh LH, Kasim NS, Ju YH (2010) Extraction and analysis of neutral lipids from activated sludge with and without sub-critical water pre-treatment. Bioresour Technol 101(22): 8891–8896
14. Rajam L, Kumar DRS, Sundaresan A, Arumughan C (2005) A novel process for physically refining rice bran oil through simultaneous degumming and dewaxing. J Am Oil Chem Soc 82(3):213–220
15. Wang H, Yao LX, Lee SL, Wang T (2017) Simultaneous texturization and extraction of phospholipids from liquid egg yolk using renewable solvents. Eur J Lipid Sci Technol 119(2)
16. Kuksis A, Ravandi A, Schneider M (2005) Covalent binding of acetone to aminophospholipids in vitro and in vivo. Maillard React 1043:417–439
17. Ando N, Ando S, Yamakawa T (1971) The structure and formation mechanism of N-acetone derivatives of phosphatidyl ethanolamine. J Biochem 70(2):341
18. Nielsen H, Shukla VKS (2004) In situ solid phase extraction of lipids from spray-dried egg yolk by ethanol with subsequent removal of triacylglycerols by cold temperature crystallization. Lebensm Wiss Technol 37(6):613–618
19. Gladkowski W, Chojnacka A, Kielbowicz G, Trziszka T, Wawrzenczyk C (2012) Isolation of pure phospholipid fraction from egg yolk. J Am Oil Chem Soc 89(1):179–182
20. Maximiano FA, da Silva MA, Daghastanli KRP, de Araujo PS, Chaimovich H, Cuccovia IM (2008) A convenient method for lecithin purification from fresh eggs. Quim Nova 31(4): 910–913
21. Le TT, Phan TTQ, Van Camp J, Dewetting K (2015) Milk and dairy polar lipids: occurrence, purification, and nutritional and technological properties. Polar lipids. Elsevier, pp 91–143
22. van Nieuwenhuyzen W (2015) Production and utilization of natural phospholipids. Ploar lipids. Elsevier, pp 245–276
23. Bonilla-Mendez JR, Hoyos-Concha JL (2018) Methods of extraction, refining and concentration of fish oil as a source of omega-3 fatty acids. Revista Corpoica-Ciencia Y Tecnologia Agropecuaria 19(3):645–668
24. Bueschelberger HG, Tirok S, Stoffels I, Schoeppe A, Norn V (2015) Lecithins. John Wiley & Sons Ltd, Chichester, pp 21–60
25. Kabri T-H (2013) Conception raisonnée d'émulsions submicroniques vectrices d'acides gras polyinsaturés n-3. Impact de la structure moléculaire et supramoléculaire des lipides sur l'oxydation. Thèse de doctorat Nantes
26. Patil VV, Galge RV, Thorat BN (2010) Extraction and purification of phosphatidylcholine from soyabean lecithin. Sep Purif Technol 75(2):138–144
27. Shah A, Nagao T, Kurihara H, Takahashi K (2017) Production of a health-beneficial food emulsifier by enzymatic partial hydrolysis of phospholipids obtained from the head of autumn chum salmon. J Oleo Sci 66(2): 147–155
28. Wang H, Yao LX, Wang T (2014) Extraction of phospholipids from structured dry egg yolk. J Am Oil Chem Soc 91(3):513–520
29. Smith LM, Jack EL (1959) Isolation of milk phospholipids and determination of their polyunsaturated fatty acids. J Dairy Sci 42(5): 767–779
30. Pimentel L, Fontes AL, Salsinha S, Machado M, Correia I, Gomes AM et al (2018) Suitable simple and fast methods for selective isolation of phospholipids as a tool for their analysis. Electrophoresis 39(15): 1835–1845
31. Kuo CH, Huang CY, Chen JW, Wang HMD, Shieh CJ (2020) Concentration of docosahexaenoic and eicosapentaenoic acid from cobia liver oil by acetone fractionation of fatty acid salts. Appl Biochem Biotechnol 192(2): 517–529

Chapter 4

Stripping of Vegetable Oils by Direct Mixing with Adsorbent Materials

Claire C. Berton-Carabin, Lucie Ribourg-Birault, and Thais Cristina Benatti Gallo

Abstract

It is often useful to conduct experiments with triacylglycerol oils, such as vegetable oils, that have been stripped from surface-active impurities and/or endogenous antioxidants. This chapter describes a rapid, simple, and solvent-free procedure to achieve such purification by the direct mixing of oil with an adsorbent powdered material (magnesium silicate or aluminum oxide), followed by gravitational separation. In particular, we give a step-by-step description of the method, as well as practical tips, and provide experimental data regarding the properties of a comprehensive set of oils obtained via this procedure.

Key words Vegetable oil, Stripping, Purification, Triacylglycerol oil, Adsorbent materials, Alumina, Silica, Surface-active impurities, Tocopherols

1 Introduction

Edible oils such as vegetable oils are mostly composed of triacylglycerols (TAGs), yet often still contain minor components that can have a noticeable effect on their stability and functional properties. Two major examples in that respect relate to (1) the presence of endogenous lipophilic antioxidants, such as tocopherols; and (2) the presence of surface-active impurities (such as free fatty acids, mono- and diacylglycerols, phospholipids, or lipid oxidation products) which, even in trace amounts, can largely alter the properties of adsorbed layers formed at the oil–water interface [1, 2].

It can therefore be relevant to strip vegetable oils from such components when conducting studies that pertain to the oxidative stability of the oil-based systems [3, 4], and/or to the characterization of structures and phenomena occurring at the oil–water interface [5]. For this, selected adsorbent materials have proven useful; they may be used either as chromatographic columns [6, 7], which implies first diluting the oil in an organic solvent, which shall then

Christelle Lopez et al. (eds.), *Multidimensional Characterization of Dietary Lipids*, Methods and Protocols in Food Science, https://doi.org/10.1007/978-1-0716-3758-6_4,

be removed after elution, or by direct mixing with the oil followed by gravimetric separation [8, 9], which is a solvent-free procedure. The choice of the adsorbent material is of high importance depending on the targeted components to remove. For instance, magnesium silicate may be used to remove surface-active polar contaminants [10, 11] and alumina to remove not only such components but also lipophilic antioxidants such as tocopherols [9]. In addition, lipophilic pigments such as carotenoids and chlorophylls can be eliminated by activated charcoal [12]. Another protocol elaborated for the purification of mineral oils/nonpolar phases has also been proposed, which requires the consecutive treatment with iron (III) chloride for the reduction of peroxides, potassium hydroxide for the extraction of alcohols, silica gel for the adsorption of polar contaminants, and finally, sodium sulfate for the removal of water [13].

This chapter focuses on a solvent-free procedure for stripping vegetable oils, either using magnesium silicate (removal of surface-active impurities) or using aluminum oxide (removal of surface-active impurities + tocopherols and tocotrienols). We strongly recommend that the stripped oils obtained with this procedure are checked for interfacial tension against water (e.g., using an automated drop tensiometer; see for instance, ref. [11]) and, when antioxidant stripping is targeted, for residual tocopherol/tocotrienol content (see Chapter 10). A comprehensive series of examples of such properties for various reference lipids/oils and stripped oils is given later (Subheading 4 *See* **Note 1**, Table 1).

2 Material

2.1 Equipment

1. 50-mL conical centrifuge tubes (polypropylene).
2. Rotative tube agitator equipped with a 50-mL tube carousel.
3. Or glass beaker and magnetic bar/stirrer.
4. Centrifuge with swing rotor adapted for 50-mL conical tubes.
5. Plastic or glass transfer pipettes.

2.2 Chemicals

1. Magnesium silicate powder (also referred to as silica; CAS: 1343-88-0; example of commercial sample: Florisil®, 100–200 mesh, Supelco).
2. Or aluminum oxide powder (also referred to as alumina; CAS: 1344-28-1; an example of commercial sample: MP Alumina N—Super I, MP Biomedical).

NB: Magnesium silicate powder (silica) should be used when the stripping procedure aims at removing surface-active impurities, but not endogenous tocopherols. Conversely, stripping with aluminum oxide (alumina) should be selected when both surface-active impurities and endogenous tocopherols/tocotrienols shall

Table 1
Interfacial tension values (oil–water interface) and tocopherol contents reported for various lipid phases, with a focus on vegetable oils

Oil or nonpolar phase	Stripping method	Interfacial tension value (mN/m)[a]	Total tocopherol content (mg/kg oil)[a]	Source/reference
Tetradecane	Stripped by consecutive treatment with iron (III) chloride, potassium hydroxide, silica gel, and sodium sulfate	51.5	NA	[13]
Hexane		50.0	NA	
1-Octanol		8.7	NA	
Medium chain triglycerides		21.0	–	
Commercial vegetable oils[b]	Nonstripped	19.5–23.5	–	[14]
Stripped commercial vegetable oils[b]	Stripped by passage through a packed column of magnesium silicate	30.0–31.5	–	
Commercial soybean oil with 0.1 wt.% lecithin	Nonstripped	~5	–	[15]
Commercial corn oil	Nonstripped	–	538.9 ± 12.8	[12] (in this article, data are also given for other (stripped) oils, namely camelina oil, chia oil, Sophia seed oil, olive oil and DHA single cell oil)
Stripped commercial corn oil	Simplified stripping method by direct mixing with hexane, activated silicic acid, and charcoal under N_2, followed by solvent evaporation	–	ND	
Stripped commercial rapeseed oil mixed with brominated oil[c] (65 wt.% – 35 wt.%)	Rapeseed oil: preliminary stripped with procedure in Subheading 3.1 with aluminum oxide, at 4 °C Brominated oil: nonstripped	25.4	–	Courtesy Sten ten Klooster, Wageningen University & Research (The Netherlands), 2021
Stripped commercial rapeseed oil mixed with brominated oil[c] (65 wt.% – 35 wt.%)	Rapeseed oil: preliminary stripped with procedure in Subheading 3.1 with aluminum oxide, at 4 °C Brominated oil: preliminary mixed with hexane (4:1, v/v), then stripped with procedure in Subheading 3.1 with aluminum oxide, at 4 °C. Hexane was then evaporated at 45 °C for 3 h under N_2 flow.	34.8	ND	
Commercial rapeseed oil[d]	Nonstripped	13.0	501.2	[16]
Stripped commercial rapeseed oil[d]	Procedure described in Subheading 3.1 with magnesium silicate, at 4 °C	28.7	513.4	
Stripped commercial rapeseed oil[d]	Procedure described in Subheading 3.1 with aluminum oxide, at 4 °C	28.7	1.8	

(continued)

Table 1 (continued)

Oil or nonpolar phase	Stripping method	Interfacial tension value (mN/m)[a]	Total tocopherol content (mg/kg oil)[a]	Source/reference
Stripped commercial rapeseed oil	Procedure described in Subheading 3.1 with a 50/50 mixture (w/w) of magnesium silicate/aluminum oxide, at 4 °C	–	30.7	
Stripped commercial rapeseed oil	Procedure described in Subheading 3.1 with magnesium silicate, at room temperature	27.8	–	Courtesy Jack Yang, Wageningen University & Research (The Netherlands), 2021
Stripped commercial rapeseed oil	Procedure described in Subheading 3.2 with aluminum oxide, at 20 °C	30.5	ND	Courtesy Jilu Feng, Wageningen University & Research (The Netherlands), 2021
Stripped commercial sunflower oil	Procedure described in Subheading 3.2 with aluminum oxide, at 20 °C	28.0 (against phosphate buffer 10 mM, pH 7.0)	–	[2]
Stripped commercial sunflower oil	Procedure described in Subheading 3.1 with aluminum oxide, at 4 °C	30.0 (against acetate buffer 50 mM, pH 4.5)	ND	[17]
Commercial rapeseed oil	Nonstripped oil	22.3	840.1 ± 5.2	Data from Thais Cristina Benatti Gallo and Lucie Birault, INRAE (France), 2021 (unpublished)
Stripped commercial rapeseed oil	Procedure described in Subheading 3.1 with magnesium silicate, at 4 °C	27.9	903.3 ± 7.8	
Stripped commercial rapeseed oil	Procedure described in Subheading 3.1 with aluminum oxide (*alumina bottle opened for >1 year*), at 4 °C	28.4	176.0 ± 3.0	
Stripped commercial rapeseed oil	Procedure described in Subheading 3.1 with aluminum oxide (*new alumina bottle*), at 4 °C	28.8	2.1 ± 0.0	

NA nonapplicable, *ND* nondetectable (below detection threshold)

[a]Interfacial tension was measured at room temperature against ultrapure water, unless otherwise stated. Tocopherols were measured by high-performance liquid chromatography (HPLC) as described in Chapter X

[b]Various oils tested (soybean, corn, canola, and olive oils)

[c]Commercial brominated vegetable oil, CAS number: 8016-94-2

[d]In these samples, the amount of hydroperoxides was also measured, and was found to be 1.1 mmol/kg oil for the nonstripped oil and 0.3 mmol/kg oil for the stripped oils (both with silica and with alumina)

be removed. After opening, it should be noted that the quality of adsorbent materials may decrease after a given storage period, which can decrease the stripping efficiency (see Subheading 4, *see* **Note 2**). Recommendations regarding handling of those materials are given in Subheading 4 (*see* **Note 3**).

3 Methods

3.1 Recommended Procedure: Agitation Using a Rotative Agitator

1. Fill in 50-mL conical centrifuge tubes with approximately 1 volume (15 mL) of adsorbent powder (silica or alumina) and 2 volumes (30 mL) of oil.
2. Mix cautiously the content of each tube with a spatula to disperse the adsorbent powder in the oil, and put immediately on rotative agitation at 15 rpm (be careful not to let the tubes stand still for too long before starting the agitation, as the adsorbent powder tends to sediment within minutes).
3. Incubate the tubes under rotative agitation at 15 rpm in the dark for at least a few hours (recommended: overnight, i.e., 10–12 h minimum). For oils that do not crystallize in refrigerated conditions (e.g., rapeseed oil), it is recommended to conduct the agitation at 4 °C. For oils with a higher melting point, a higher incubation temperature may be chosen (it is important that lipids are completely melted upon the stripping procedure).
4. At the end of the agitation period, pick the tubes and centrifuge them (centrifuge equipped with a swing rotor, 2000 × *g*, 20 min, 20 °C).
5. Collect the supernatant (oil layer) with a transfer pipette (stop collecting when reaching about 0.5 cm above the bottom adsorbent layer), and transfer it to a new 50-mL conical centrifuge tube.
6. Perform a second centrifugation in the same conditions as previously.
7. Collect the stripped oil with a transfer pipette (stop collecting when reaching about 1 cm above the bottom of the tube, i.e., the conical part. Generally, residual adsorbent particles are hardly visible for the second centrifugation). It is recommended to store the stripped oil in clean, glass tubes, under a nitrogen blanket and away from light, ideally at −20 °C or below, until further use.

3.2 Alternative Procedure: Agitation Using a Magnetic Stirrer

When a rotative agitator is not available, the aforementioned procedure can be modified to perform the mixing between oil and the adsorbent material directly in a beaker. The following steps should then be taken:

1. In a glass beaker of appropriate volume, put 1 volume of the adsorbent powder (silica or alumina) and 2 volumes of oil.
2. Mix cautiously with a spatula to disperse the adsorbent powder in the oil, and put immediately on magnetic stirring using a large enough magnetic bar. The stirring intensity should be high enough to keep the mixture well dispersed and avoid the formation of an oil layer depleted from the adsorbent material on top. It should be noted that the mixture is quite viscous, which may (1) cause the magnetic bar to slow down/stop at some point and/or (2) induce an increase in the temperature of the magnetic stirrer, which then heats the mixture. This may, in practice, not be resolved by putting the stirrer + mixture system in a refrigerated room because this would typically increase the viscosity of the mixture, resulting in stopping the magnetic bar from spinning.
3. Incubate the beaker under magnetic stirring at appropriate speed (see comment in the previous point), protecting it from light for at least a few hours (recommended: overnight, i.e., 10–12 h minimum) at room temperature.
4. At the end of the agitation period, rapidly transfer the content of the beaker into 50-mL conical centrifuge tubes.
5. Follow then **steps 4–7** as described in Subheading 3.1.

4 Notes

1. *Examples of typical properties of various reference lipids/oils and of stripped oils*: Table 1 summarizes typical interfacial tension values and tocopherol contents for different lipid phases, with an emphasis on vegetable oils.
2. *Useful remarks*: as can be seen from the last two lines of Table 1, the quality of adsorbent materials can decrease after a given storage period. In fact, it can be noticed that alumina loses its adsorbing properties toward tocopherols (only ~80% of tocopherols removed for an alumina bottle opened for >1 year, versus almost full removal for a freshly opened bottle). This can be solved by regenerating alumina, which can be done by heating (see conditions advised by the manufacturer). Nevertheless, this points out the importance of checking the quality of the stripped oil with regard to the important features targeted (interfacial tension and/or the removal of tocopherols or of other lipophilic components) for every batch.
3. *Safety*: Although the adsorbent materials used for the present procedure are not classified for specific hazards, please note that because of the very fine size of the powder particles and their use in a dry state, they may easily be suspended in the air. We

thus recommend that **steps 1** and **2** are conducted under a fume hood.

Acknowledgments

Anja Schröder, Jack Yang, Sten ten Klooster, and Jilu Feng (Wageningen University & Research, the Netherlands) are gratefully acknowledged for kindly providing experimental values measured on stripped oils.

References

1. Abousalham A, Fotiadu F, Buono G, Verger R (2000) Surface properties of unsaturated non-oxidized and oxidized free fatty acids spread as monomolecular films at an argon/water interface. Chem Phys Lipids 104:93–99
2. Berton-Carabin CC, Schröder A, Rovalino Cordova A et al (2016) Protein and lipid oxidation affect the viscoelasticity of whey protein layers at the oil-water interface. Eur J Lipid Sci Technol 118:1630–1643. https://doi.org/10.1002/ejlt.201600066
3. Chaiyasit W, Elias RJ, McClements DJ, Decker EA (2007) Role of physical structures in bulk oils on lipid oxidation. Crit Rev Food Sci Nutr 47:299–317. https://doi.org/10.1080/10408390600754248
4. Hoppenreijs LJG, Berton-Carabin CC, Dubbelboer A, Hennebelle M (2021) Evaluation of oxygen partial pressure, temperature and stripping of antioxidants for accelerated shelf-life testing of oil blends using 1H NMR. Food Res Int 147:110555. https://doi.org/10.1016/j.foodres.2021.110555
5. Bergfreund J, Bertsch P, Fischer P (2021) Effect of the hydrophobic phase on interfacial phenomena of surfactants, proteins, and particles at fluid interfaces. Curr Opin Colloid Interface Sci 101509. https://doi.org/10.1016/j.cocis.2021.101509
6. Boon CS, Xu Z, Yue X et al (2008) Factors affecting lycopene oxidation in oil-in-water emulsions. J Agric Food Chem 56:1408–1414. https://doi.org/10.1021/jf072929+
7. Villiere A, Viau M, Bronnec I et al (2005) Oxidative stability of bovine serum albumin- and sodium caseinate-stabilized emulsions depends on metal availability. J Agric Food Chem 53:1514–1520. https://doi.org/10.1021/jf0486951
8. Maldonado-Valderrama J, Woodward NC, Patrick Gunning A et al (2008) Interfacial characterization of β-lactoglobulin networks: Displacement by bile salts. Langmuir 24:6759–6767. https://doi.org/10.1021/la800551u
9. Berton C, Genot C, Ropers M-H (2011) Quantification of unadsorbed protein and surfactant emulsifiers in oil-in-water emulsions. J Colloid Interface Sci 354:739–748
10. Mishra K, Bergfreund J, Bertsch P et al (2020) Crystallization-induced network formation of tri- and monopalmitin at the middle-chain triglyceride oil/air interface. Langmuir 36:7566–7572. https://doi.org/10.1021/acs.langmuir.0c01195
11. Hinderink E, Sagis L, Schroën K, Berton-Carabin CC (2020) Behavior of plant-dairy protein blends at air-water and oil-water interfaces. Colloids Surfaces B Biointerfaces 192:111015. https://doi.org/10.1016/j.colsurfb.2020.111015
12. Abad A, Shahidi F (2020) A robust stripping method for the removal of minor components from edible oils. Food Prod Process Nutr 2:1–9. https://doi.org/10.1186/s43014-019-0015-2
13. Bergfreund J, Bertsch P, Kuster S, Fischer P (2018) Effect of oil hydrophobicity on the adsorption and rheology of β -lactoglobulin at oil–water interfaces. Langmuir 34:4929–4936. https://doi.org/10.1021/acs.langmuir.8b00458
14. Gaonkar AG (1989) Interfacial tensions of vegetable oil/water systems: Effect of oil purification. J Am Oil Chem Soc 66:1090–1092
15. Johansson D, Bergenståhl B (1995) Lecithins in oil-continuous emulsions. Fat crystal wetting and interfacial tension. J Am Oil Chem Soc 72:205–211. https://doi.org/10.1007/BF02638901
16. Berton C (2011) Construction raisonnée d'interfaces pour protéger les lipides émulsionnés contre l'oxydation. Université de Nantes
17. Schröder A, Laguerre M, Tenon M et al (2021) Natural particles can armor emulsions against lipid oxidation and coalescence. Food Chem 347:129003. https://doi.org/10.1016/j.foodchem.2021.129003

Part II

Analysis of Lipid Composition

Chapter 5

Preparation of Fatty Acid Derivatives for their Further Analysis and Quantification by Gas Chromatography

Michelle Viau and Anne Meynier

Abstract

Fatty acids are the major component of lipids, and the physical, chemical, and physiological properties of a lipid class depend primarily on its fatty acid composition. The fatty acid composition is determined as the methyl esters of fatty acids by gas chromatography. There are literally dozens of methods for preparing fatty acid derivatives. The generation of fatty acid methyl esters (FAMEs) can be done in acidic or alkaline conditions. The main procedures generally used will be presented in this chapter. The choice of the (trans)-methylation method is highly dependent on the lipid classes present in the sample and on the fatty acid's nature, i.e., the chain length, the double-bound configuration, and the unsaturation degree. As no single method is suitable for all types of lipids, neither could be use in any circumstances, the objective of this chapter is to give rational information to guide the user in the choice of the appropriate methods for obtaining an accurate fatty acid composition.

For further information, see W.W. Christie, in Advances in Lipid Methodology—Two, pp. 69–111 (1993) (Ed. W.W. Christie, Oily Press, Dundee).

Key words Fatty acid composition, FAMEs

1 Introduction

Gas chromatography (GC) is the routine procedure for the analysis of long-chain fatty acids (C > 12) and this topic was largely reviewed [1–5]. However, separating carboxylated compounds by GC is complicated due to their relatively high polarity and low volatility. Therefore, nonpolar derivatives of fatty acids, more volatile than the corresponding free form, were prepared and analyzed. In this sense, fatty acid methyl esters derivatives of fatty acids (FAMEs) are used almost universally for GC analysis. However, a case can be made for isopropyl or butyl esters in some circumstances (dairy products, for example, or lipids that contain short-chain fatty acids). Esterification of free fatty acids (FFA) and transesterification of esterified lipids generally occur in a one-step reaction in the presence of acid catalysts (FFA and esterified lipids) or basic

Christelle Lopez et al. (eds.), *Multidimensional Characterization of Dietary Lipids*, Methods and Protocols in Food Science,
https://doi.org/10.1007/978-1-0716-3758-6_5,

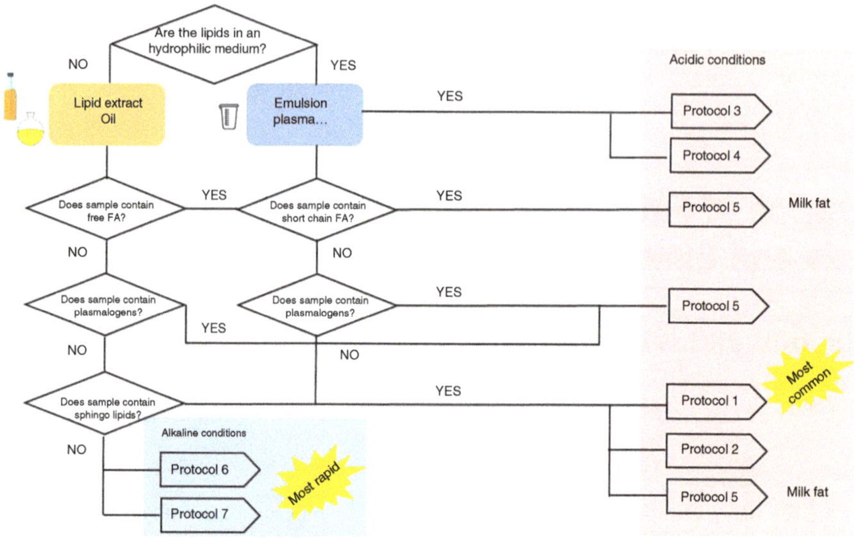

Fig. 1 Decision tree to select the appropriate protocol

catalysts (esterified lipids only) and an alcoholic phase in excess, generally methanol. There is no need to hydrolyze lipids before the derivation step to obtain the FFA before preparing esters, as most lipids can be transesterified directly. The huge majority of dietary lipids consist of triacylglycerols and phospholipids; thus, the main derivation procedure is transesterification.

One of the first considerations when deciding which procedure to adopt for preparing ester derivatives for GC is the lipid composition of the samples to be analyzed (Fig. 1). If these are FFA alone, mixtures containing significant amounts of free acids, or mixtures of unknown composition that might contain free acids, then acid-catalyzed procedures are generally preferred. FFAs are esterified and O-acyl lipids transesterified by heating them with a large excess of anhydrous methanol in the presence of an acidic catalyst (Eq. 1). The literature describes a multitude of methods using various acid catalysts. The mildest acidic reagent is 5% (w/v) anhydrous hydrogen chloride in methanol, but its transmethylation capacity is low, thus requiring reaction times higher than 30 min [6]. The most common is the method using boron trifluoride (BF_3) as a catalyst. BF_3 in methanol (12–14%, w/v) has revealed a high methylation power requiring a short time period to react.

Equation 1

1. Free fatty acids in media with acid pH. The reaction is called esterification or methylation (with methanol)

$$\mathrm{RCOOH} + \mathrm{CH_3OH} \rightarrow (+H^+) \rightarrow \mathrm{RCOOCH_3} + \mathrm{H_2O}$$

2. Fatty acids included in complex lipids by O-acyl bonds. The reaction is called transesterification or transmethylation (with methanol)

$$RCOOR' + CH_3OH \rightarrow (+H^+) \rightarrow RCOOCH_3 + R'OH$$

The preparation of FAMEs using methanol sulfuric acid (1–2%, v/v) is described and it transesterifies lipids in the same manner and rate similar to that of methanolic HCl. H_2SO_4 is a very corrosive viscous liquid and must be handled with care.

During the derivatization in the presence of basic catalysts, the fatty esters form an anionic intermediate, which is transformed into a new ester in the presence of a large excess of alcohol. O-acyl lipids are transesterified rapidly in anhydrous methanol in the presence of a basic catalyst (Eq. 2). Free fatty acids are not sensitive to nucleophilic attack by alcohols or bases and thus are not esterified in these conditions.

$$RCOOR' + CH_3OH \rightarrow (+OH^-) \rightarrow RCOOCH_3 + R'OH \quad (2)$$

Potassium hydroxide, sodium hydroxide, and sodium methoxide are the most common catalyst used in basic catalysis. Compared to acid catalysis, basic catalysis has the advantages of speed and mild heating conditions. Thus, this type of catalysis is recommended in samples with short-chain fatty acids or labile fatty acids (polyunsaturated, cyclopropane rings, conjugated unsaturation, etc.). However, alkali catalyst reactions require strict anhydrous conditions, which may be difficult to fulfill in the case of hydrated samples (emulsions). They must be considered for mixed lipid samples that contain little or no unesterified fatty acids or for single lipid classes containing ester-bound fatty acids.

The most useful basic transesterifying agents are 0.5–2 M sodium or potassium methoxide in anhydrous methanol, prepared by dissolving the clean metals in anhydrous methanol (the reaction is strongly exothermic).

If a single method is required for use with all routine lipid samples, BF_3–methanol, despite the comparatively long reaction times needed, is probably the best general-purpose reagent available. It is convenient to have more than one method for routine use in the laboratory, i.e., BF3–methanol for samples containing free fatty acids and sodium methoxide in methanol for all other lipid classes except sphingolipids for which special procedures are in any case necessary (see protocols 1, 2, and 5).

A further consideration when deciding which procedure will be applied is the fatty acid composition of the samples to be esterified. If short-chain fatty acids (protocol 5) or others of unusual structure (protocol 8 for plasmalogens) are present, appropriate methods must be adopted. Polyunsaturated fatty acids must always be handled with care and should not be subjected to more vigorous conditions than are necessary. It should be recognized that reagents can be destructive to fatty acids if used carelessly.

Because of the high sensitivity of chromatographic analysis procedures, small amounts of lipids (usually less than 1 mg) are required, and the procedures described below are on this scale. The efficiency of a selected catalyst/procedure should be verified (e.g., by thin-layer chromatography) to ensure the complete transmethylation of fatty acids (*see* Chapter 7).

2 Materials

2.1 Chemicals

2.2 Mandatory Equipment

1. Analytical precision balance, sensitivity: 0.01 mg.
2. Positive displacement pipettes and disposable capillary piston tips (1000 μL, 100 μL).
3. Pasteur pipette.
4. Magnetic stirrer and bar magnets.
5. Vortex mixer.
6. Centrifuge.
7. Water bath or thermoblock reactor for 10 mL tubes.
8. Nitrogen generator or nitrogen cylinder.
9. 10 mL screw-capped (Teflon-lined) tubes.
10. 50 mL glass bottle.
11. Weighing shoes.

3 Preparation of Derivatives of Fatty Acids.

3.1 Acid-Catalyzed Esterification and Transesterification

3.1.1 Protocol 1: Methylation and Ttrans-methylation with BF_3 Methanol

Boron trifluoride (BF_3) in methanol (12–14%, w/v) has been much used as a transesterification catalyst and in particular as a rapid means of esterifying free fatty acids [7].

Note that this protocol allows the derivatization of fatty aldehydes, but for accurate determination, protocol 8 will be preferred. This reagent is quickly destroyed in the presence of water and therefore *must be applied to water-free samples* (Table 2).

Reagents:

Cyclohexane, toluene, boron trifluoride (14%)–methanol solution, chloroform, and heptadecanoic acid (Fig. 2).

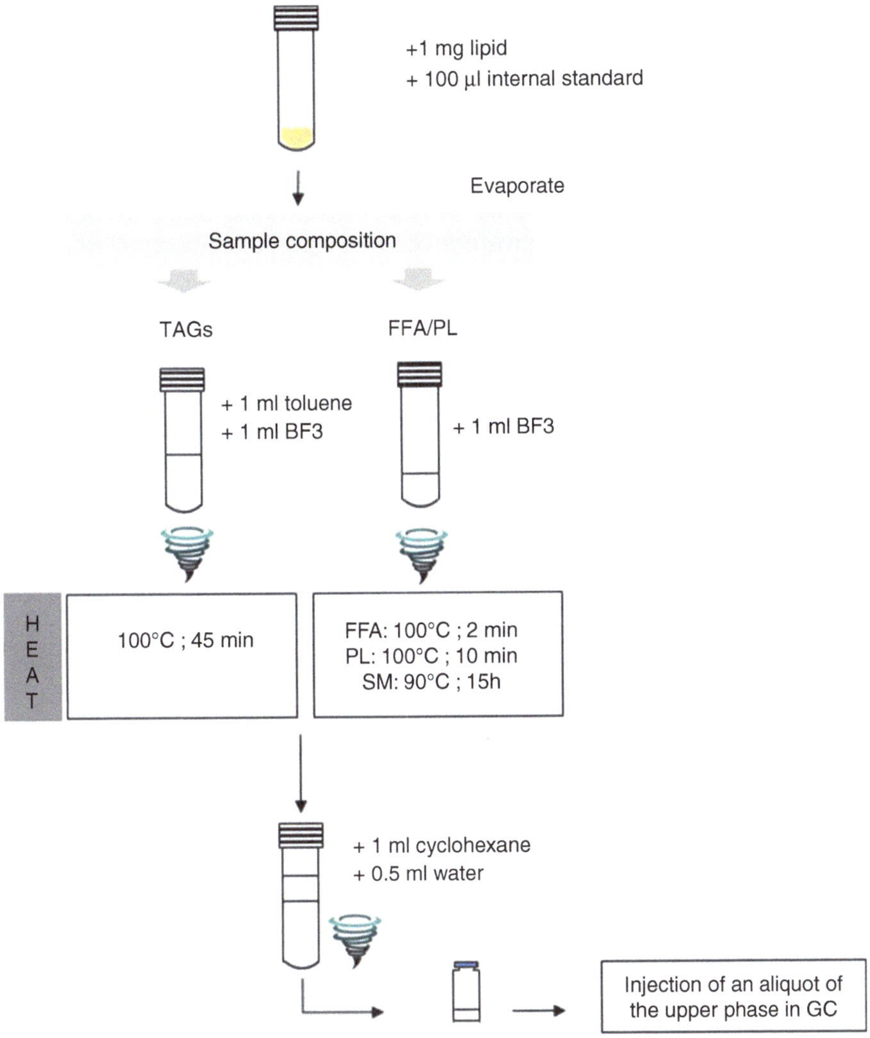

Fig. 2 Schematic diagram of procedure 1

Derivatization Procedure

1. If quantification is required, prepare an internal standard solution. Accurately weigh (± 0.1 mg) approximately 25 mg of $C_{17:0}$ into a 25 mL volumetric flask and make it to volume with chloroform.
2. Dissolve the lipid extract or the oil in chloroform or any solvent of the lipids that can be easily evaporated (e.g., 10 mg/mL).
3. Add around 1 mg of lipids in a 10 mL capped glass tube.
4. If quantification is required, add 100 µL (= 100 µg) of C_{17} as an internal standard (10% of total lipids).
5. Evaporate under nitrogen.

6. If triacylglycerols or sterol esters are analyzed alone or are abundant in the extract, add 1 mL of toluene to solubilize them.
7. Add 1 mL of the BF_3 reagent.
8. Close the tube under nitrogen.
9. Incubate the samples at time and temperature depending on the composition of the lipids as indicated in Table 1. Note particular attention to samples containing sphingomyelins which require to be incubated 15 h at 90 °C.
10. When quantitative analysis is required, take care that no leak occurs during the heating step (especially at the beginning).
11. To extract the FAMEs, add 1 mL of cyclohexane and 0.5 mL of water.
12. Vortex the tube.
13. Centrifuge 5 min at 1000 × g.
14. Transfer a clear portion of the top layer with a Pasteur pipette into a 2 mL autosampler vial (without any trace of the lower layer). The sample is ready for chromatographic separation and quantification.

Notes

The BF_3–methanol reagent has a limited shelf life, even when refrigerated, and the use of old solutions can result in the production of artifacts and losses of appreciable amounts of polyunsaturated fatty acids. Keep the BF_3–methanol reagent refrigerated under nitrogen and discard it after 3 months or when solids appear at the bottom of the flask.

The proportion of BF_3 (10–14%) and brand did not significantly affect the qualitative and quantitative results on common vegetable or fish oil.

3.1.2 Protocols 2 and 3: Methylation and Transmethylation with HCl Methanol

Preparation of fatty acid methyl esters from various sources using commercial aqueous HCl was also described [8]. Yields of FAME are the same as those obtained with the boron trifluoride method. Furthermore, the reagent is very convenient, safe, and cheap.

Protocol 2

Reagents:

Cyclohexane, toluene, methanol, HCl (37%, w/w), chloroform, and heptadecanoic acid (Fig. 3).

Derivatization Procedure

1. If quantification is required, prepare a standard solution. Accurately weigh (±0.1 mg) approximately 25 mg of $C_{17:0}$ into a 25 mL volumetric flask and make to volume with toluene.
2. Prepare a reagent solution from 9.7 mL commercial concentrated HCl (37%, w/w) diluted with 41.5 mL of methanol to make 50 mL of 8% (w/v) HCl. The reagent solution is stored at

Table 1
Chemical products

Chemical	Chemical formula	CAS number	Molecular weight (g/mol)	Quality	Safety instructions	Protocol number
Boron trifluoride-methanol solution	BF_3CH_4O	373-57-9	99.85	14% in methanol		1
Cyclohexane	C_6H_{12}	110-82-7	84.16	Pesti-S		1; 2; 3; 4
Toluene	C_7H_8	108-88-3	92.14	For GC		1; 2; 3; 6
Chloroform	$CHCl_3$	67-66-3	119.38	For GC		1; 2; 3
Acetone	C_3H_6O	67-64-1	58.08	For GC		4
Ethanol	C_2H_5OH	64-17-5	46.07	99% absolute		4
Methanol	CH_4O	67-56-1	32.04	For GC		4;7
Potassium hydroxide	KOH	1310-58-3	56,11	Reagent grade, 90%, flakes		7
Sulfuric acid	H_2SO_4	7664-93-9	98.08	95–97%		4; 5

(continued)

Table 1 (continued)

Chemical	Chemical formula	CAS number	Molecular weight (g/mol)	Quality	Safety instructions	Protocol number
Glacial acetic acid	$C_2H_4O_2$	64-19-7	60.05	≥99.0%		6
Heptadecanoic acid	$C_{17}H_{34}O_2$	506-12-7	270.45	≥99.0%		1; 2; 3; 4;5
Triheptadecanoin	$C_{54}H_{104}O_6$	2438-40-6	849.4	≥99.0%		6; 7
Sodium methoxide	CH_3ONa	124-41-4	54.02	0.5 M, for GC derivatization		6
Methanolic HCl (3 N)	CH_4O –HCl	7647-01-0	36.46	for GC derivatization		3
2,6-Di-tert-butyl-p-cresol	$C_{15}H_{24}O$	128-37-0	220.35	≥99.0%		3
Hydrochloric acid	HCl	7647-01-0	36.46	ACS reagent,37%		2
Hexane	C_6H_{14}	110-54-3	86.18	For GC >95%		5; 6;7
Isopropanol	C_3H_8O	67-63-0	60.10	ACS reagent, ≥99.5%		5

Table 2
Conditions for the preparation of methyl esters from various lipids with BF_3

Class of Lipid	Reagent (mL) BF$_3$-MeOH	Reagent (mL) Toluene	Reaction time (min)	Temperature °C
Total Fat	1	1	45	100
Neutral lipids	1	1	45	100
Free Fatty acids	1		2	100
Polar lipids	1		10	100
Sphingomyelins	1		900	90

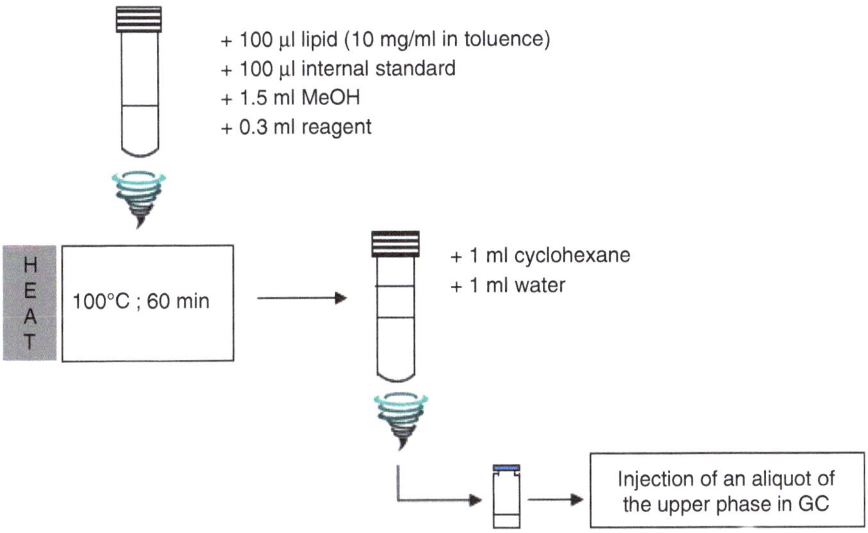

Fig. 3 Schematic diagram of procedure 2

4–7 °C. Commercial products of methanolic HCl are available. However, anhydrous methanolic HCl is unstable and may contain considerable amount of water that are probably formed during storage.

3. Dissolve the lipid extract or the oil in toluene to obtain a 10 mg/mL concentration. For example, accurately weigh (±0.1 mg) approximately 50 mg lipid into a 5 mL volumetric flask and make to volume with toluene.
4. Add 100 µL (about 1 mg of lipids) in a 10 mL capped glass tube.
5. Add 100 µL (= 100 µg) of the internal standard (10% of total lipids).

6. Add 1.5 mL methanol.
7. Add 0.3 mL of the reagent solution.
8. Close the tube.
9. Shake the sample for 30 s and heat it at 100 °C for 60 min. The reaction time at 100 °C was extended up to 1.5 h for samples containing sterol esters.
10. Cool the tube to room temperature and then add 1 mL of water and 1 mL of cyclohexane.
11. Shake the tube for 30 s.
12. Transfer a clear portion of the top layer with a transfer pipette into a 2 mL autosampler vial (without any trace of the lower layer).

Protocol 3

Protocol 3 is another direct transmethylation procedure without prior lipid extraction. This protocol is particularly suitable for samples with low lipid content or small sample sizes available. It can be easily applied to accurately determine fatty acids (C6–C24) in emulsions, plasma samples, and digesta.

Reagents

Cyclohexane, toluene, methanolic HCl 3 N, 2,6-di-tert-butyl-p-cresol, chloroform, and heptadecanoic acid (Fig. 4).

Derivatization Procedure

The total FA from 100 μL (for lipid content around 150–200 μg) of the liquid sample as digesta, plasma, etc. is derived in situ into methyl esters, avoiding initial extraction and drying steps. The fatty

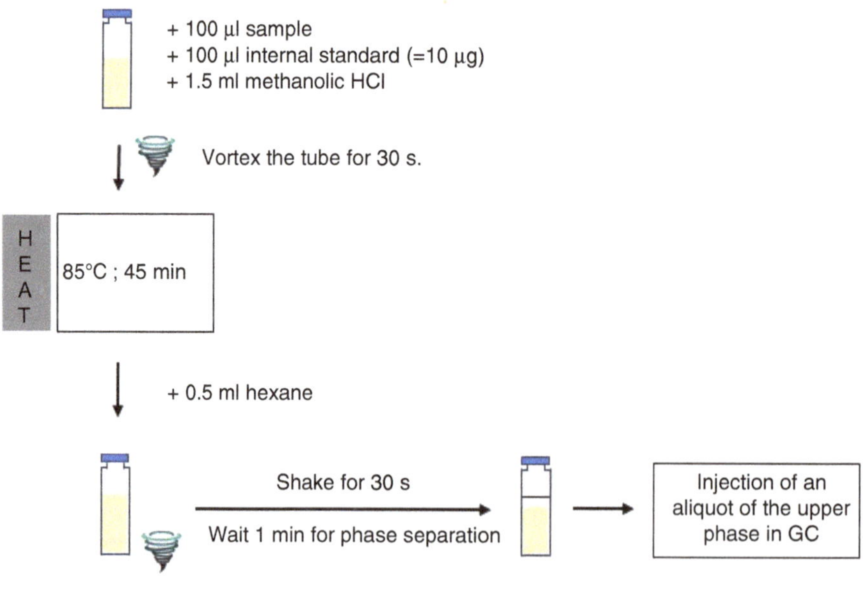

Fig. 4 Schematic diagram of procedure 3

acid methyl esters are extracted once and analyzed by gas chromatography. For details, see ref. [9]:

1. Add 100 μL of sample into a closed glass tube.
2. Prepare internal standard solution: accurately weigh (±0.1 mg) approximately 10 mg of $C_{17:0}$ into a 100 mL volumetric flask and make up to volume with methanol/chloroform (70/30; v/v).
3. To prevent FA oxidation, prepare methanolic HCl 3N and hexane each containing 2 g/L 2,6-di-tert-butyl-*p*-cresol.
4. Add 100 μL of internal standard (=10 μg) and 1.5 mL methanolic HCl 3N, containing 2 g/L 2,6-di-tert-butyl-*p*-cresol.
5. Shake the sample for 30 s, close the tube, and heat to 85 °C for 45 min.
6. Cool the tube to room temperature, add 0.5 mL hexane (containing 2 g/L 2,6-di-tert-butyl-p-cresol), and shake the tubes for 30 s.
7. Let it stand for approximately 1 min at room temperature to allow phase separation.
8. Transfer an aliquot of the upper hexane phase into a 2 mL vial for storage and further GC analysis.

3.1.3 Protocol 4: Methylation and Transmethylation with H_2SO_4 Methanol (Fig. 5)

The following protocol is another direct transmethylation without prior extraction for hydrophilic samples using H_2SO_4 as an acid catalyst [10].

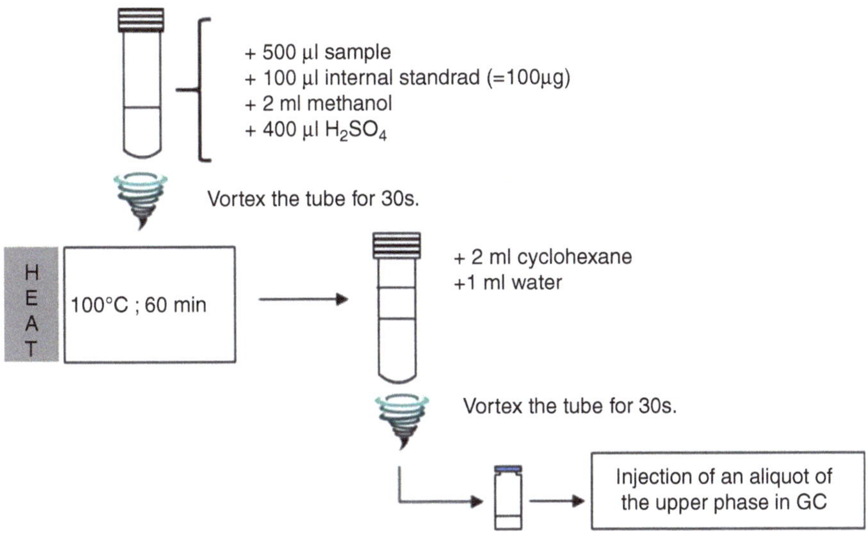

Fig. 5 Schematic diagram of procedure 4

Reagents

Acetone, heptadecanoic acid, methanol, H_2SO_4 (95–97%), and cyclohexane.

Derivatization Procedure.

1. Add 500 μL (≈1 mg of lipid) of the sample in a glass-capped tube.
2. Prepare the internal standard solution: accurately weigh (±0.1 mg) approximately 50 mg of $C_{17:0}$ into a 50 mL volumetric flask and make to volume with acetone/methanol (2/1).
3. Add 100 μL of internal standard (=100 μg).
4. Add 2 mL methanol.
5. Add slowly 400 μL H_2SO_4, take care to avoid splashing.
6. Shake the sample for 30 s.
7. Heat to 100 °C for 60 min.
8. Cool to room temperature, add 2 mL cyclohexane and 1 mL water.
9. Shake the tubes for 30 s.
10. Centrifuge (2 min, 1000× g, 20 °C).
11. Transfer an aliquot of the upper cyclohexane phase into 2 mL vials for storage and gas chromatographic (GC) analysis.

3.1.4 Protocol 5: Preparation of Fatty Acid Isopropyl Esters (FAIPE)

This procedure is recommended to analyze short-chain fatty acids (milk fat, coconut oil, etc.) because quantitative recovery of the methyl esters can be difficult due to their high volatility and partial solubility in water (<C8). The best esterification procedure has no aqueous extraction or solvent removal steps. Wolff (1987) first

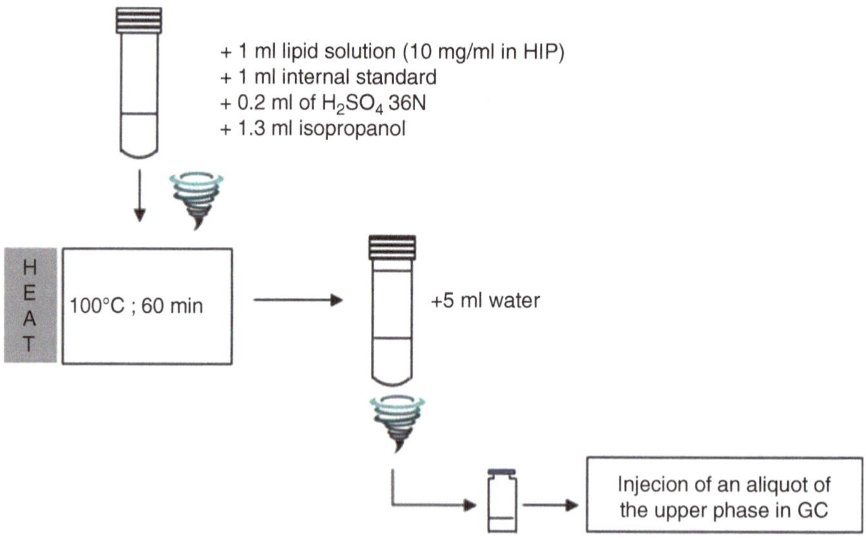

Fig. 6 Schematic diagram of procedure 5

explored the use of propan-2-ol esters as part of a one-step procedure for extracting and simultaneously transesterifying lipids (Fig. 6).

Reagents

Hexane, isopropanol, sulfuric acid (95–97%), and heptadecanoic acid.

Derivatization Procedure

1. If quantification is required, prepare a standard solution. Accurately weigh (±0.1 mg) approximately 25 mg of $C_{17:0}$ (IS) into a 25 mL volumetric flask and make it to volume with hexane/isopropanol (3/2; v/v).
2. Dissolve the lipid extract or oil to obtain a concentration of 10 mg/mL in hexane/isopropanol (3/2; v/v).
3. Add 1 mL (10 mg of lipids) into a 10 mL capped glass tube.
4. Add 1 mL (= 1 mg) of C_{17} internal standard (10% of total lipids).
5. Add 0.2 mL of H_2SO_4 36N and 1.3 mL of isopropanol.
6. Close the tube under nitrogen.
7. Vortex the tube (5s × 2).
8. Incubate samples for 1 h at 100 °C.
9. Cool the tubes under tap water to room temperature.
10. Add 5 mL of water.
11. Vortex vigorously the tubes.
12. After standing for about 1 min, withdraw 2 mL of the upper layer containing FAIPE with a transfer pipette. To obtain a 1 mg/ml solution dilute by 5 into a 2 mL autosampler vial. The sample is ready for chromatographic separation and quantification.

3.2 Base-Catalyzed Transesterification

3.2.1 Protocol 6: Methylation and Transmethylation with Sodium Methoxide

Sodium methoxide in methanol is a valuable reagent for the rapid transesterification of fatty acids linked by ester bonds to alcohols (e.g., cholesterol, glycerol). It will not esterify free fatty acids or transesterified amide-bound fatty acids such as in sphingolipids. Also, unlike acidic catalysts, it will not liberate aldehydes from plasmalogens (Fig. 7).

Reagents

Hexane, sodium methoxide, toluene, acetic acid glacial, and triheptadecanoin.

Derivatization Procedure

1. If quantification is required, prepare a standard solution. Accurately weigh (± 0.1 mg) approximately 25 mg of TAG $_{C17}$ into a 25 mL volumetric flask and make it to volume with toluene.

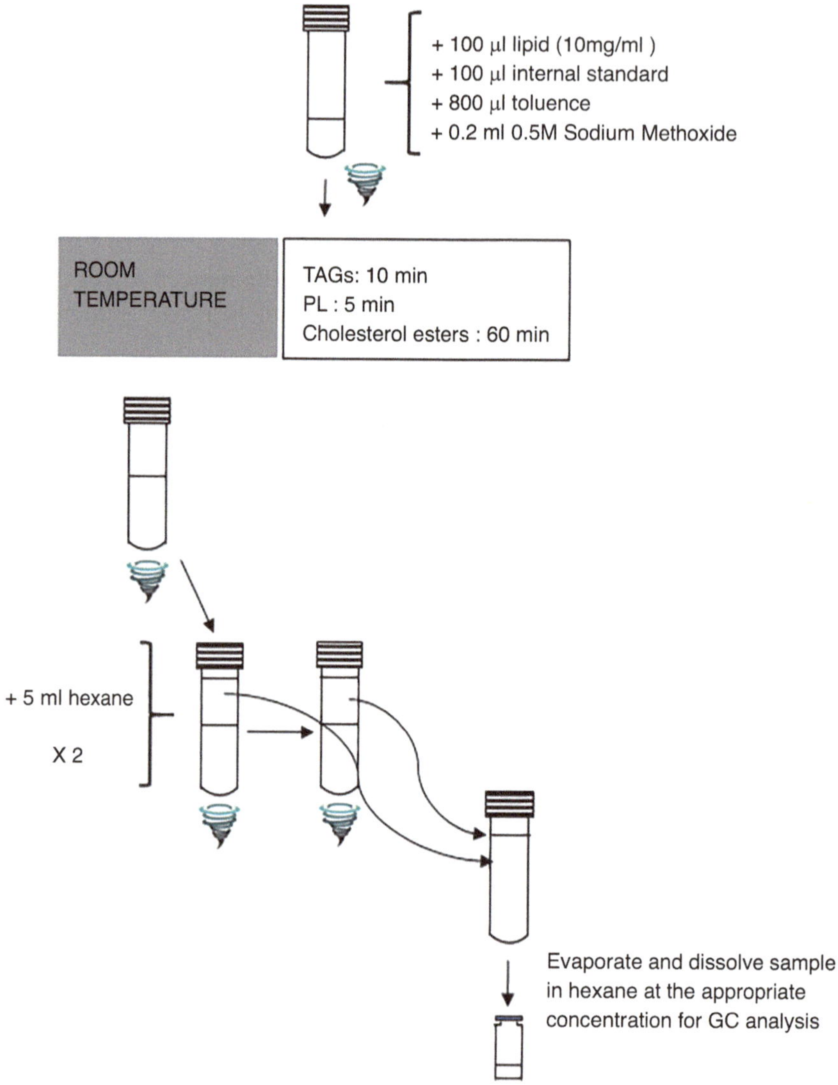

Fig. 7 Schematic diagram of procedure 6

2. Dissolve the lipid extract or oil to obtain a concentration of 10 mg/mL in toluene.
3. Add 100 μL (1 mg of lipids) into a 10 mL capped glass tube.
4. Add 100 μL (= 100 μg) of C_{17} internal standard (10% of total lipids).
5. Add 800 μL of toluene (final volume = 1 mL toluene).
6. Add 2 mL of 0.5 M sodium methoxide in anhydrous methanol, *see* **Note 1**.
7. Close the tube under nitrogen.
8. Vortex the tube (5s × 2).

Table 3
Conditions for preparation of FAMEs from lipid classes using basic-catalyst

Lipids	Heating time (min)
Triacylglycerols	10
Glycerophospholipids	5
Cholesterol esters	60

9. Incubate samples at room temperature for the time indicated in Table 3 for the respective lipid classes.
10. Cool the tubes under tap water to room temperature.
11. To neutralize the sodium methoxide and to minimize the risk of hydrolysis occurring, add 0.1 mL of glacial acetic acid followed by 5 mL of water.
12. Vortex vigorously the tubes.
13. Extract the required esters into hexane (add 5 mL) and repeat one time.
14. Vortex vigorously the tubes.
15. After standing for about 1 min, withdraw the upper layer containing esters with a transfer pipette.
16. Remove the solvent under a rotary evaporator and dissolve the sample in hexane at the appropriate concentration for GC analysis.

Notes

1. The reagent is stable for several months at 4 °C until a white precipitate of bicarbonate salt is formed.

3.2.2 Protocol 7: Methylation and Transmethylation with 2 M Methanolic KOH

An improved rapid procedure to analyze fatty acid esters from triacylglycerols and phospholipids is described below by Ichihara et al. [11]. In this protocol, the transesterification is complete in a matter of minutes which is the primary advantage of the method. It must be pointed out that sterol esters and waxes do not react under these conditions (Fig. 8).

Reagents

Hexane, 2 M methanolic KOH, and tri-heptadecanoin.

Derivatization Procedure

1. Prepare 2 M methanolic KOH. Potassium hydroxide can be dissolved in methanol if you have time to wait, if not, you can add a few drops of water. Weigh 1.122 g of KOH into a small beaker. Then add a few drops of distilled/deionized water and dissolve all the KOH by gentle shaking. Pour the solution into a 10 mL volumetric flask. Rinse carefully the beaker with methanol to ensure that all the KOH was transferred. Fill the

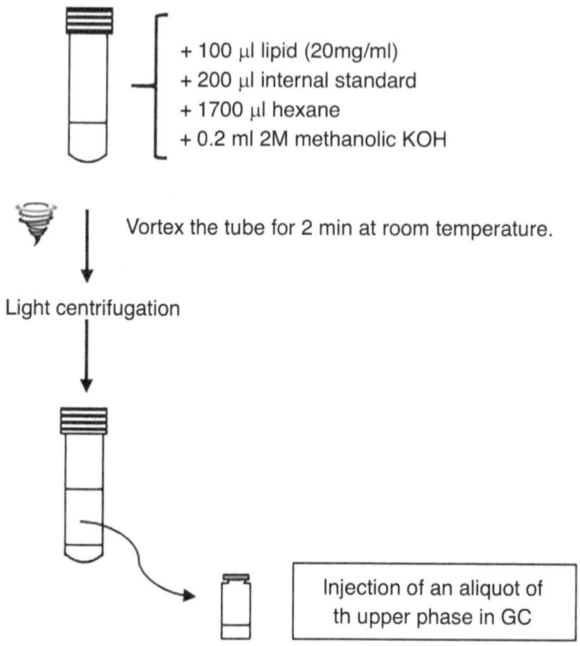

Fig. 8 Schematic diagram of procedure 7

flask to about 80–90% as the solution might contract or expand still. Let the bottle stand until the solution equilibrates to room temperature. Then, carefully adjust the volume by adding methanol until the bottom of the meniscus. This solution is not stable and it is recommended to prepare it every week (*see* **Note 1**).

2. If quantification is required, prepare a standard solution. Accurately weigh (±0.1 mg) approximately 25 mg of TAG $_{C17}$ (IS) into a 25 mL volumetric flask and make up to volume with hexane.
3. Dissolve the lipid extract or the oil to obtain a concentration of 20 mg/mL in hexane.
4. Introduce 100 μL (2 mg of lipids) in a 10 mL capped glass tube.
5. Add 200 μL (= 200 μg) of IS (10% of total lipids).
6. Add 1700 μL of hexane (final volume = 2 mL hexane).
7. Add 0.2 mL of 2 M methanolic KOH.
8. Vortex the tube for 2 min at room temperature.
9. Centrifuge (2 min, 1000 × g) and collect an aliquot of the hexane layer for GC analysis.

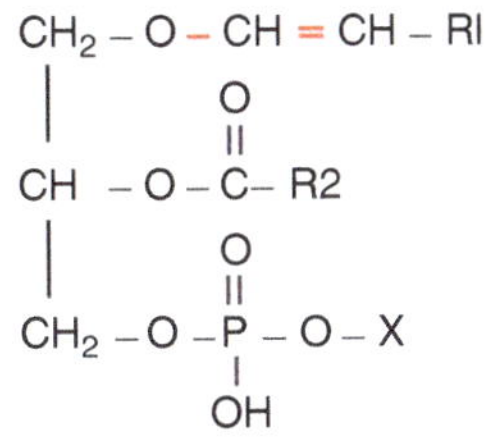

Fig. 9 The simultaneous compositions of fatty aldehydes and acids can be achieved after the appropriate derivatization of phospholipids with hydrochloric acid and 2,2-dimethoxypropane in methanol. The aldehydes originating from the plasmalogens are thus derivatized into dimethylacetals (DMA) while fatty acids are derivatized into FAME

Notes

1. Artifact formation is minimized and the shelf life of the reagent is improved if oxygen-free methanol is used in its preparation.

3.3 Protocol 8: A Special Case, Dimethylacetals (DMA)

Application field: Measurement of plasmalogens and associated fatty acids (mainly polyunsaturated fatty acids).

Plasmalogens are essential components of various membranes and structures such as the heart, brain, and muscle. Plasmalogens are a class of membrane glycerophospholipids containing a fatty aldehyde with a vinyl-ether bond at the sn-1 position and enriched in polyunsaturated fatty acids at the sn-2 position of the glycerol backbone (Fig. 9).

A procedure to analyze fatty acid esters and plasmalogens from phospholipids is described below [12] (Fig. 10).

3.3.1 Reagents

Hexane, methanol, HCl 37%, and 2,2-dimethoxypropane.

3.3.2 Derivatization Procedure

1. Prepare the reagent solution: methanol/HCl/2.2 dimethoxypropane (100/10/4 v/v/v)).
2. If quantification is required, prepare a standard solution. Accurately weigh (±0.1 mg) approximately 25 mg of $C_{17:0}$ (IS) into a 25 mL volumetric flask and make it to volume with chloroform.
3. Dissolve the lipid sample in chloroform (e.g., 10 mg/mL).
4. Add around 1 mg of lipids in a 10 mL capped tube.
5. Add 100 μL (= 100 μg) of the internal standard (10% of total lipids).
6. Evaporate under nitrogen.
7. Add 5 mL of reagent solution.
8. Close the tube.
9. Vortex the tube.
10. Incubate the samples for 4 h at 65 °C.

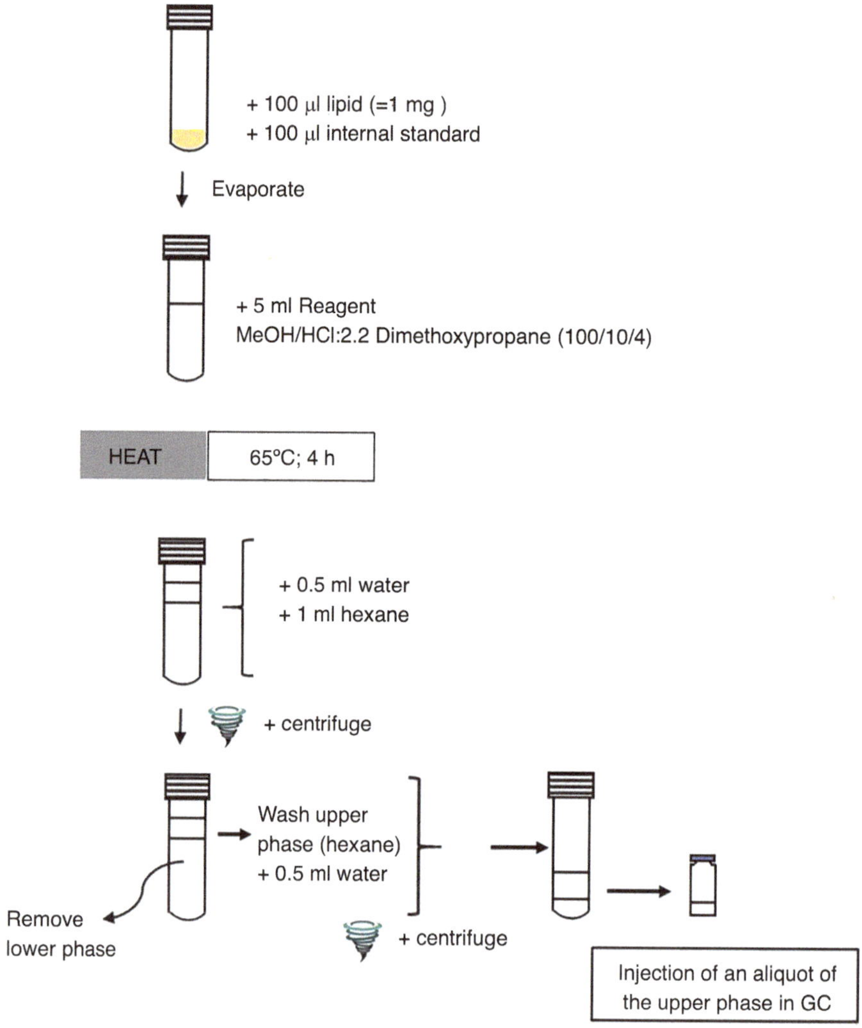

Fig. 10 Schematic diagram of procedure 8

11. Cool the tubes under tap water to room temperature.
12. Add 0.5 mL of water and 1 mL of hexane.
13. Vortex the tubes.
14. Centrifuge 5 min at 1000 × g.
15. Remove the lower layer with Pasteur pipette and add 0.5 mL of water.
16. Vortex and centrifuge.
17. Repeat the washing phase of the hexane layer once.
18. Transfer a clear portion of the top layer with a transfer pipette into a 2 mL autosampler vial.

References

1. Cyberlipid. http://cyberlipid.gerli.com/
2. Carrapiso AI, García C (2000) Development in lipid analysis: Some new extraction techniques and in situ transesterification. Lipids 35:1167–1177. https://doi.org/10.1007/s11745-000-0633-8
3. Christie WW (1993) Advances in lipid methodology-Two. Oily Press, Dundee
4. Perkins EG (1993) Analyses of fats, oils and derivatives. American Oils Chemist's Society, Champaign
5. Ruiz-Rodriguez A, Reglero G, Ibañez E (2010) Recent trends in the advanced analysis of bioactive fatty acids. J Pharm Biomed Anal 51:305–326. https://doi.org/10.1016/j.jpba.2009.05.012
6. Stoffel W, Chu F, Ahrens EH (1959) Analysis of long-chain fatty acids by gas-liquid chromatography: micromethod for preparation of methyl esters. Anal Chem 31:307–308. https://doi.org/10.1021/ac60146a047
7. Morrison WR, Smith LM (1964) Preparation of fatty acid methyl esters and dimethylacetals from lipids. J Lipid Res 5:600–608. https://doi.org/10.1016/S0022-2275(20)40190-7
8. Ichihara K, Fukubayashi Y (2010) Preparation of fatty acid methyl esters for gas-liquid chromatography. J Lipid Res 51:635–640. https://doi.org/10.1194/jlr.D001065
9. Glaser C, Demmelmair H, Koletzko B (2010) High-throughput analysis of total plasma fatty acid composition with direct in situ transesterification. PLoS One 5. https://doi.org/10.1371/journal.pone.0012045
10. Berton C, Genot C, Ropers MH (2011) Quantification of unadsorbed protein and surfactant emulsifiers in oil-in-water emulsions. J Colloid Interface Sci 354:739–748. https://doi.org/10.1016/j.jcis.2010.11.055
11. Ichihara K, Shibahara A, Yamamoto K, Nakayama T (1996) Erratum: An improved method for rapid analysis of the fatty acids of glycerolipids (Lipids) (May 1996) 31:5 (535-539). Lipids 31:889. https://doi.org/10.1007/bf02522986
12. Berry JF, Cevallos WH, Wade RR (1965) Lipid class and fatty acid composition of intact peripheral nerve and during wallerian degeneration. J Am Oil Chem Soc 42:492–500. https://doi.org/10.1007/BF02540090

Chapter 6

Gas Chromatography Analysis of Fatty Acid Derivatives

Michelle Viau and Anne Meynier

Abstract

The analysis of fatty acid methyl esters (FAMEs), derived from food, is a very important characterization procedure. Capillary gas chromatography is the traditionally used technique for the analysis of FAMEs in a wide number of research areas. FAME analysis demands high chromatographic resolution, especially to provide evidence of positional and geometrical isomers of unsaturated fatty acids in complex mixtures. To meet these requirements, polar stationary phases such as polyethylene glycols or cyanopropyl silicones are normally used for the separation of complex FAME mixtures since they enable separation according to unsaturation and carbon number. In recent reviews, the esterification methods, injection techniques, and analytical columns, as well as the analysis and identification of fatty acid isomers, have been described in detail (Gutnikov, J Chromatogr B Biomed Sci Appl 671:71–89, 1995; Seppänen-Laakso et al., Anal Chim Acta 465:39–62, 2002; Ruiz-Rodriguez et al., J Pharm Biomed Anal 51:305–326, 2010). Nowadays, gas chromatography (GC) analysis of fatty acids (as methyl esters) in oils (fish oils and vegetable oils) and fat (e.g., milk) is a well-established technique (AOCS, Off Method Ce 1e-91:1–4; Int Stand ISO 15304, 2003).

For routine fatty acid analysis, a medium polar cyanopropyl DB-225 column provides excellent separation for complex FAME mixtures. This application is described in this chapter.

Key words Gas Chromatography, Fatty acid composition, FAMEs

1 Introduction

The analysis of FAMEs is used for the characterization of the lipid fraction in foods and is one of the most important analyses for food. Most edible fats and oils contain mainly fatty acids ranging from lauric acid (C12:0) to arachidic acid (C20:0). Besides the linear saturated fatty acids, branched fatty acids, mono-unsaturated, di-unsaturated, and higher unsaturated fatty acids can also occur (Tables 1 and 2).

After the preparation of the FAMEs (see Chapter 5), they are separated according to the carbon number (number of carbon atoms in the fatty acid chain, excluding the methyl ester carbon) and the degree of unsaturation. Moreover, the position of the

Christelle Lopez et al. (eds.), *Multidimensional Characterization of Dietary Lipids*, Methods and Protocols in Food Science,
https://doi.org/10.1007/978-1-0716-3758-6_6,

Table 1
Systematic and common name of fatty acids

No.	Fatty acid	Cas-Number	Common name	Abbreviation
1	Butanoic acid	107-92-6	Butyric acid	C4: 0
2	Hexanoic acid	142-62-1	Caproic acid	C6: 0
3	Octanoic acid	124-07-2	Caprylic acid	C8: 0
4	Decanoic acid	334-48-5	Capric acid	C10: 0
5	Undecanoic acid	112-37-8		C11: 0
6	Dodecanoic acid	143-07-7	Lauric acid	C12: 0
7	Tridecanoic acid	638-53-9		C13: 0
8	Tetradecanoic acid	544-63-8	Myristic acid	C14: 0
9	Tetradecenoic acid	544-64-9	Myristoleic acid	C14: 1n-5
10	Pentadecanoic acid	1002-84-2		C15: 0
11	Pentadecenoic acid	84743-29-3		C15: 1
12	Hexadecanoic acid	57-10-3	Palmitic acid	C16: 0
13	Hexadecenoic acid	373-49-9	Palmitoleic acid	C16: 1
14	Heptadecanoic Acid	506-12-7	Margaric acid	C17: 0
15	Heptadecenoic acid	75190-82-8		C17: 1
16	Octadecanoic acid	57-11-4	Stearic acid	C18: 0
17	*trans*-9-Octadecenoic acid	112-79-8	Elaidic acid	C18: 1n-9 t
18	*cis*-9-Octadecenoic acid	112-80-1	Oleic acid	C18: 1n-9
19	*cis-11*-Octadecenoic acid	506-17-2	Vaccenic acid	C18: 1n-7
20	*trans*-9,12-Octadecadienoic acid	506-21-8	Linolelaidic acid	C18: 2n-6 t
21	*cis*-9,12-Octadecadienoic acid	60-33-3	Linoleic acid (LA)	C18: 2n-6
22	cis-6,9,12-Octadecatrienoic acid	506-26-3	γ-Linolenic acid (GLA)	C18: 3n-6
23	*cis*-9,12,15-Octadecatrienoic acid	463-40-1	α-Linolenic acid (ALA)	C18: 3n-3
24	*cis*-6,9,12,15-Octadecatetraenoic acid	20290-75-9	Stearidonic acid (SDA)	C18: 4n-3
25	Eicosanoic acid	506-30-9	Arachidic acid	C20: 0
26	*cis*-11-Eicosenoic acid	5561-99-9		C20: 1n-9
27	*cis*-11,14- Eicosadienoic acid	5598-38-9		C20: 2n-6
28	cis-8,11,14- Eicosatrienoic acid	1783-84-2	Dihomo-γ-linolenic acid	C20: 3n-6
29	*cis*-5,8,11,14- Eicosatetraenoic acid	506-32-1	Arachidonic acid	C20: 4n-6
30	Heneicosanoic acid	2363-71-5	HEA	C21: 0
31	*cis*-11,14,17- Eicosatrienoic acid	17046-59-2		C20: 3n-3

(continued)

Table 1
(continued)

No.	Fatty acid	Cas-Number	Common name	Abbreviation
32	*cis*-8,11,14,17-Eicosatetraenoic acid	24880-40-8	w-3-Arachidonic acid	C20: 4n-3
33	*cis*-5,8,11,14,17-Eicosapentenoic acid	10417-94-4	EPA	C20: 5n-3
34	Docosanoic acid	112-85-6	Behenic acid	C22: 0
35	*cis*-13-Docosenoic acid	112-86-7	Erucic acid	C22: 1n-9
36	*cis*-13,16-Docosadienoic acid	17735-98-7		C22: 2n-6
37	Tricosanoic Acid	2433-96-7		C23: 0
38	*cis*-7,10,13,16-Docosatetraenoic acid	28874-58-0	DTA	C22: 4n-6
39	*cis*-7,10,13,16,19-docosapentaenoic acid	24880-45-3	DPA	C22: 5n-3
40	*cis*-4,7,10,13,16,19-Docosahexaenoic acid	6217-54-5	DHA	C22: 6n-3
41	Tetracosanoic acid	557-59-5	Lignoceric acid	C24: 0
42	*cis*-15-Tetracosenoic acid	506-37-6	Nervonic acid	C24: 1n-9

double bonds and the geometric configuration (*cis/trans*) are also important parameters and their determination adds additional information to the characterization of the lipid fraction in food. Separation of FAMEs can be carried out with different types of capillary columns, polar and highly polar stationary phases. The polarity of the stationary phase influences the retention times of FAMEs, especially those of polyunsaturated types. The column should be chosen according to the sample to be analyzed and is summarized in Fig. 1.

The separation of FAMEs on polyethylene glycol columns is widely used and is applied to the characterization of classical samples, such as vegetable oils and animal fat. For the analysis of complex samples, such as fish oils, additional resolution is needed and is obtained using a capillary column coated with a cyanopropyl stationary phase, such as a DB-23, or cyanopropylphenyl, such as DB-225. On these columns, highly unsaturated fatty acids such as EPA and DHA are separated from other FAMEs. Separation of the *cis*- and *trans*-isomers is also possible. Due to the stronger interaction of the *cis*-isomer with the cyano-dipole, the *trans*-isomers elute before the *cis*-isomers. However, the polarity of the stationary phase is not sufficient to fully separate *cis–trans* mixtures. So, for a complex FAME mixture with a large amount of *trans*-fatty acids, a highly polar HP-88 column is generally preferred [6].

However, very polar phases have a shorter lifetime than polar phases. Because the DB-225 column provides sufficient separation of FAMEs for simple and complex mixtures, we recommend a 30 m column of DB225 type for routine fatty acid analysis and only this application is described below.

Table 2
Molecular weight of fatty acid and methyl esters

No.	Abbreviation	Formula FA	Molecular weight FA	Formula FAME	Molecular weight FAME
1	C4: 0	$C_4H_8O_2$	88.11	$C_5H_{10}O_2$	102.13
2	C6: 0	$C_6H_{12}O_2$	116.16	$C_7H_{14}O_2$	130.19
3	C8: 0	$C_8H_{16}O_2$	144.21	$C_9H_{18}O_2$	158.24
4	C10: 0	$C_{10}H_{20}O_2$	172.26	$C_{11}H_{22}O_2$	186.29
5	C11: 0	$C_{11}H_{22}O_2$	186.29	$C_{12}H_{24}O_2$	200.32
6	C12: 0	$C_{12}H_{24}O_2$	200.32	$C_{13}H_{26}O_2$	214.34
7	C13: 0	$C_{13}H_{26}O_2$	241.35	$Cl_4H_{28}O_2$	228.37
8	C14: 0	$C_{14}H_{28}O_2$	228.37	$C_{15}H_{30}O_2$	242.40
9	C14: 1n-5	$C_{14}H_{26}O_2$	226.36	$C_{15}H_{28}O_2$	240.38
10	C15: 0	$C_{15}H_{30}O_2$	242.40	$C_{16}H_{32}O_2$	256.42
11	C15: 1	$C_{15}H_{28}O_2$	240.38	$C_{16}H_{30}O$	254.41
12	C16: 0	$C_{16}H_{32}O_2$	256.40	$C_{17}H_{34}O_2$	270.45
13	C16: 1	$C_{16}H_{30}O_2$	254.41	$C_{17}H_{32}O_2$	268.40
14	C17: 0	$C_{17}H_{34}O_2$	270.45	$C_{18}H_{36}O_2$	284.50
15	C17: 1	$C_{17}H_{32}O_2$	268.40	$C_{18}H_{34}O_2$	282.50
16	C18: 0	$C_{18}H_{36}O_2$	284.48	$C_{19}H_{38}O_2$	298.50
17	C18: 1n-9t	$C_{18}H_{34}O_2$	282.47	$C_{19}H_{36}O_2$	296.49
18	C18: 1n-9	$C_{18}H_{34}O_2$	282.47	$C_{19}H_{36}O_2$	296.49
19	C18: 1n-7	$C_{18}H_{34}O_2$	282.47	$C_{19}H_{36}O_2$	296.49
20	C18: 2n-6t	$C_{18}H_{32}O_2$	280.40	$C_{19}H_{34}O_2$	294.50
21	C18: 2n-6	$C_{18}H_{32}O_2$	280.40	$C_{19}H_{34}O_2$	294.50
22	C18: 3n-6	$C_{18}H_{30}O_2$	278.40	$C_{19}H_{32}O_2$	292.50
23	C18: 3n-3	$C_{18}H_{30}O_2$	278.40	$C_{19}H_{32}O_2$	292.50
24	C18: 4n-3	$C_{18}H_{28}O_2$	276.40	$C_{19}H_{30}O_2$	290.50
25	C20: 0	$C_{20}H_{40}O_2$	312.53	$C_{21}H_{42}O_2$	326.60
26	C20: 1n-9	$C_{20}H_{38}O_2$	310.51	$C_{21}H_{40}O_2$	324.50
27	C20: 2n-6	$C_{20}H_{36}O_2$	308.50	$C_{21}H_{38}O_2$	322.53
28	C20: 3n-6	$C_{20}H_{34}O_2$	306.50	$C_{21}H_{36}O_2$	320.51
29	C20: 4n-6	$C_{20}H_{32}O_2$	304.50	$C_{21}H_{34}O_2$	318.50
30	C21: 0	$C_{21}H_{42}O_2$	326.56	$C_{22}H_{44}O_2$	340.60
31	C20: 3n-3	$C_{20}H_{34}O_2$	306.50	$C_{21}H_{36}O_2$	320.51

(continued)

Table 2 (continued)

No.	Abbreviation	Formula FA	Molecular weight FA	Formula FAME	Molecular weight FAME
32	C20: 4n-3	$C_{20}H_{32}O_2$	304.50	$C_{21}H_{34}O_2$	318.50
33	C20: 5n-3	$C_{20}H_{30}O_2$	302.45	$C_{21}H_{32}O_2$	316.50
34	C22: 0	$C_{22}H_{44}O_2$	340.58	$C_{23}H_{46}O_2$	354.60
35	C22: 1n-9	$C_{22}H_{42}O_2$	338.57	$C_{23}H_{44}O_2$	352.59
36	C22: 2n-6	$C_{22}H_{40}O_2$	336.60	$C_{23}H_{42}O_2$	350.60
37	C23: 0	$C_{23}H_{46}O_2$	354.61	$C_{24}H_{48}O_2$	368.60
38	C22: 4n-6	$C_{22}H_{36}O_2$	332.50	$C_{23}H_{38}O_2$	346.60
39	C22: 5n-3	$C_{22}H_{34}O_2$	330.50	$C_{23}H_{36}O_2$	344.53
40	C22: 6n-3	$C_{22}H_{32}O_2$	328.49	$C_{23}H_{34}O_2$	342.51
41	C24: 0	$C_{24}H_{48}O_2$	368.63	$C_{25}H_{50}O_2$	382.70
42	C24: 1n-9	$C_{24}H_{46}O_2$	366.62	$C_{25}H_{48}O_2$	380.70

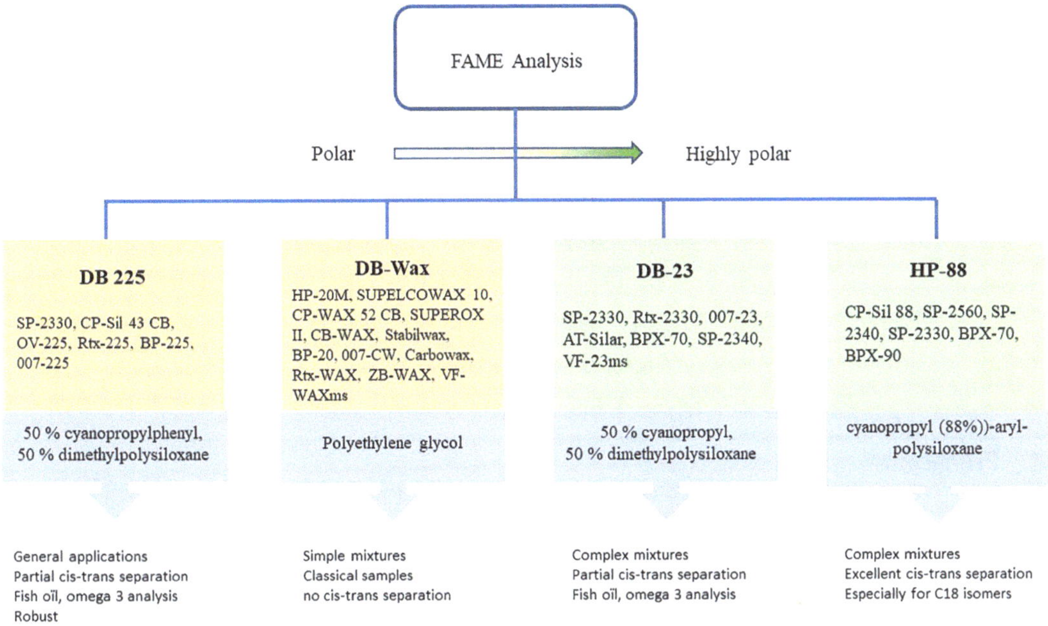

Fig. 1 Overview of column selection for FAMEs analysis

Analytical conditions must be adapted from published values to obtain reliable and precise results. The temperature gradient program is the main parameter to be modified according to the nature of the column and the complexity of the fatty acid mixture.

2 Mandatory Equipment

1. Gas chromatograph with split/spitless capillary injection system and flame-ionization detector (FID).
2. Carrier gas (hydrogen or helium), dried air, and hydrogen for FID.
3. The capillary GLC column should be a flexible fused silica, 30 m or more in length and 0.25–0.32 mm i.d. coated with DB 225 ((cyanopropylphenyl (50%) and dimethylpolysiloxane (50%)) or an equivalent phase. The upper temperature limit of the column should be noted and verified.

3 Reagents

1. Gases.
2. Carrier gas, helium or hydrogen minimum purity 99.995%, dried, $H_20 < 1$ ppm, and $O_2 < 0{,}5$ ppm.
3. FID, hydrogen, min purity 99.995%; air dry and hydrocarbon-free.
4. Reference standards. A mixture of methyl esters, preferably similar to that of the fatty acid matter to be analyzed. Reference mixtures simulating most fats and oils may be obtained from Supelco Inc. (Supelco Park, Bellefonte, PA, USA. Nu-Check-Prep, Inc., P.O. Box 295 Elysian, MN 56028, USA. Larodan, PO Box 156, Monroe, MI 48161, USA)
5. F.A.M.E. Mix, C4-C24 (Supelco, 18919-1AMP).
6. 37 Component FAME Mix (Supelco, CRM47885).
7. Qualmix Fish (Larodan, 89-5550).
8. GLC reference standards (Nu-Check-Prep, GLC37).

4 Procedure

1. Condition a new column. To get the best performance, be sure to condition 6–8 h (or overnight) at the isothermal temperature limit (240 °C for a DB-225).
2. Set up the gas chromatograph:
   (a) Inlet: 250 °C, split/spitless mode, split ratio = 25:1.
   (b) Carrier gas: Hydrogen, 2 ml/min, constant flow mode.
   (c) Detector FID: 250 °C, Hydrogen 40 mL/min, Air 400 mL/min (can be adjusted according to the manufacturer specifications).

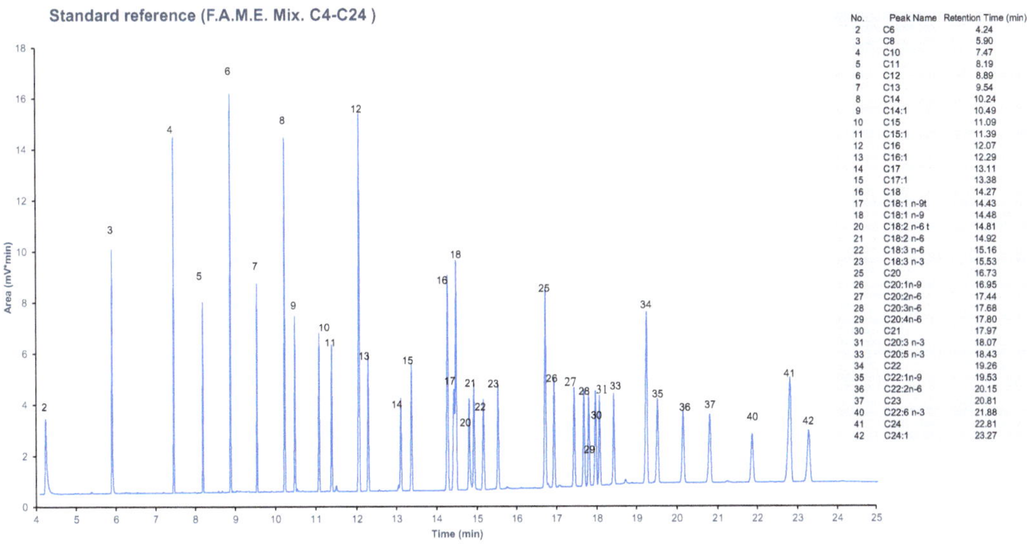

Fig. 2 Typical chromatogram of FAME standard mix C4-C24) on a DB225 column

(d) Oven temperature profile: 50 °C (1 min), 180 °C@ 15 °C/min (1 min), 220 °C @ 5 °C/min (10 min).

3. Prepare the methyl esters according to Chapter 5.
4. Inject 1 μL of the methyl esters from the reference standard into the gas chromatograph. Identify each component from its retention time (Figs. 2 and 3). *See* **Note 1**.
5. Inject 1 μL of the methyl esters from the sample into the gas chromatograph. Make the proper assignments for the peaks by comparing the retention times with those of standards, using the chromatogram shown in Fig. 2 as a guideline.

4.1 Expression of Result and Quantitative Estimation of Fatty Acid

When flame ionization detectors are used and high precision is required, small correction factors can be applied to compensate for the fact that the carboxyl carbon atom in each ester is not ionized appreciably during combustion (Table 3), [7], *see* **Note 2**.

Results are frequently expressed as weight percent (area normalization), this is currently used in nutrition works, and the use of an internal standard (generally C17 or C19) is recommended.

1. Without Internet standard

   Calculate the relative percentage of each component X_i by determining the corrected area of the corresponding peak ($A_i \times TRF_i$) to the sum of the corrected areas of all the peaks ($\Sigma Ai \times TRFi$), as follows:

   Area percent of the component i expressed as percent of the total fatty acid methyl esters:

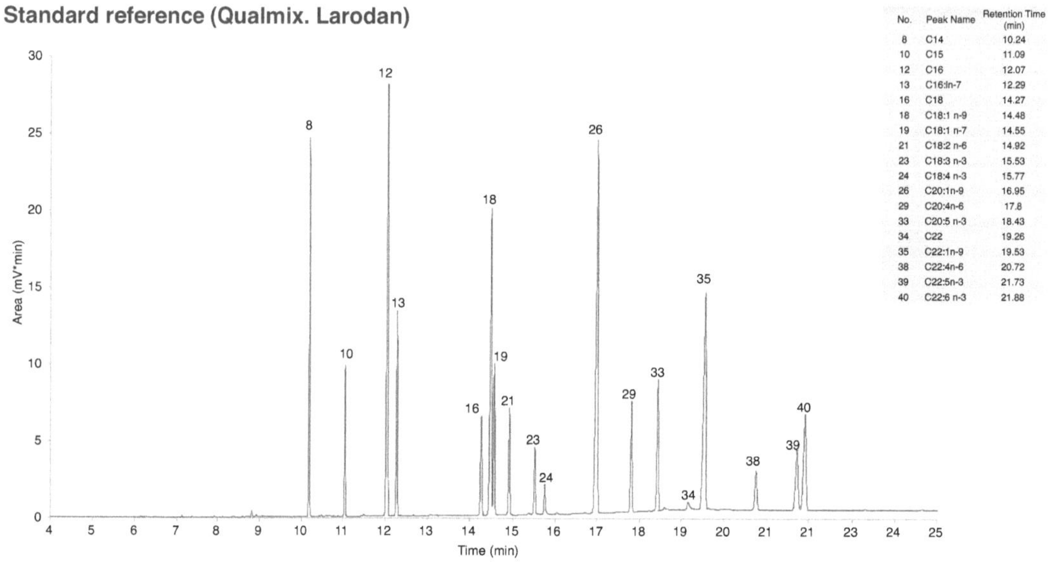

Fig. 3 Typical chromatogram of FAME standard (Qualmix Larodan) on a DB225 column

Table 3
Theoretical relative FID response corrections (TRFs) for FAMEs

FAME	TRF	MWF	FAME	TRF	MWF	FAME	TRF	MWF
4:0	1.5396	2.647	*16:0*	1.0193	1.000	*20:2*	0.9724	0.838
5:0	1.4009	2.328	*16:1*	1.0117	1.007	*20:3*	0.9663	0.843
6:0	1.3084	2.077	*16:2*	1.0041	1.015	*20:4*	0.9603	0.849
7:0	1.2423	1.875	*16:3*	0.9965	1.023	*20:5*	0.9452	0.854
8:0	1.1927	1.709	*16:4*	0.9889	1.031	*21:0*	0.9780	0.794
9:0	1.1542	1.570	*17:0*	1.0091	0.951	*22:0*	0.9720	0.763
10:0	1.1233	1.452	*17:1*	1.0019	0.957	*22:1*	0.9664	0.767
11:0	1.0981	1.350	*18:0*	1.0000	0.906	*22:2*	0.9609	0.771
12:0	1.0771	1.262	*18:1*	0.9932	0.912	*22:3*	0.9554	0.776
12:1	1.0670	1.274	*18:2*	0.9865	0.918	*22:4*	0.9499	0.780
13:0	1.0593	1.184	*18:3*	0.9797	0.925	*22:5*	0.9443	0.784
14:0	1.0440	1.116	*18:4*	0.9730	0.931	*22:6*	0.9388	0.789
14:1	1.0354	1.125	*19:0*	0.9919	0.865	*23:0*	0.9665	0.734
15:0	1.0308	1.055	*20:0*	0.9846	0.827	*24:0*	0.9614	0.707
15:1	1.0227	1.055	*20:1*	0.9785	0.833	*24:1*	0.9564	0.711

$$Xi\% = \frac{Ai \times TRFi}{\Sigma Ai \times TRFi} \times 100$$

where

- Ai = area of the peak corresponding to component i.
- TRF_i = theoretical relative FID response corrections (Table 3).
- $\Sigma A_i \times TRF_i$ = sum of the corrected areas under all the peaks.

2. With Internet standard

   Calculate the amount of each component with an area normalization method with unity response factors for each component as follows:

   Calculate the area percent for fatty acids by the formula:

$$\text{Area\%fatty acids} = \frac{Ai \times TRFi}{\Sigma Ai \times TRFi - (AIS \times TRFIS)} \times 100$$

where

- A_i = area of the peak corresponding to component *i*.
- TRF_i = theoretical relative FID response corrections (Table 3).
- A_{IS} = area of the peak corresponding to internal standard.
- TRF_{IS} = internal standard theoretical relative FID response correction.
- $\Sigma A_i \times TRF_i$ = sum of the corrected areas under all the peaks.

Calculation of mg of FAs per g of sample:

$$\text{Corrected area} = A_{ic} = A_i \times TRF_i$$

where

- Ai = area of the peak corresponding to component i
- $TRFi$ = theoretical relative FID response corrections (Table 3)

$$WFAi\ (\mu g) = \frac{WIS \text{ x } Aic}{A\ ISC}$$

where

- A_{ic} = corrected area of the peak corresponding to component i
- A_{ISC} = corrected area of the peak corresponding to internal standard
- W_{IS} = mass of internal standard added to the sample in μg

$$WFAi\ (mg/g) = \frac{WFAi}{WS}$$

where

- W_{FAi} = mass of FAi in μg.
- W_S = mass of sample in mg.

The theoretical FID correction factors of Table 3, calculated by Craske and Bannon [8], were included in the AOAC International method and also in the AOCS [4, 5, 9].

Another alternative way to express results is to calculate the molar amounts of each acid. This is performed simply by multiplying the area of each peak by a molecular weight factor (MWF), obtained by dividing the weight of a selected standard ester (say 16: 0) by the molecular weight of the component, followed by re-normalizing to 100% (see Table 3).

5 Notes

1. If possible, identification must be confirmed by other investigations such as GC-MS.
2. The theoretical detector response factors do not always work for short-chain fatty acids, and it appears that empirical response factors are necessary in the analysis of milk fats. Careful calibration with standards (purity>99) is then necessary.

References

1. Gutnikov G (1995) Fatty acid profiles of lipid samples. J Chromatogr B Biomed Sci Appl 671: 71–89. https://doi.org/10.1016/0378-4347(95)00116-Z
2. Seppänen-Laakso T, Laakso I, Hiltunen R (2002) Analysis of fatty acids by gas chromatography, and its relevance to research on health and nutrition. Anal Chim Acta 465:39–62
3. Ruiz-Rodriguez A, Reglero G, Ibañez E (2010) Recent trends in the advanced analysis of bioactive fatty acids. J Pharm Biomed Anal 51:305–326. https://doi.org/10.1016/j.jpba.2009.05.012
4. AOCS Détermination of Fatty Acids in Edible Oils and Fats by Capillary GLC. Off Method Ce 1e-91:1–4
5. (2003) Animal and vegetable fats and oils-Determination of the content of trans fatty acid isomers of vegetable fats and oils-Gas chromatographic method. Int Stand ISO 15304
6. AOCS Official Method Ce 1c-89 (1997) Fatty Acid Composition by GLC cis and trans Isomers. Off Methods Recomm Pract Am Oil Chem Soc 2–5
7. Ackman RG (1972) The analysis of fatty acids and related materials by gas-liquid chromatography. Prog Chem Fats Other Lipids 12:165–284. https://doi.org/10.1016/0079-6832(72)90003-1
8. Craske JD, Bannon CD (1988) Letter to the editor. J Am Oil Chem Soc 65:1190–1191. https://doi.org/10.1007/BF02660581
9. (1997) AOCS Official Method Ce 1b-89. Fatty Acid Composition of Marine Oils by GLC

Chapter 7

Analysis of Neutral Lipids, Phospholipids, Glycolipids, and Total Lipids by High-Performance Thin Layer Chromatography

Brigitte Laillet, Michelle Viau, and Ha Phuong Ta

Abstract

High-performance thin layer chromatography (HPTLC) is a technique of planar chromatography that is routinely used to separate and identify lipid classes. The solutes in the samples applied on the lower edge of an HPTLC plate and carried by the mobile phase traveling through the plate are separated according to their polarity and their interactions with the stationary phase. The separated solutes are derivatized and visualized by specific or nonspecific staining reagents and finally identified and if possible quantified by comparison with standard compounds. In this chapter, we describe the first materials, chemicals, preparation of solvent mixtures (eluents), and derivatization solutions needed for HPTLC analysis of lipids. Then, we present the common steps for HPTLC analysis whatever the specific application while emphasizing lipid analysis. Finally, we detail five protocols, requiring one-step, two-step, or three-step development for the analysis of neutral lipids (2 protocols), phospholipids (1 protocol), glycolipids (1 protocol), and neutral lipids–phospholipids (1 protocol) from various food materials.

Key words HPTLC, Neutral lipids, Phospholipids, Glycolipids, Total lipids, Silica gel

1 Introduction

High-performance thin layer chromatography (HPTLC) is a planar chromatography technique to separate the compounds in a sample. The involved theoretical principles are the basic principles of chromatography, namely, the separation of compounds according to their interactions with two immiscible phases: the stationary phase and the mobile phase. The compounds trained by the mobile phase driven by capillary action travel through the stationary phase. Each compound is subjected to its own unique adsorption retention force and mobile phase entrainment force (Annex 1: Calculated eluotropic forces of solvents [1]). The resulting pseudo-equilibrium results in a differential migration of the compounds; those with higher affinity with the stationary phase move slowly,

Christelle Lopez et al. (eds.), *Multidimensional Characterization of Dietary Lipids*, Methods and Protocols in Food Science, https://doi.org/10.1007/978-1-0716-3758-6_7,

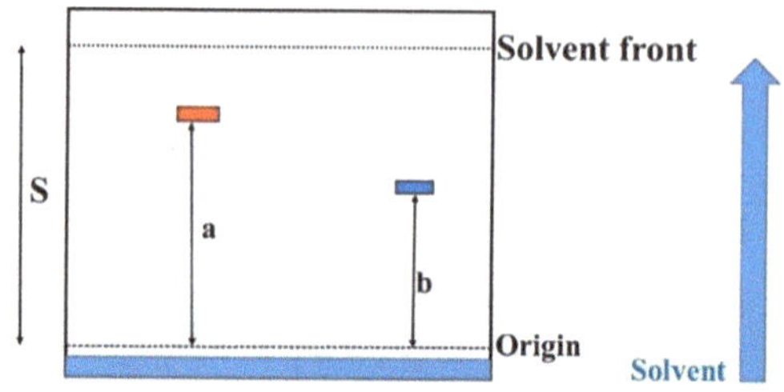

Fig. 1 Schema of HPTLC plate after development and derivatization. The distance traveled by the compounds from the origin (where the samples were put onto the plate) (**a** or **b**) divided by the total distance traveled by the solvent (S) is called the retention factor (R_f) of the compound. For red compound, $R_f = a/S$

while the others travel faster, which allows their separation. When the separation is completed, individual compounds (or groups of compounds with the same relative affinity toward the stationary and the mobile phase, i.e., lipid classes in our analysis) appear as vertically separated spots or bands (Fig. 1). Each spot has a retardation or retention factor (R_f), which is equal to the distance migrated by the aforesaid compound over the total distance covered by the solvent [2–4].

In the first instance, this chapter aims to give a general description of the application of HPTLC in the analysis of lipid classes. We will then focus on giving full details about our routine protocols for the separation and characterization of different lipid classes by HPTLC. The coupling of HPTLC to other techniques of detection such as mass spectroscopy will not be broached in this chapter. For further reading, more exhaustive references are available [5–7].

Figure 2 illustrates the general process of lipid analysis from sample extraction to the determination of lipid classes by HPTLC and, when possible, the quantification of these lipids from the HPTLC plate.

In this chapter:

- We first present the chemical materials, the preparation of solvent and derivatization solutions, and the equipment used for the HPTLC analysis.
- Then we describe a general HPTLC procedure for lipid analysis.
- Finally, we present our specific analysis of neutral lipids, phospholipids, glycolipids, and total lipids applied to different food materials.

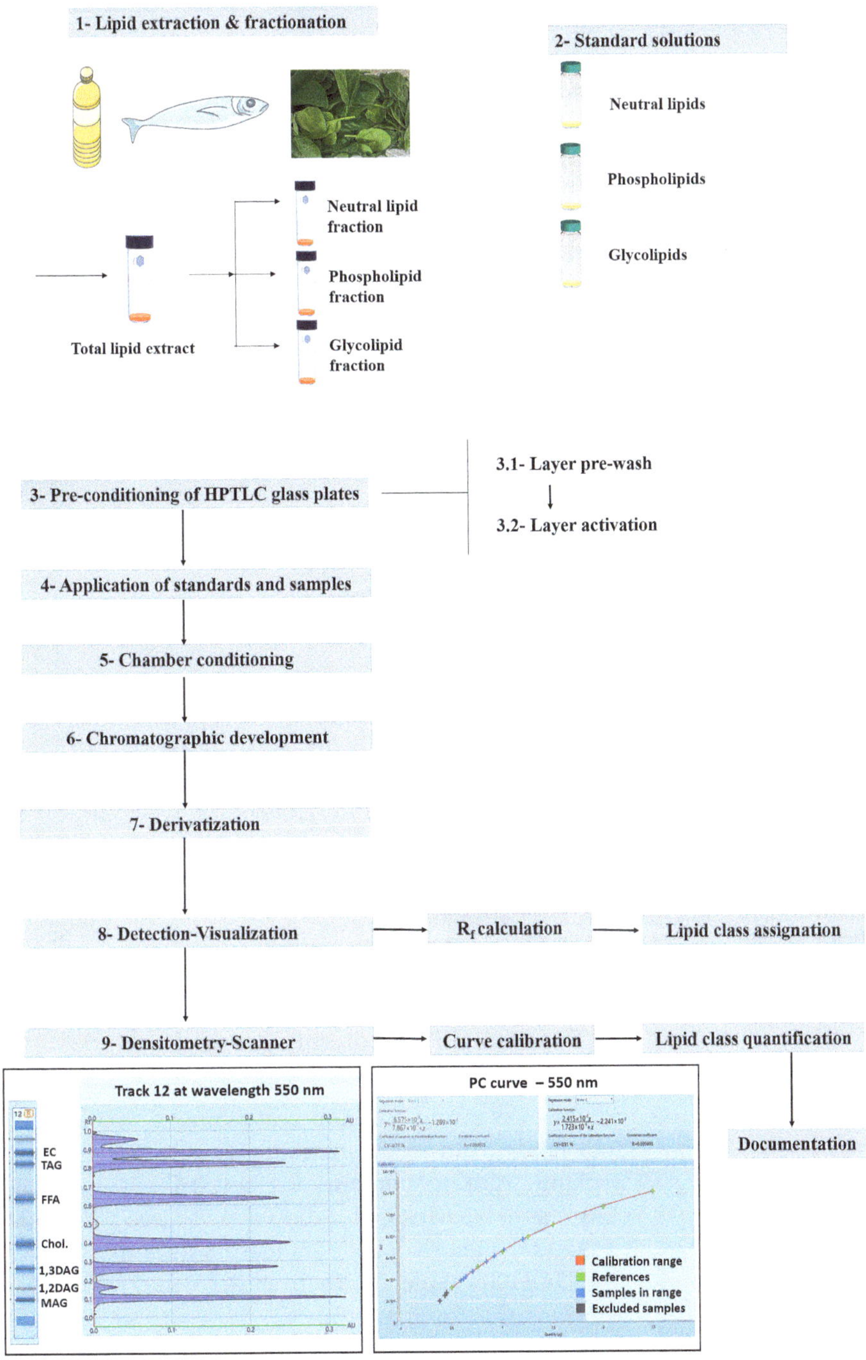

Fig. 2 Schema illustrating the general HPTLC method of lipid analysis

2 Materials

2.1 Mandatory Materials

1. Vortex agitator.
2. Glass weighs shoes.
3. Analytical precision balance, sensitivity: 0.01 mg.
4. 20 mL glass volumetric flasks.
5. 20, 50, 100, and 250 mL glass graduated cylinders.
6. Positive displacement pipettes and disposable capillary piston tips compatible with organic solvents (1 mL, 100, and 250 μL).
7. Pasteur pipettes.
8. 1.5 mL screw top vials.
9. TLC tank.
10. HPTLC plates.
11. Bottle of nitrogen gas.
12. −20 °C and −80 °C freezers.
13. Cold room at 4 °C.

2.2 Chemical Products

Water is of ultrapure quality (18.2 MΩ/cm at 25 °C). Without any further specification, the organic solvents are of HPLC grade. The chemical products other than lipid standards necessary for the analysis of lipids by HPTLC following different protocols are given in Table 1 (*see* **Notes 1** and **2**). The lipid standards and their references are given in Table 2.

2.3 Preparation of Standards and Samples

1. Prepare stock solutions of individual lipids by dissolving dry lipid standards generally at a concentration of 10 mg/mL in chloroform. The concentration of glycolipid standard solutions is 0.5 mg/mL in chloroform. The stock solutions are stored at −80 °C.
2. Prepare a working solution containing a mixture of the standard lipids from stock solutions. For instance, a neutral lipid working solution contains 0.1 mg/mL of each target lipid class, namely, MAG, 1,2-DAG, 1,3-DAG, FFA, and TAG (Table 2). The glycolipid standard solution contains MGDG and DGDG (Table 2), each at 0.25 mg/mL. The working solutions are stored at −20 °C.
3. Dissolve the lipid samples to analyze in chloroform to obtain a concentration between 0.1 and 1.5 mg/mL.

2.4 Staining Solutions

2.4.1 Common Staining Reagents

The detection of lipids is based on their visualization after binding/reacting to a dye. If the interaction is not destructive, the lipid spots can be recovered for further analysis (by mass spectrometry for example).

Table 1
Chemical products

Chemical	Chemical formula	CAS number	Molecular weight (g/mol)	Quality	Safety instructions	Protocol number
Acetone	C_3H_6O	67-64-1	58.08	99.9%		2-3-4-5
Alpha-naphthol	$C_{10}H_7OH$	90-15-3	144.17	99%		4
Ammonium hydroxyde	NH_4OH	1336-21-6	35.05	28–30%, analysis		4
Chloroform (stabilized with ethanol)	$CHCl_3$	67-66-3	119.38	99%		2-3-4-5
Copper sulfate pentahydrate	$CuSO_4.5H_2O$	7758-99-8	249.69	≥99.5%		1-2-3-5
Diethyl ether	$C_4H_{10}O$	60-29-7	74.12	For analysis		1-2-5
Ethanol absolute	C_2H_5OH	64-17-5	46.07	RPE for Analysis		3-4-5
Ethyl acetate	$C_4H_8O_2$	141-78-6	88.11	HPLC grade		3-5
Glacial acetic acid	$C_2H_4O_2$	64-19-7	60.05	≥99.0%		1-3-5
Hexane	C_6H_{14}	110-54-3	86.18	HPLC grade		1-5

(continued)

Table 1
(continued)

Chemical	Chemical formula	CAS number	Molecular weight (g/mol)	Quality	Safety instructions	Protocol number
Isopropanol	C_3H_8O	67-63-0	60.10	ACS reagent, ≥99.5%		1-2-3-4-5
Magnesium chloride hexahydrate	$MgCl_2.6H_2O$	7791-18-6	203.30	≥99%		1-4
Methanol	CH_3OH	67-56-1	32.04	99%		2-3-4-5
Methylene chloride	CH_2Cl_2	75-09-2	84.93	HPLC grade		3-5
n-Pentane	C_5H_{12}	109-66-0	72.15	HPLC grade		2
Ortho-phosphoric acid	H_3PO_4	7664-38-2	98.00	HPLC grade, 85%		2-3-4
Primuline	$C_{21}H_{14}N_3NaO_3S_3$	8064-60-6	475.54	99%		1-2-3-5
Sulfuric acid	H_2SO_4	7664-93-9	98.08	95–98%		4

Table 2
Lipid standards

Lipid classes	Type of lipid	Suggested supplier and reference	Composition	CAS number	Quality
Neutral lipid	Cholesterol	Sigma C8667	Cholesterol	57-88-5	≥99%
	Cholesterol ester	Sigma C9253	Cholesteryl oleate.	303-43-5	≥98%
	Mono-, di-, and triglyceride mix	Supelco 1787-1AMP	1,3-Diolein (1,3-DAG) 10 mg 1,2-Dioleoyl-*rac*-glycerol (1,2-DAG) 10 mg Glyceryl trioleate (TAG) 10 mg. Monoolein (MAG) 10 mg.		MQ-100
	Oleic acid	Sigma O1008	*cis*-9-Octadecenoic acid (standard for free fatty acids (FFA))	112-80-1	>99%
Phospholipids	Soy PA	Avanti 840074C	L-α-phosphatidic acid (Soy) (sodium salt)	475995-54-1	>99%
	Egg PC	Avanti 840051C	L-α-phosphatidylcholine (Egg, Chicken)	97281-44-2	>99%
	Soy Lyso PC (LPC)	Avanti 840072P	L-α-lysophosphatidylcholine (Soy)	97281-38-4	>99%
	Egg PE	Avanti 840021P	L-α-phosphatidylethanolamine (Egg, Chicken)	39382-08-6	>99%
	Egg Lyso PE	Avanti 860081P	L-α-lysophosphatidylethanolamine (Egg, Chicken)	97281-40-8	>99%
	Egg PG	Avanti 841138P	L-α-phosphatidylglycerol (Egg, chicken) (sodium salt)	383907-64-0	>99%
	Soy PI	Avanti 840044C	L-α-phosphatidylinositol (Soy) (sodium salt)	383907-36-6	>99%
	Soy PS	Avanti 870336P	L-α-phosphatidylserine (sodium salt)	383908-63-2	>99%
	Milk SM	Avanti 860063P	Sphingomyelin (Milk, Bovine)	475662-40-9	>99%
Glycolipids	MGDG	Avanti 840523P	Monogalactosyldiacylglycerol (Plant)	63142-69-8	>99%
	DGDG	Avanti 840524P	Digalactosyldiacylglycerol (Plant)	63142-69-8	>99%

Table 3
Commonly used reagents for the specific staining of lipids. Please note that this list is not exhaustive

Lipid class	Reagent	Results/remarks
Cholesterol and cholesteryl esters	Acid ferric chloride	Cholesterol: Red to violet spots within 2–3 min. Cholesteryl esters: violets spots (the color appears later than for cholesterol).
Free fatty acids	2′,7′-Dichlorofluorescein/ $AlCl_3/FeCl_3$	Pink color after a few minutes
Phospholipids	Molybdic oxide/ molybdenum	Blue spots on a white background within a few minutes without heating, detection threshold: about 10 μg.
Phospholipids containing choline	Dragendorff reagent (bismuth nitrate+KI)	Choline-positive spots (such as PC, LPC, and SM) become detectable as orange spots
Phospholipids containing free amino groups	Ninhydrin	Purple-mauve spots
Glycolipids	Alpha-naphthol	Pink or purple spots.
Glycolipids	Orcinol	Blue-purple spots on a white background
Glycolipids	5-Hydroxy-1-tetralone	Glycolipids give yellow spots easily distinguishable from the light blue spots of phospholipids

Adapted from refs. [5, 7]

For instance, we use primuline for nonspecific and nondestructive detection of all lipid classes [6], copper sulfate for nonspecific revelation of neutral lipids and phospholipids [8–10], and the alpha-naphthol reagent for the specific detection of glycolipids [11, 12]. Note that the derivatization by copper sulfate and the alpha-naphthol reagent is destructive. After primuline derivatization, other reagents can be used to determine the specific functional groups (see Table 3). For the preparation of specific staining reagents mentioned in Table 3, one can find instructions in the book: Techniques of Lipidology [5].

2.4.2 Preparation of Staining Solutions

Prepare all solutions with ultrapure water (resistivity = 18.2 MΩ/cm at 25 °C).

Primuline Reagent

62.5 μg/mL primuline in acetone/water (80/20, v/v)):

1. For a total of 200 mL, mix together 160 mL acetone and 40 mL water.
2. Weight 12.5 mg of primuline in a glass weighing shoe.
3. Dissolve the weighed primuline in the acetone/water mixture.
4. Store the primuline reagent in an air-tight flask at 4 °C until use.

Copper Sulfate Reagent [8, 10]

$CuSO_4$ 10% and H_3PO_4 8% in water/methanol (19/1; v/v):

1. Weigh 25 g copper sulfate pentahydrate in a glass weighing shoe.
2. Dissolve copper sulfate in 200 mL ultrapure water.
3. Add 20 g ortho-phosphoric acid 85%.
4. Add 12.5 mL cold methanol.
5. Fill up the mixture to 250 mL with fridge cold water. The temperature of the mixture should not exceed 20 °C.
6. Store the reagent at 4 °C.

Alpha-Naphthol Reagent [12]

About 2.4% alpha-naphthol in sulfuric acid/ethanol/water solution (1/8/1, v/v/v):

1. Weigh 4.8 g alpha-naphthol into a 250 mL glass bottle in a fume cupboard.
2. Add 200 mL sulfuric acid/ethanol/water solution (1/8/1, v/v/v).
3. Close the bottle and shake gently to dissolve the crystals completely.
4. Store the reagent at 4 °C (up to 6 months, the reagent always works).

2.5 Preparation of HPTLC Silica Gel Glass Plates

We routinely use HPTLC silica gel 60 glass plates (size 20 × 10 cm, 0.20 mm, Merck, Darmstadt, Germany) for the separation of neutral lipid, phospholipid, or glycolipid mixtures:

1. Wash the HPTLC plate to remove eventual impurities. For this, the plate is put in a clean TLC tank containing 40 mL of 2-propanol until the migration of the solvent to the top of the plate (*see* **Note 3**).
2. Dry the plate in air, under a fume hood for at least 1 h.
3. Activate the HPTLC plate just before spotting, on a plate heater for 30 min at 120 °C. (*see* **Note 4**).

2.6 Elution Systems

The elution systems used for different protocols are indicated in Table 4 (*see* **Note 5**).

2.7 Preparation of Saturated Salt solution

The chromatographic separation on silica gel can be influenced by the activity of the stationary phase and thus depending on the relative humidity in the laboratory. With the "humidity control," the activity of the stationary phase is adjusted and the development is standardized. The desired humidity is adjusted in the developing chamber of the HPTLC system using a saturated salt solution.

Table 4
Elution systems used for the different protocols

Protocol number	Eluant	Run number	Lipids class separation	Solvents	Height (mm)
1	A	1	Neutral lipids	Hexane/diethyl ether/acetic acid, (68/32/1, v/v/v)	80
2	B	1	Neutral lipids	Chloroform/n-Pentane/Methanol, (45/52/3, v/v/v)	60
	C	2		n-Pentane/Diethyl ether, (97/3 v/v)	80
3	D	1	Phospholipids	Methylene chloride/Ethyl Acetate/Acetone, (80/16/4, v/v/v)	80
	E	2		Chloroform/ethyl acetate/acetone/isopropanol/ethanol/methanol/water/acetic acid, (30/6/6/6/16/28/6/2, v/v/v/v/v/v/v/v)	80
4	G	1	Glycolipids	Chloroform/Methanol/30% ammonium hydroxide (65/35/5, v/v/v)	85
5	D	1	Neutral lipids + Phospholipids	Methylene chloride/ethyl acetate/acetone, (80/16/4v/v/v)	55
	E	2		Chloroform/ethyl acetate/acetone/isopropanol/ethanol/methanol/water/acetic acid, (30/6/6/6/16/28/6/2, v/v/v/v/v/v/v/v)	55
	F	3		Hexane/diethyl ether/acetic acid, (80/20/2, v/v/v)	85

We used magnesium chloride hexahydrate for a 33% relative humidity. The preparation is as follows:

1. Add 900 g $MgCl_2$ $6H_2O$ into a beaker.
2. Add 200 mL distilled water and heat it up to 40–50 °C under stirring.
3. Cool down after 10 min to room temperature.

The final solution should contain about 1 cm of undissolved salt at the bottom to ensure the saturation of the salt solution.

2.8 HPTLC Equipment

HPLTC analyses were performed using a Camag® (Muttenz, Switzerland) HPTLC system equipped with an automatic HPTLC sampler (ATS 4), an Automatic Developing Chamber (ADC2) with humidity control, an Automatic Multiple Development (AMD2) systems, a TLC plate heater, a plate immersion device, a visualizer (CAMAG® TLC Visualizer 2), and a TLC scanner 4 controlled with visionCATS software.

3 General Procedures for HPTLC Analysis of Lipids

3.1 Application of Samples and Standards on the HPTLC Plate

A working solution of standards and sample solutions contained in 1.5 mL glass vials are put in the sample holder of the ATS4 system. They are then applied automatically by the ATS4 autosampler as 5.0-mm-long bands at 8.0 mm from the bottom of the plate and at a constant application rate of 250 nL/s, under continuous drying with a stream of nitrogen at 5 bars (*see* **Note 6**). For quantification, a mixture of lipid standards (working solutions) is spotted at increasing volumes to load 100–1000 ng per spot (*see* **Notes 6–9**).

3.2 Chromatogram Development

Depending on the composition of the lipid samples, the development of the HPTLC plate can require one- or multiple-step process. With the CAMAG system (https://www.camag.com/products/hptlc), the ADC2 system is designed for one-step development, while the AMD2 system is more suitable for the multiple step development process.

3.2.1 One-Step Development Process

Plates are developed at room temperature in the ADC2 system controlled in humidity. Prior to each experiment, a pad of filter paper and the activated HTPLC plate are introduced in the twin-trough glass chamber.

All the following parameters are set before experiments and are applied automatically by the ADC2 system:

1. The relative humidity of the developing chamber is equilibrated at 33% with the saturated solution of $MgCl_2$ (refer to Subheading 2.7 for the preparation) for 10 min.
2. 25 mL of the mobile phase is used to saturate the twin-trough glass chamber for 20 min.
3. 10 mL of the mobile phase is used for plate development.
4. The HPTLC plate is developed to a predefined level (refer to Subheading 2.6, Table 4).
5. After development, the plate is automatically dried under vacuum for 2 min.

3.2.2 Multi-step Development Process

Multi-step development of lipids on the silica HPTLC layer is based on an n-step gradient of solvents performed over an increasing distance of solvent migration. Between runs, the solvent is completely removed from the developing chamber and the silica layer is dried. For neutral lipids, solvents of decreasing polarity are applied (Table 4, protocol 2). With the first, most polar solvent, all substances migrate and concentrate as narrow bands in order of decreasing polarity, the most polar concentrating near the solvent front. As solvent polarity decreases, the most polar compounds almost remain in position while the less polar migrate much more and become separated.

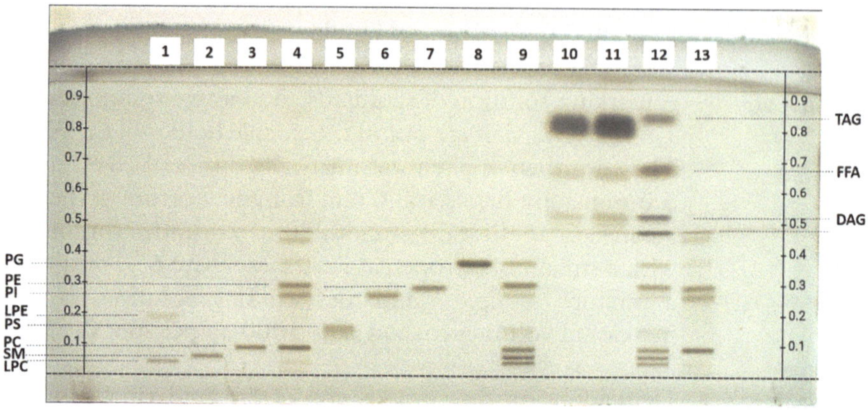

Fig. 3 Example of separation of total lipids in chia oil (*Salvia hispanica* L.) by three-step development, realized with the AMD2 system (see Subheading 2.6, Table 4, protocol 5 for solvent systems). The chia oil dissolved in chloroform was deposited at tracks 10 and 11 at 10 and 20 μg, respectively. The amounts of standards deposited were as follows: tracks 1, 2, 3, 5, 6, 7, 8: individual phospholipids and LPE-LPC at about 1 μg; tracks 4 and 13: 10.5 μg of soy lecithin (Lecigran™ 1000P; Cargill); track 9: LPC 0.27 μg, SM 0.26 μg, PC 0.25 μg, PS 0.06 μg, PI 0.06 μg, PE 0.25 μg, PG 0.09 μg; and track 12: mixture of standards of phospholipids and neutral lipids. Development in three steps with (1) methylene chloride/ethyl acetate/acetone (80/16/4 v/v/v) (eluant D, Table 4, Subheading 2.6) to 45 mm; (2) chloroform/ethyl acetate/acetone/propanol/ethanol/methanol/water/acetic acid (30/6/6/6/16/28/6/2, v/v/v/v/v/v/v/v) (eluant E, Table 4, Subheading 2.6) to 45 mm; and (3) hexane/diethyl ether/acetic acid (80/20/2, v/v/v) (eluant F, Table 4, Subheading 2.6) to 85 mm. After each development step, the plate was dried by the cool air stream from a hair dryer for 5 min. Finally, the plate was derivatized with the copper sulfate reagent (see Subheading 3.3.2)

The combination of the focusing effect and gradient elution results in extremely narrow bands (Fig. 3). We used the AMD2 system for automated multiple development of the HPTLC plates. The same saturation conditions as for the ADC2 system are applied. The HPTLC plate is automatically developed to a predefined distance with each eluent system (refer to Subheading 2.6, Table 4). After each development, the plate is automatically dried under vacuum for 2 min. Figure 3 illustrates the separation of individual phospholipids and neutral lipids in a three-step development procedure (see Subheading 2.6, Table 4, protocol 5). TAG was shown to be the major class in chia oil (*Salvia hispanica* L.).

3.3 Staining of the Lipid Classes

The compounds separated on the HPTLC plate are derivatized in the plate immersion device. For proper execution of the dipping technique, the plate must be immersed and withdrawn at a controlled uniform speed. By maintaining a well-defined vertical speed and immersion time, the derivatization conditions are standardized, avoiding "tide marks," which can interfere with densitometric evaluation. With our CAMAG derivatization system, uniform vertical speed is available between 30 mm/s and 50 mm/s and the immersion time can be chosen between 1 and 8 s.

3.3.1 Primuline Staining

1. Immerse the dried plate for 2 s in a solution of primuline; immersion speed is 30 mm/s.
2. Dry the plate again for 2 min at room temperature in a fume hood.
3. Complete the drying of the plate with a stream of warm air from a hair dryer in a fume hood.
4. Visualize individual lipid spots under UV fluorescence at 366 nm.

3.3.2 Copper Sulfate Staining

Detection and quantification of neutral lipids and phospholipids by charring in the presence of copper salts are well-established analytical techniques [8–10]. When using densitometric detection with copper sulfate salt, this method is relatively insensitive to the degree of unsaturation [9]:

1. Immerse the dried plate for 15 s in the copper sulfate reagent; immersion speed is 30 mm/s. Remove the excess of the charring reagent if any.
2. Dry the plate at room temperature in a fume hood (in air or with the cool air stream from a hair dryer).
3. Heat the plate at 150 °C for 15 min then at 170 °C for 5 min on a plate heater. Individual lipid classes become detectable as black-brown spots.

3.3.3 Alpha-Naphthol Staining

1. Immerse the dried plate for 5 s in the alpha-naphthol solution (or spray the plate to wet the surface). Remove the excess reagent on the back of the plate if any.
2. Dry the plate (in air or with a stream of warm air from a hair dryer) in a fume hood.
3. Heat the plate at 120 °C on a heating plate for 1–2 min until glycolipid bands are stained pink or purple. Overheating the plate will lead to dark blue spots of glycolipids then charred spots of all lipids.

3.4 Visualizer and Scanner

The plate is photographed immediately after derivatization, using a visualizer (CAMAG® TLC Visualizer 2), under white or an appropriate wavelength:

- After primuline derivatization, a digital image of the plate is captured under lighting at 366 nm in the fluorescence mode.
- After copper sulfate derivatization, a digital image of the plate is captured under 550 nm in absorbance mode with a deuterium-tungsten lamp.
- After alpha-naphthol derivatization, a digital image of the plate is captured either under white light (for specific staining of

glycolipids) or under 550 nm in absorbance mode with a deuterium-tungsten lamp (if charring).

The identification of lipid classes is established from their R_f in comparison with the corresponding standards.

For quantification, densitograms (CAMAG® TLC Scanner 4) and image profiles can be generated (Fig. 4). The data are processed by visionCATS software. Spot intensities are integrated and peak surfaces expressed in arbitrary units. The quantity of the lipids present in a sample is calculated from a calibration curve of pure standards carried out on the same HPTLC plate. The range of concentration of calibration solution needs to be adapted according to the amount in the sample (or vice versa). The percentage by weight of lipids in each sample is calculated. A good correlation ($r^2 > 0.999$) of the applied lipid mass and scanned intensity is obtained for at least the range of 20–1000 ng of lipid (Fig. 5).

4 Applications to Specific Analysis

4.1 Separation of Neutral Lipids with One-Step Development (Protocol 1)

This method is applied for the rapid analysis of neutral lipids. It involves one-step development for the separation of different neutral lipid classes by the solvent system hexane/diethyl ether/acetic acid (68/32/1, v/v/v) (see Subheading 2.6, Table 4, protocol 1). The derivatization is done with the primuline reagent (see Subheading 3.3.1). The plate is visualized under 366 nm wavelength. The identification of neutral lipid classes is established from their R_f values compared to those of standards.

Figure 6 shows the separation of neutral lipids of edible caterpillars from different batches.

4.2 Separation of Neutral Lipids with Two-Step-Development (Protocol 2)

In this method, the HPTLC plate undergoes two-step development with drying after each development. The second solvent system travels through the layer, re-concentrating the spots, and producing narrow bands. Multiple developments can be performed over different separation distances, using the same solvent or different solvents of varying polarities [13].

In order to separate different neutral lipid classes, we use these solvent systems: firstly chloroform/n-pentane/methanol (45/52/3, v/v/v) with a distance of migration of 60 mm and then n-pentane/diethyl ether (97/3, v/v) to 80 mm (see Subheading 2.6, Table 4, protocol 2). The derivatization is done with the copper sulfate reagent (see Subheading 3.3.2). The plate is visualized under 550 nm wavelength. The identification of neutral lipid classes is established from their R_f values compared to those of standards (Fig. 7).

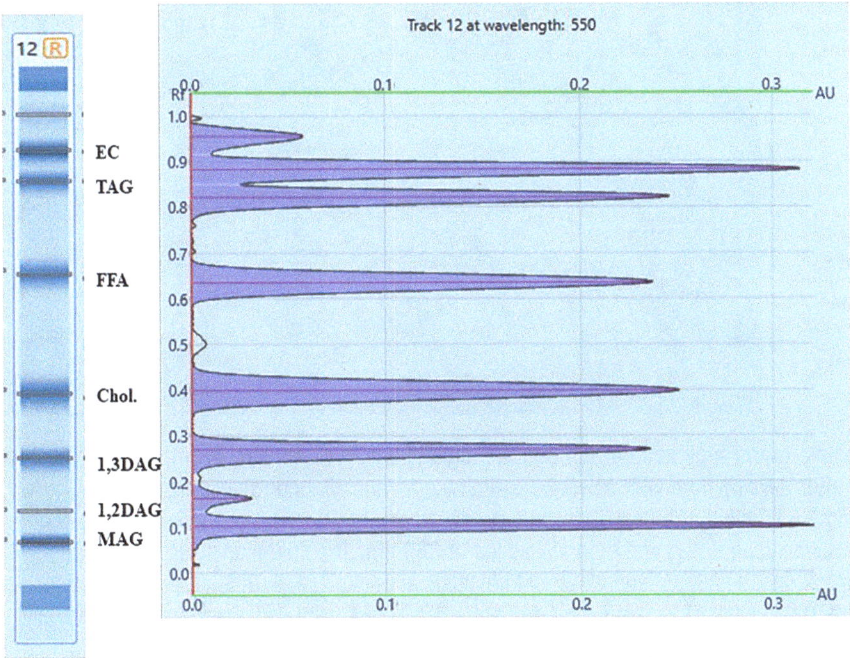

Fig. 4 Densitometric chromatogram of a standard mixture of neutral lipids separated by HPTLC in a two-step development (see Subheading 2.6, Table 4, protocol 2). All lipids were deposited at 0.8 μg. The plate was derivatized with the copper sulfate reagent (see Subheading 3.3.2)

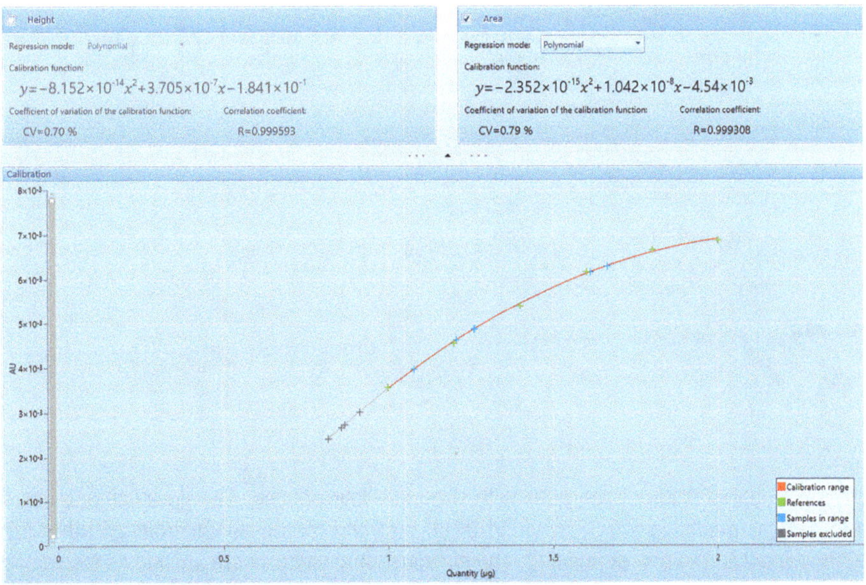

Fig. 5 Example of a calibration curve of TAG obtained from densitometric integration by the visionCATS software. The curve was calculated from the peak areas of six amounts (1–2 μg) of the standard. The correlation coefficient was 0.999308

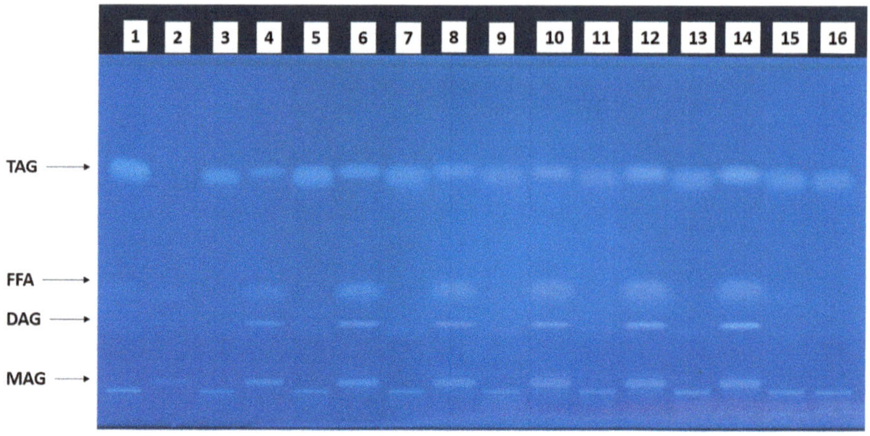

Fig. 6 Example of the separation of the neutral lipids of edible caterpillars. Samples were compared to lipid standards after development with solvent system hexane/diethyl ether/acetic acid (68/32/1, v/v/v) to 80 mm (Subheading 2.6, Table 4, protocol 1). Total time development was 40 min. Lipid spots were derivatized with primuline (Subheading 3.3.1) and then visualized at 366 nm. The lipid standards were deposited at different quantities: 2, 9, 16, 21, 28, 35, and 42 μg of total lipids at tracks 2, 4, 6, 8, 10, 12, and 14, respectively. The R_f values were 0.04, 0.21, 0.31, and 0.67 for MAG, DAG, FFA, and TAG, respectively. The lipid extracts of caterpillar (about 0.25 mg/mL in chloroform) of different batches were deposited at tracks 1, 3, 5, 7, 9, 11, 13, 15, and 16

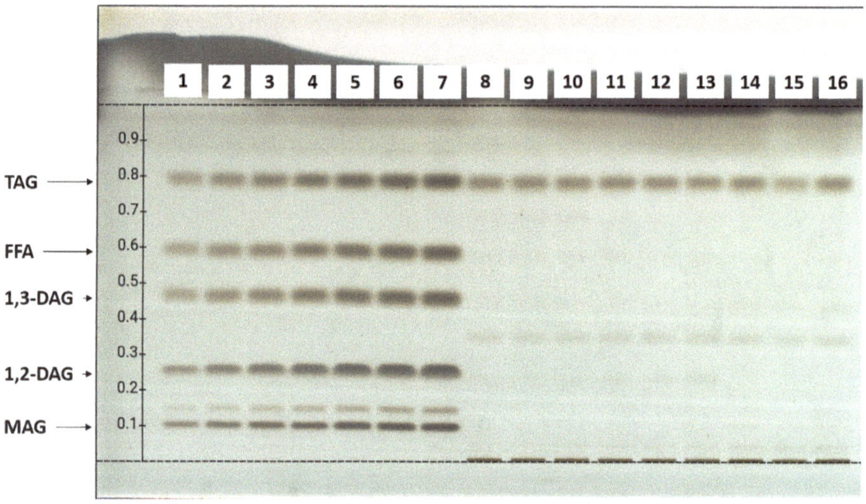

Fig. 7 Analysis of neutral lipids of pea flour with two-step development. The development was firstly with chloroform/n-pentane/methanol (45/52/3, v/v/v) to 60 mm and then n-pentane/diethyl ether (97/3 v/v) to 80 mm (Subheading 2.6, Table 4, protocol 2). The standards and samples are applied as follows: tracks 1–7: mixtures of standards of TAG; FFA; 1, 3-DAG; 1,2-DAG; and MAG (refer to Subheading 2.2, Table 2: Lipid standards) deposited at 0.4, 0.6, 0.8, 1.0, 1.2, 1.4, 1.6 μg of each standard; tracks 8–16: lipid extracts from three samples of pea flour deposited at 1 μg per track, in triplicate successively. Lipid spots are visualized by derivatization with the copper sulfate reagent (see Subheading 3.3.2). The R_f values were 0.70, 0.59, 0.46, 0.25, and 0.09 for TAG, FFA, 1, 3-DAG, 1,2-DAG, and MAG, respectively

4.3 Separation of Phospholipids with Two-Step Development (Protocol 3)

Phospholipids are separated following the procedure proposed by Weerheim et al. [14] with modifications to separate individual phospholipid classes depending on their polar head groups. During the optimization of the protocol, emphasis was set on achieving the best possible separation of the PC, PS, PI, and PE [7, 14–16].

Figure 8 is an illustration of the separation of mixture of phospholipid standards by two successive developpement steps, using methylene chloride/ethyl acetate/acetone (80/16/4, v/v/v) and then chloroform/ethyl acetate/acetone/isopropanol/ethanol/methanol/water/acetic acid (30/6/6/6/16/28/6/2, v/v/v/v/v/v/v/v) (refer to Subheading 2.6, Table 4, protocol 3).

4.4 Separation and Identification of Glycolipids (Protocol 4)

For the separation of glycolipids, several solvent systems are suggested: chloroform/methanol/30% ammonium hydroxide (65/35/5 or 40/10/1, v/v/v), chloroform/methanol/water (65/35/5 or 65/25/4, v/v/v), chloroform/acetone/water (30/60/2, v/v/v), diisobutyl ketone/acetic acid/water (40/25/5, v/v/v), chloroform/methanol/acetic acid/water (25/15/4/2, v/v/v/v), or chloroform/acetone/methanol/acetic acid/water (10/4/2/2/1, v/v/v/v/v) [6, 11]. We tested successfully chloroform/methanol/30% ammonium hydroxide (65/35/5, v/v/v)

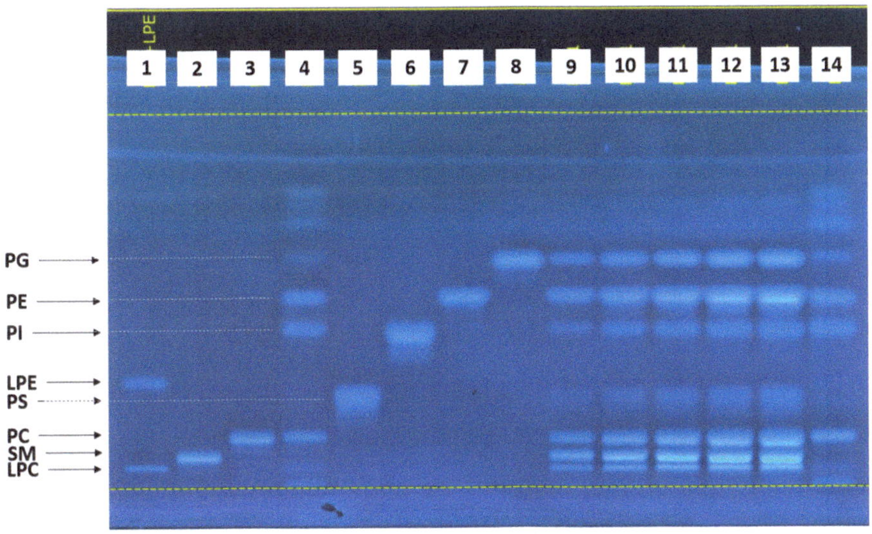

Fig. 8 Separation of mixtures of phospholipid standards. Separation was achieved with a two-step development procedure with the following solvent systems: methylene chloride/ethyl acetate/acetone (80/16/4 v/v/v) to 80 mm then chloroform/ethyl acetate/acetone/isopropanol/ethanol/methanol/water/acetic acid (30/6/6/6/16/28/6/2, v/v/v/v/v/v/v/v) to 70 mm. The plate was derivatized by primuline and then visualized at 366 nm. The lipid solutions were applied as follows: the individual phospholipids (tracks 1,2,3,5,6,7,8, about 1 μg); soy lecithin (Lecigran™ 1000P, Cargill) at 2.626 μg/μL (tracks 4 and 14, 5.2, and 10.5 μg), a mixture of individual phospholipids at increasing amounts were applied on tracks 9–13 (1–5 times, respectively). Track 9: 0.05, 0.12, 0.03, 0.03, 0.13, 0.13, 0.13 μg for PG, PE, PI, PS, PC, SM, and LPC, respectively. The Rf values were 0.61, 0.51, 0.40, 0.28, 0.23, 0.14, 0.08, and 0.05 for PG, PE, PI, PS, PC, SM, and LPC, respectively

for the separation of galactolipids (Fig. 9) and bacterial glycolipids (data not shown). This solvent system is also efficient for the separation of some phospholipids (PG, PS, PE) (Fig. 9). Other separation systems for one- or two-dimensional TLC can be found in previously mentioned refs. [6, 11].

After development (see Subheading 3.2. Development), the plate is dried with the hot air stream from a hair dryer. To stain and highlight all lipids in the samples, we use the primuline reagent (Fig. 9a) (see Subheading "Primuline Reagent" for the preparation and Subheading 3.3.1 for the staining).

For the specific revelation of glycolipids, an orcinol–sulfuric acid mixture, naphthol, and diphenylamine can be used [6]. We use alpha-naphthol following the protocol of Wang and Benning [12] which is in our opinion easier to handle as compared to the procedure mentioned elsewhere [6, 11] (Fig. 9b). It involves the application of the reagent in only one step followed by a heating step. See Subheading "Alpha-Naphthol Reagent" for the preparation and Subheading 3.3.3 for the staining.

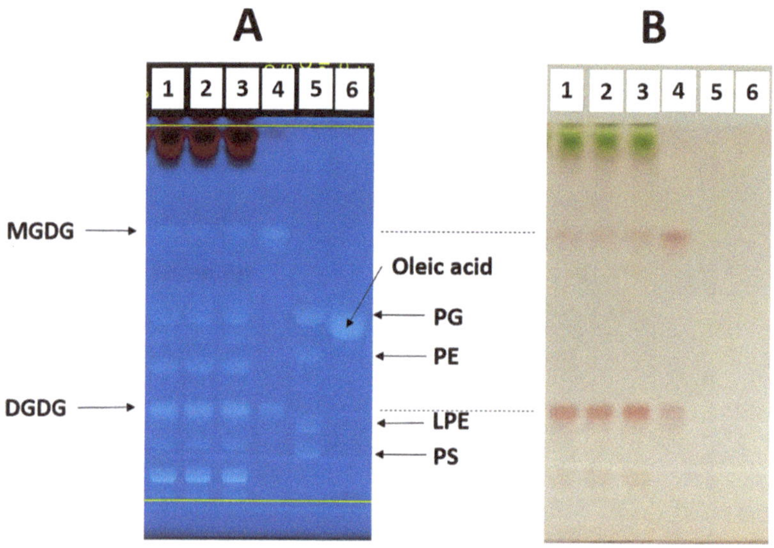

Fig. 9 Separation of total lipids and specific detection of glycolipids in spinach *(Spinacia oleracea)*. The development solvent was chloroform/methanol/30% ammonium hydroxide (65/35/5) (Subheading 2.6, Table 4, protocol 4). The plate was first derivatized by primuline (Subheading 3.3.1), then visualized at 366 nm (**a**), then derivatized by the alpha-naphthol reagent (Subheading 3.3.3), and finally observed under white light (**b**). The lipid standard's positions were marked. The glycolipids were stained purple. The lipid solutions were applied as follows: spinach sample as triplicate (tracks 1, 2, 3, about ~70 μg); mixture of MGDG and DGDG standards (track 4: MGDG 5 μg, DGDG 2.5 μg); mixture of phospholipid standards (track 5, 2.5 μg/phospholipid); and oleic acid (track 6, 10 μg). The R_f values were 0.12, 0.20, 0.25, 0.38, 0.45, 0.49, and 0.70 for PS, LPE, DGDG, PE, oleic acid, PG, and MGDG, respectively

4.5 Separation of Total Lipids (Phospholipids and Neutral Lipids) with Three Development (Protocol 5)

We have optimized a protocol for the qualitative determination of different phospholipids and neutral lipids eventually present in a lipid sample. Three successive developments with the following solvents: 1/(methylene chloride/ethyl acetate/acetone) (80/16/4 v/v/v) then 2/(chloroform/ethyl acetate/acetone/isopropanol/ethanol/methanol/water/acetic acid) (30/6/6/6/16/28/6/2, v/v/v/v/v/v/v/v), and finally 3/(hexane/diethyl ether/acetic acid) (80/20/2, v/v/v) (refer to Subheading 2.6, Table 4, protocol 5) give a good separation of phospholipids (LPC, SM, PC, PS, PI, PE, PG) and neutral lipids (TAG, FFA, 1,2-DAG, 1,3-DAG, MAG). This protocol was used for the separation of lipid classes in different oil samples (Fig. 10).

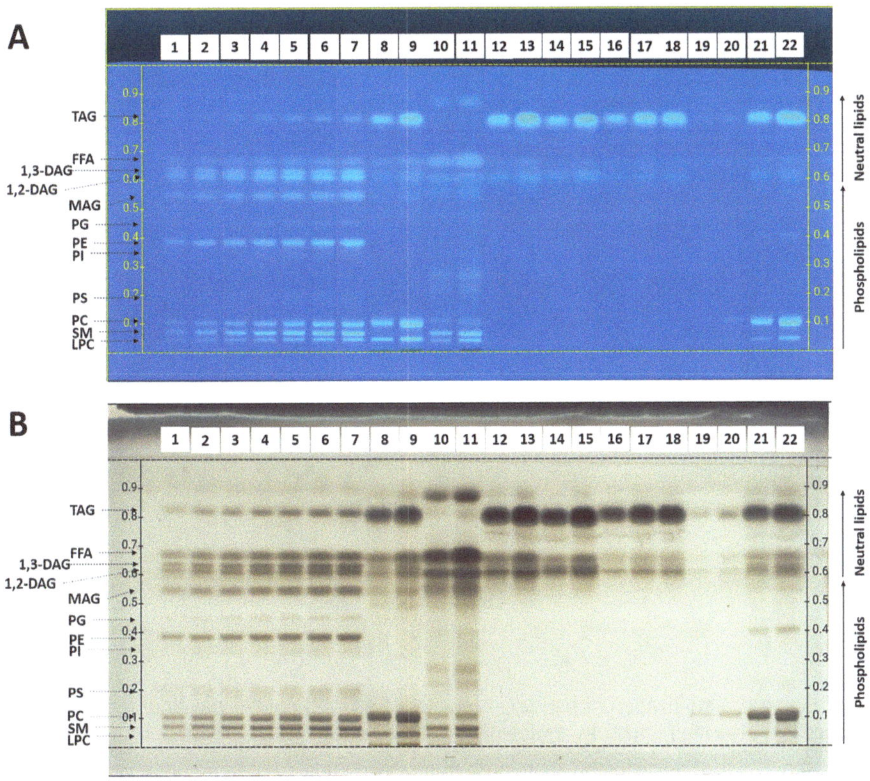

Fig. 10 Separation of mixtures of phospholipid and neutral standards. Separation was achieved with a three-step development procedure with the following solvents: (methylene chloride /ethyl acetate/acetone) (80/16/4 v/v/v) to 55 mm then (chloroform/ethyl acetate /acetone/propanol/ethanol/methanol/water/acetic acid) (30/6/6/6/16/28/6/2, v/v/v/v/v/v/v/v) to 55 mm and finally (hexane/diethyl ether/acetic acid) (80/20/2, v/v/v) to 85 mm (refer to Subheading 2.6, Table 4, Protocol number 5). The plate was first derivatized by primuline (Subheading 3.3.1), then visualized at 366 nm (**a**), then derivatized by the copper sulfate reagent (Subheading 3.3.2), and visualized at 550 nm (**b**). The lipid solutions were applied as follows: tracks 1–7: a mixture of standards lipids at 4, 6, 8, 10, 12, 14, and 16 μL of a standard solution of different neutral lipids, each at 0.1 μg/μL and phospholipids (PG, PE, PI, PS, PC, SM, LPC at 0.016, 0.123, 0.006, 0.031, 0.125, 0.101, 0.080 μg/μL, respectively). The lipid samples were different marine products (squid, tuna, salmon, cod liver oil) at 1 μg/μL, applied at 2 and 4 μL successively

4.6 Quantification of Phospholipids

We also use HPTLC for the quantification of phospholipids. The lipid samples are deposited at different amounts on the HPTLC plate in parallel with a series of increasing quantities of a mixture of lipid standards. The lipid nature is determined by R_f values in comparison with the standards. For each type of phospholipids, the quantity is calculated based on the calibration curve of the corresponding standard (refer to Subheading 3.4). Fig. 11 is an example of an analysis of phospholipids from pea flour.

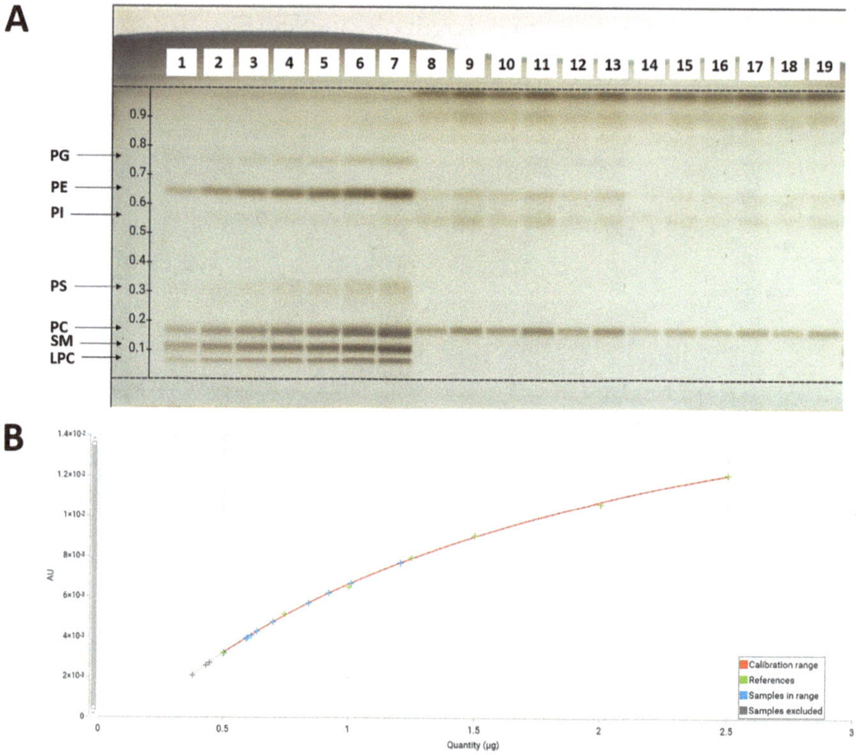

Fig. 11 Separation and visualization of pea flour phospholipids with double-development and copper sulfate charring (**a**) and PC curve calibration (**b**). The standards and samples are applied as follows: tracks 1–7: 4, 6, 8, 10, 12, 16, and 20 µL of a mixture of standards of phospholipids (standard concentrations in µg/mL: LPC 318; SM 405; PC 500; PS 125; PI 25; PE 492; and PG 63); tracks 8–19: lipid extracts from different pea flour samples deposited at 2 and 3 µg successively. Separation was achieved with a two-step development procedure with the following solvent systems (methylene chloride/ethyl acetate/acetone (80/16/4 v/v/v) to 75 mm, then chloroform/ethyl acetate/acetone/propanol/ethanol/methanol/water/acetic acid (30/6/6/6/16/28/6/2, v/v/v/v/v/v/v/v) to 80 mm. The plate was derivatized by the copper sulfate reagent (Subheading 3.3.2) and visualized at 550 nm. For the curve calibration, PC was applied at 0.5–2.5 µg. Standard curve determined after charring and photo densimetric analysis in regression mode Michaelis–Menten (r^2 0.996, CV%)

5 Notes

1. All solvents and other reagents are analytical grade. The solvents recommended are chemically inert with the samples and the stationary phase and easily evaporate.
2. Safety: The use of solvents requires working under a fume hood and working with gloves.
3. The HPTLC plate can also be washed with methanol or chloroform/methanol (1/1, v/v) or the elution system. Note the direction of migration with a soft pencil in the upper right corner of the plate, not with ink. Plates can be prepared in advance and be kept in the dark, in a desiccator.
4. The silica gel is hygroscopic; hydratation can alter chromatography and interactions of the solvent system with the sample. This step makes it possible to "re-initialize" the silica because the water content of the silica influences the separation of the molecules.
5. To prepare the multi-component mobile phases, accurate volumetric measurements of the components of the mobile phase must be performed separately and precisely in adequate volumetric glassware. Final shaking should ensure proper mixing of the content.
6. The sample extracts are applied as bands onto the plate with a software-controlled applicator. The precision of the applied volume, exact positioning, and compactness of the application zone determine the quality of the final result. The finer the deposits, the better the resolution of the separation. Deposits applied as bands give more reproductible results than spots. The obtained bands are finer if the vitesse of deposition is slower (e.g., 50 or 100 nL/s).
7. Derivatization can be performed either by immersing or by spraying the plate with a reagent; for better reproducibility, immersion is the preferred technique.
8. The intensity of staining of each of the four classes of neutral lipids (monoacylglycerols, diacylglycerols, triacylglycerols, and free fatty acids) depends on the number or amount of carbon–carbon double bonds (C=C) with 3% cupric acetate in 8% phosphoric acid. In contrast, with 10% cupric sulfate in 8% phosphoric acid, the staining intensity is relatively independent of the number or amount of C=C [9].
9. For quantification and establishment of calibration curves, different equations can be used: polynomial, linear, or Michaelis–Menten. The equation of Michaelis–Menten $y = ax/(b + x)$ is

recognized as the most adapted to the phenomenon of saturation. In the fluorescence mode, saturation does not occur and a linear response is currently and easily obtained.

Acknowledgments

The authors would like to thank Dr. C. Genot (BIA, INRAE, France) for her comments and advice to improve this manuscript.

Annex 1 Calculated eluotropic forces of solvents [1]

	Eluent	Eluent strength $\varepsilon°$
Nonpolar	n-Heptane	0.01
	Cyclohexane	0.03
	Diisobutylene	0.05
	Carbon tetrachloride	0.11
	Carbon disulfide	0.2
	Ethyl benzene	0.2
	Isopropyl ether	0.21
	Isopropyl chloride	0.22
	Chlorobenzene	0.23
	Toluene	0.23
	Benzene	0.25
	Chloroform	0.26
	Butyl acetate	0.33
	Ethyl ether	0.38
	Ethyl acetate	0.38
	Nitrobenzene	0.4
	Acetone	0.47
	Aniline	0.48
	Dioxane	0.49
	Butanol	0.56
	Ethanol	0.68
Very polar	Methanol	0.73

References

1. Snyder LR (1966) Linear elution adsorption chromatography. XIII. Further studies on a narrow pore silica. J Chromatogr A 25:274–293. https://doi.org/10.1016/s0021-9673(01)98244-5
2. Johnston AV (1971) Basic lipid methodology. College of Agriculture, University of Illinois at Urbana-Champaign
3. Sherma J (1992) Planar Chromatography. Anal Chem 64:134–147. https://doi.org/10.1021/ac00036a007
4. Touchstone JC (1995) Thin-layer chromatographic procedures for lipid separation. J Chromatogr B Biomed Sci Appl 671:169–195. https://doi.org/10.1016/0378-4347(95)00232-8
5. Kates M (1982) Techniques for separation of lipid mixtures. In: Techniques of lipidology: isolation, analysis and identification of lipids. North-Holland Publishing company, pp 393–469
6. Christie WW, Han X (2010) Lipid analysis. The Oily Press
7. Fuchs B, Süß R, Teuber K et al (2011) Lipid analysis by thin-layer chromatography-A review of the current state. J Chromatogr A 1218:2754–2774. https://doi.org/10.1016/j.chroma.2010.11.066
8. Churchward MA, Brandman DM, Rogasevskaia T, Coorssen JR (2008) Copper (II) sulfate charring for high sensitivity on-plate fluorescent detection of lipids and sterols: quantitative analyses of the composition of functional secretory vesicles. J Chem Biol 1:79–87. https://doi.org/10.1007/s12154-008-0007-1
9. Baron CB, Coburn RF (1984) Comparison of two copper reagents for detection of saturated and unsaturated neutral lipids by charring densitometry. J Liq Chromatogr 7:2793–2801. https://doi.org/10.1080/01483918408067046
10. Handloser D, Widmer V, Reich E (2008) Separation of phospholipids by HPTLC - An investigation of important parameters. J Liq Chromatogr Relat Technol 31:1857–1870. https://doi.org/10.1080/10826070802188940
11. Kates M (1982) Lipid extraction procedures. In: Techniques of lipidology. Elsevier Science & Technology Books, pp 347–353
12. Wang Z, Benning C (2011) Arabidopsis thaliana polar glycerolipid profiling by thin layer chromatography (TLC) coupled with gas-liquid chromatography (GLC). J Vis Exp 2–7. https://doi.org/10.3791/2518
13. Olsen RE, Henderson RJ (1989) The rapid analysis of neutral and polar marine lipids using double-development HPTLC and scanning densitometry. J Exp Mar Bio Ecol 129: 189–197. https://doi.org/10.1016/0022-0981(89)90056-7
14. Weerheim AM, Kolb AM, Sturk A, Nieuwland R (2002) Phospholipid composition of cell-derived microparticles determined by one-dimensional high-performance thin-layer chromatography. Anal Biochem 302:191–198. https://doi.org/10.1006/abio.2001.5552
15. Pinault M, Guimaraes C, Dumas JF et al (2020) A 1D high performance thin layer chromatography method validated to quantify phospholipids including cardiolipin and monolysocardiolipin from biological samples. Eur J Lipid Sci Technol 122. https://doi.org/10.1002/ejlt.201900240
16. Parchem K, Kusznierewicz B, Chmiel T et al (2019) Profiling and qualitative assessment of enzymatically and thermally oxidized egg yolk phospholipids using a two-step high-performance liquid chromatography protocol. J Am Oil Chem Soc 96:693–706. https://doi.org/10.1002/aocs.12218

Chapter 8

HPLC Analysis of Triacylglycerol Molecular Species

Alain Riaublanc and Michelle Viau

Abstract

Triacylglycerols (TAGs) make up the major part of naturally occurring fats and oils. The composition and fine structure of TAGs determine to a large extent the functionality of fats and oils as food ingredients and the physiological effects of fats and oils as components of the human diet.

Reversed-phase C_{18} high-performance liquid chromatography (RP-HPLC) is the method of choice for the separation of molecular species of TAGs. It provides a good resolution of individual TAG molecules. Thus, TAG molecules are separated according to both chain length and degree of unsaturation of the fatty acids (FAs). The complexity of natural TAG mixtures would, however, require gradient elution to fully utilize the resolution power of the HPLC system. In this chapter, we describe two methods, one for the determination of profiles of fats and vegetable oils and the second one for milk products. Both of them are performed using a binary gradient of acetonitrile (ACN) – chloroform and an evaporative light scattering detection. Indeed, the method can easily be adjusted for use with different oils by a simple modification of the gradient and run time.

Key words RP-HPLC, Triacylglycerols (TAGs), TAGs composition, TAGs molecular species

1 Introduction

To characterize the nutritional properties of food lipids, the characterization of their fatty acid composition by gas chromatography after transformation into methyl esters is often sufficient, but to understand their physical and thermal properties, it is necessary to analyze their composition in molecular species of triacylglycerols (TAGs). Indeed, the crystalline organization and the melting temperature of TAG crystals strongly depend on the nature of the three fatty acids associated within a TAG and their distribution on the three positions of glycerol (see Chapters 23 and 24 from Multidimensional Characterization of Dietary Lipids). Depending on the number of different fatty acids present in dietary lipids, the number of three-by-three combinations can be very important as in the case of dairy lipids.

Christelle Lopez et al. (eds.), *Multidimensional Characterization of Dietary Lipids*, Methods and Protocols in Food Science,
https://doi.org/10.1007/978-1-0716-3758-6_8,

Table 1
Fatty acids name and symbol

Fatty acid name		Abbreviation
C4:0	Butyric acid	B (or Bu)
C6:0	Caproic acid	Ca
C8:0	Caprylic acid	Cy
C10:0	Capric acid	C
C12:0	Lauric acid	La
C14:0	Myristic acid	M (or My)
C14:1	Myristoleic acid	Mo
C15:0	Pentadecanoic acid	Pd
C16:0	Palmitic acid	P
C16:1	Palmitoleic acid	Po
C18:0	Stearic acid	S
C18:1	Oleic acid	O
C18:2	Linoleic acid	L
C18:3	Linolenic acid	Ln
C20:0	Arachidic acid	A
C20:1	Gadoleic acid	G

TAGs are classically named using three letters indicating the diversity of the acylated fatty acids (see Table 1), for example, the homogeneous LLL, OOO, PPP, SSS, or the heterogeneous OOL, SOO, or PSO (L: linoleic acid C18:2 n-6, O: oleic acid 18:1 n-9, P: palmitic acid C16:0, and S: stearic acid C18:0).

Gradient reversed-phase C_{18} high-performance liquid chromatography (RP-HPLC) is widely used and is the method of choice for the separation of the molecular species of TAGs in natural and restructured lipid sources. Numerous practical applications have been described [1–7]. Under these elution conditions, the molecular species of TAGs are separated according to their hydrophobicity. The separation increases with the length of the carbon chains of the fatty acids but decreases with the number of unsaturations present in these fatty chains. In the first approximation, this separation depends on a parameter called equivalent carbon number (ECN), which corresponds to the total number of carbon atoms (noted C) of the three fatty chains minus two times the total number of unsaturation of these three chains (noted C=C). Hence, the retention in RP-HPLC increases with increasing equivalent carbon number:

$$ECN = \Sigma C - 2 * \Sigma C = C$$

For compounds with the same ECN, the retention time decreases with the number of double bonds present on fatty chains. For example, the group of OOO, OOP, OPP, and PPP with the ECN equal to 48 migrate in the order below:

$$OOO < OOP < OPP < PPP$$

In practice, chromatograms are highly complex owing to the broad variety of TAGs in dietary fats and oils.

2 Materials

2.1 Mandatory Equipment

1. High-performance liquid chromatograph equipped with two pumps for solvent gradients, thermostatic control of column temperature, and an evaporative light scattering detector (ELSD).
2. Two reverse phase columns C_{18} (Nucleoshell RP18 HPLC column, 90A, 2.7 μm, 4.6 × 150 mm) must be connected in series to improve species separation, and the temperature of the column must be controlled for a better reproducibility of retention times.
3. Analytical precision balance, sensitivity: 0.01 mg.
4. Positive displacement pipettes and disposable capillary piston tips (1000–50 μL).
5. 5 or 10 mL volumetric flask.
6. HPLC glass vial.

2.2 Reagents

1. Chloroform (HPLC grade).
2. Acetonitrile (HPLC grade).
3. Methanol (HPLC grade).
4. Trilinolein (LLL; glyceryl trilinoleate; purity 99%).
5. Standard oils (coco, sunflower, palm, flax seed, macadamia).
6. Anhydrous milk fat from cow (AMF).

3 Methods

3.1 Procedure

3.1.1 Sample Preparation

In a 2 mL vial, weigh accurately about 10 mg of the neutral lipid sample and add the volume of chloroform/methanol (1:1, v/v) necessary to obtain a final concentration of 10 mg/mL. The use of this chloroform/methanol mixture is essential to optimize the separation of TAG species.

3.1.2 Standard Oils or Milk Fat Preparation

In a 5 mL volumetric flask, weigh accurately about 50 mg of standard vegetable oils or AMF, previously melted at 60 °C, and make up to volume with chloroform/methanol (1:1,v/v) to obtain a final concentration around 10 mg/mL.

3.1.3 Standard Trilinolein Preparation

1. Prepare chloroform/methanol mixture (1/1, v/v).
2. Prepare standard solution: accurately weigh (± 0.1 mg) approximately 50 mg of trilinolein into a 5 mL volumetric flask and make up to volume with chloroform/methanol (1/1; v/v) to obtain a final concentration around 10 mg/mL.
3. Then perform a 10-fold dilution by pipetting 100 μL of the standard solution into a vial and adding 900 μL of chloroform/methanol 1:1, v/v. Always use positive displacement pipettes to pipet chloroform/methanol or standard solutions.
4. Inject increasing volumes of the diluted standard solution (5, 10, 15, 20, and 25 μL), corresponding to 5, 10, 15, 20, and 25 μg of trilinolein.

3.2 Chromatographic Separation

The sample dissolved in chloroform/methanol (1:1; v/v) (10 mg/mL) was eluted using a gradient elution from solvent A: acetonitrile to solvent B: chloroform.

For neutral lipids other than those of mammalian milk, the number of fatty acids and TAGs is quite small, which allows a fast separation that corresponds to method 1. However, for ruminant milk lipids, which contain a large variety of fatty acids, from C4 to C20, and therefore an even larger variety of TAGs, a different elution gradient corresponding to method 2 was necessary for better separation.

3.2.1 Method 1

1. Before analysis, the column was allowed to equilibrate with 85% A and 15% B. The analytical columns were maintained at 30 °C.
2. Inject 10 μL of the sample or standard oils solution or variable volumes of the standard trilinolein solution.
3. After injection, the gradient went from 15% to 45% B in 65 min.
4. The mobile phase was brought back to initial conditions and the column was allowed to equilibrate for 4 min until the next injection.
5. The mobile phase flow rate was 1 mL/min.
6. The TAGs were detected by an ELSD Sedex 85 (Sedere S.A., Alfortville, France). The nebulizer gas pressure was maintained at 3.5 bar, the drift tube temperature was set at 40 °C, and the gain at 6.

3.2.2 Method 2

1. Before the analysis, the column was allowed to equilibrate with 90% A and 10% B.
2. Inject 10 μL of the sample or standard AMF solution.
3. After injection, the gradient went from 10% to 45% B in 190 min, then from 45% to 60% in 200 min, and maintained at 60% during 5 min.
4. The mobile phase was brought back to initial conditions and the column was allowed to equilibrate for 4 min until the next injection.
5. The mobile phase flow rate was 1 mL/min.
6. The TAGs were detected by an ELSD Sedex 85 (Sedere S.A., Alfortville, France). The nebulizer gas pressure was maintained at 3.5 bar, the drift tube temperature was set at 40 °C, and the gain at 6.

3.3 Identification of the TAG Peaks

Individual peaks observed on the chromatogram were identified by comparing their retention times with those of oils of known TAG compositions (coco, sunflower, palm, flax seed, and macadamia) shown in Fig. 1.

In the case of ruminant milk lipids, the use of vegetable oils as standards is not sufficient and it is preferable to use an anhydrous milk fat (AMF), from cow's milk, to perform peak identification. A chromatogram from neutral lipids of AMF is presented in Fig. 2. The peaks have been numbered for easy reading, and the TAGs' composition of each peak is detailed in Table 2.

Matrix-assisted laser desorption/ionization time of flight mass spectrometry (MALDI TOF) can be performed and is useful for

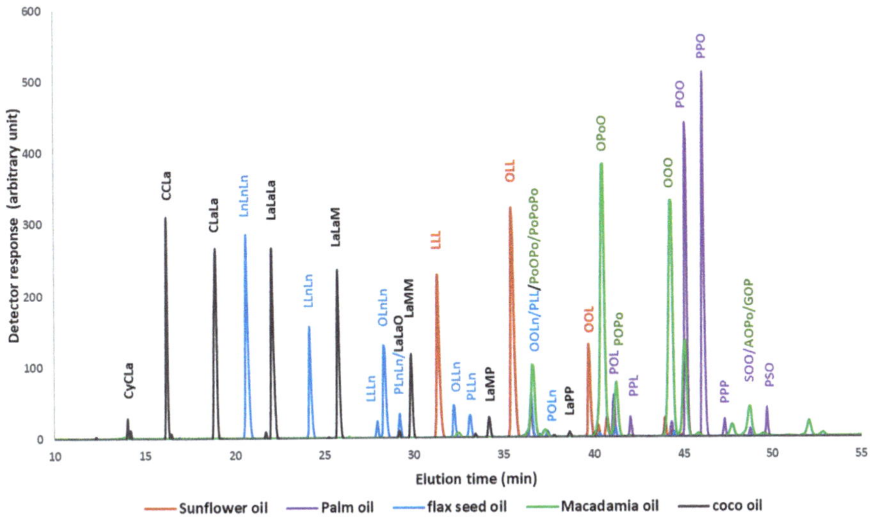

Fig. 1 Chromatograms obtained for different standard oils. The labels above each peak correspond to the TAGs species identified by matrix-assisted laser desorption/ionization time of flight mass spectrometry (MALDI TOF)

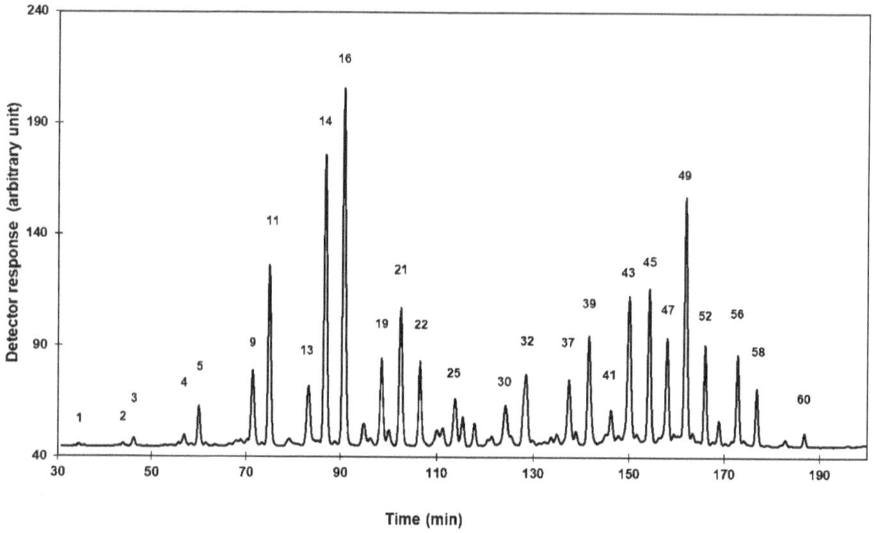

Fig. 2 Chromatograms obtained for AMF from cow. The number above each peak corresponds to the TAGs species identified in Table 2

TAG identification. The MALDI-TOF mass spectrometer is a spectrometer using a matrix-assisted laser ionization source (MALDI = matrix-assisted laser desorption/ionization) and a time-of-flight (TOF) analyzer. This gentle technique allows to ionize lipids without degrading them. The lipid sample is mixed with the matrix (2,5 dihydroxybenzoic acid) and placed on a slide and evaporated so that they co-crystallize. The deposit formed is called a target. A laser source is directed onto the target to ionize the molecules of the sample. TAGs were ionized as sodium adducts, *m/z* corresponding to [M + Na]+. The ions are then detected by measuring the time it takes for the different particles to reach the detector (*m/z* ratio). The analysis of the spectra coupled with the composition determined by HPLC and the results of structure simulation obtained by LIPID MAPS (http://www.lipidmaps.org/tools/ms) allowed to identify the major glyceride species but does not specify the position of the fatty acids on the glycerol [8–10].

3.4 Expression of Result and Quantitative Estimation of Triacylglycerol Molecular Species

The relative amount of each TAG species identified is usually expressed as a percentage by dividing the area of the corresponding peak by the sum of the areas of all peaks [11]. However, this method of calculation tends to overestimate the concentration of small peaks. For a more precise quantification of the mass of each TAG species, it is necessary to use calibration curves, but the diversity of TAG species in each oil would make it virtually impossible to construct a calibration curve for each TAG species. For simplicity, we have chosen trilinolein (LLL) to determine an

Table 2
Identification of TAGs' molecular species present in chromatograms of lipid samples from mammalian milk. The numbers correspond to those identifying the peaks in Fig. 2

Peak number	Molecular species	Peak number	Molecular species	Peak number	Molecular species	Peak number	Molecular species
1	**BCM**	18	CaOO	35	MPL	**52**	**PPP/MPS**
2	**BCO**	**19**	**CaPO/CLaO**	36	LaOO/MoPO	53	PdOS/PdPS
3	**BCP**	20	CyMP	**37**	**MMO/LaPO**	54	SOO
4	**BMPo/BLaO**	**21**	**CaPP/BOS/CaMS**	38	COS/MPPo	55	SSL
5	**BLa/BMM**	22	**BPS**	**39**	**MMP/LaPP/CPS**	56	**PSO**
6	BPoO	23	CyOO	40	MPdO/MPdP	57	PoSS
7	BMPd	24	CMO	**41**	**MOO/PPoO**	**58**	**PPS/MSS**
8	BPPo	**25**	**CyOP/CMP**	42	MSL/PPL	59	SSO
9	**BMO/CaLaP**	26	CyPP	**43**	**MPO/PPPo**	**60**	**PSS**
10	BMoS	27	CaPS	44	LaOS	61	SSS
11	**BMP**	28	LaOL	**45**	**MMS/MPP/LaPS**		
12	BPdO/BPdP	29	COO/MMoO	46	MMP/PdPP/OOO		
13	**BOO/CaMO**	**30**	**CPO/LaMO**	**47**	**POO**		
14	**BPO/CaMP**	31	MMoP/CyOS	48	PoSO		
15	CCP	32	LaMP/CPP/CMS	**49**	**PPO/MSO**		
16	**BPP/BMS**	33	CyPS	50	PPoS		
17	CaPdP	34	LaPdO/LaPdP/LMO	51	PdPM		

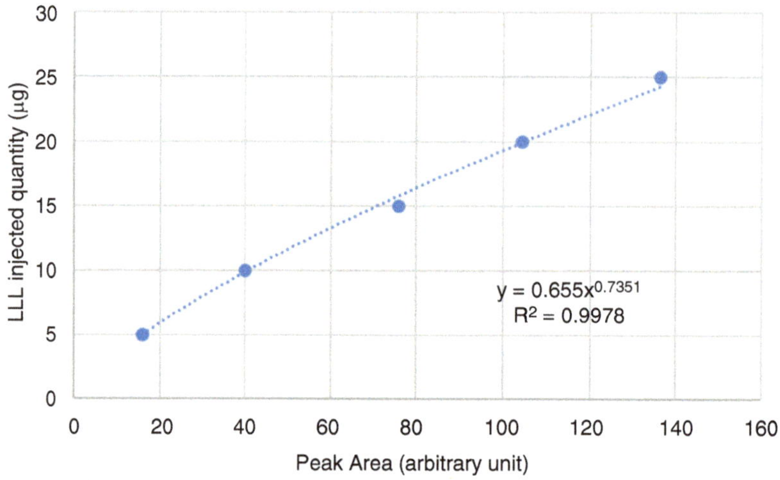

Fig. 3 Calibration curve linking the trilinolein injected quantity to the peak area on the chromatogram. The equation of the power law correlation is plotted on the graph

average response curve by plotting the peak areas given by 5–25 µg trilinolein (LLL) (*see* **Note 1**). After elution, the injected amount is plotted against the integrated area of the trilinolein peak as in Fig. 3 and a power-law regression is calculated using Excel or another spreadsheet. This power law is then used to quantify each TAG identified in a sample based on the integrated area of its elution peak.

TAG (µg) = α (peak area) $^{\beta}$ with α and β, the coefficients obtained from the power law regression.

4 Notes

1. The response of the light scattering detector depends on the composition of the solvent to be evaporated, which varies in time according to the gradient used to elute the TAGs. When ELSD is used, it is necessary to work out the optimum conditions. The operational parameters for the instrument, such as gas pressure, flow rate, evaporator temperature, and gain, must be standardized rigorously. Of course, the response differs with different commercial ELSDs. The response factor of the different TAGs will depend on their elution time (nature of the mobile phase). Trilinolein (LLL) eluted in the gradient medium is a standard TAG, which is a good compromise, but depending on the TAG composition of the lipid samples to be analyzed, it is possible to choose another one that is eluted in the beginning such as trilinolenin (LnLnLn) or at the end such as triolein (OOO) as represented in Fig. 1.

References

1. Ruiz-Gutierrez V, Barron LJR (1995) Methods for the analysis of triacylglycerols. J Chromatogr B 671:133–168
2. Morera Pons S, Castellote Bargallo AI, Lopez Sabater MC (1998) Analysis of human milk triacylglycerols by high-performance liquid chromatography with light-scattering detection. J Chromatogr A 823:475–482
3. Buchgraber M, Ulberth F, Emons H, Anklam E (2004) Triacylglycerol profiling by using chromatographic techniques. Eur J Lipid Sci Technol 106:621–648
4. Rombaut R, De Clercq N, Foubert I, Dewettinck K (2009) Triacylglycerol analysis of fats and oils by evaporative light scattering detection. J Am Oil Chem Soc 86:19–25
5. Ross KL, Hansen SL, Tu T (2011) Reversed-phase analysis of triacylglycerols by ultra-performance liquid chromatography evaporative light scattering detection (UPLC-ELSD). Lipid Technol 23:14–16
6. Endo Y, Ohta A, Kido H et al (2011) Determination of triacylglycerol composition in vegetable oils using high-performance liquid chromatography: a collaborative study. J Oleo Sci 60:369–373
7. Nikolova-Damyanova B (2012) Reversed-phase high-performance liquid chromatography: general principles and application to the analysis of fatty acids and triacylglycerols. Adv Lipid Methodol 4:193–251
8. Bail S, Stuebiger G, Unterweger H et al (2009) Characterization of volatile compounds and triacylglycerol profiles of nut oils using SPME-GC-MS and MALDI-TOF-MS. Eur J Lipid Sci Technol 11:170–182
9. Fuchs B, Suss R, Schiller J (2010) An update of MALDI-TOF mass spectrometry in lipid research. Prog Lipid Res 49:450–475
10. Fogang Mbaa AR, Kansci G, Viau M et al (2017) Lipid and amino acid profiles support the potential of Rhynchophorus phoenicis larvae for human nutrition. J Food Compos Anal 60:64–73
11. AOCS Official method Ce 5c-93, Individual Triglycerides in Oils and Fat by HPLC

Chapter 9

Identification and Quantification of Phospholipids by HPLC/ELSD: Application to Bovine Milk Phospholipids

Christelle Lopez

Abstract

The quantification of phospholipids, including glycerophospholipids and sphingolipids (mainly sphingomyelin in bovine milk), requires an extraction of total lipids from samples and then their analysis by high-performance liquid chromatography (HPLC) combined with an evaporative light scattering detector (ELSD). The sum of glycerophospholipids (PE, PI, PS, and PC) and sphingomyelin concentration is regarded as total phospholipid concentration.

Key words Phospholipid, Polar lipid, Sphingomyelin, HPLC

1 Introduction

Phospholipids, including glycerophospholipids and sphingolipids, are important nutritional and functional components of dietary lipids.

In natural oil-in-water emulsions, lipid droplets are covered by a layer of phospholipids, also called polar lipids. For example, plant oil bodies (also called oleosomes or plant lipid droplets) are covered by a monolayer of phospholipids and proteins (oleosins, caleosins) and milk fat globules are covered by a biological membrane composed of a trilayer of phospholipids and proteins (e.g., butyrophilin, mucins). Phospholipids are the main components of lecithins that are used as emulsifiers in the food industry (soya lecithin, egg lecithin). They are also found in other ingredients, such as protein powders originating from a technological process leading to the recovery of phospholipids (e.g., the by-product of the cold pressing of oil seeds containing proteins, fibers, and phospholipids).

The protocol detailed below was first published by Lopez et al. [2] to identify and quantify milk phospholipids in the biological membrane surrounding fat globules in milk, including glycerophospholipids and sphingomyelin.

Christelle Lopez et al. (eds.), *Multidimensional Characterization of Dietary Lipids*, Methods and Protocols in Food Science,
https://doi.org/10.1007/978-1-0716-3758-6_9,

2 Chemicals

Chloroform stabilized with ethanol (HPLC grade).

Methanol (HPLC grade).

Triethylamine (purity >99%).

Formic acid (purity >98%).

The phospholipid standards:

- PE (L-R-phosphatidylethanolamine dipalmitoyl, N,N-dimethyl (C16:0); purity 99%).
- PI (L-R-phosphatidylinositol ammonium salt, from soybean; purity 98%).
- PS (1,2-diacyl-sn-glycero-3-phosphoserine; purity 98%).
- PC (1,2-dipalmitoyl-sn-glycero-3-phosphocholine; purity 99%).
- Sphingomyelin (SM from bovine milk; purity 99%).

3 Extraction of Total Lipids

Extraction of total lipids is performed by an adapted protocol of the cold extraction procedure developed by Folch et al. [1], in order to recover the phospholipids (polar lipids) together with the other lipids, mainly the triacylglycerols (non polar neutral lipids).

Two grams of the sample are homogenized with chloroform/methanol (2:1, vol/vol) in a mixer (IKA, Fisher Bioblock).

The extract is shaken and equilibrated with one-fourth its volume of saline solution (NaCl 0.73%, w/w).

The upper phase is washed three times with a mixture of 40 mL of chloroform/methanol (2:1, vol/vol) and 10 mL of NaCl 0.58% (w/w).

The solvent phases containing total lipids are filtered (Whatman filter paper, 2.5 μm, Grosseron, France) through a pinch of anhydrous sodium sulfate and evaporated under vacuum.

The extraction of total lipids is performed in triplicate to obtain a coefficient of variation <5%.

Total lipids extracted are stored at −20 °C until further HPLC analysis.

4 Analysis of Phospholipids: Glycerophospholipids and Sphingolipids

4.1 Phospholipids: Concentration and Individual Classes

The quantification of the glycerophospholipids (PE, PI, PS, PC) and sphingolipids (sphingomyelin) and the determination of the phospholipid classes are performed using high-performance liquid chromatography (HPLC) combined with an evaporative light scattering detector (ELSD).

The chromatographic method used for the separation of the phospholipids extracted from the samples is adapted from the method of Rombaut et al. [3].

Glycerophospholipids and sphingomyelin separations are performed on an Agilent HPLC chain (HP 1100, Agilent, Massy, France) with four solvent lines, a degasser, and a quaternary pump.

The detector is an evaporative light scattering detector (ELSD; PL-ELS1000, Polymer Laboratories, Marseille, France).

HPChem software (Agilent, Massy, France) permits the acquisition of data from the ELSD detector.

A silica column, 150 × 3 mm with 3 μm particle diameter (AIT, Houilles, France) and a precolumn in silica with the same packing and internal diameter are used.

Dried and filtered compressed air is used as the nebulizing gas at a flow rate of 1.7 L/min and a temperature of 50 °C.

The evaporating temperature is 85 °C.

The elution program is as follows:

(i) Isocratic conditions with chloroform/methanol/buffer (1 M formic acid, neutralized to pH 3 with triethylamine) 87.5:12:0.5 (vol/vol/vol) from $t = 0$ min to $t = 7$ min.

(ii) A linear gradient of chloroform/methanol/buffer 87.5:12:0.5 (vol/vol/vol) to 28:60:12 (vol/vol/vol) from $t = 7$ min to $t = 27$ min.

The mobile phase is brought back to the initial conditions, i.e., 87.5:12:0.5 (vol/vol/vol), from $t = 27$ min to $t = 29$ min, and the column is allowed to equilibrate until the next injection at $t = 36$ min.

The total chromatographic run time is 36 min per sample, which consist of (i) the 7 min in isocratic conditions to elute the nonpolar (neutral) lipids, (ii) the linear gradient to elute the polar lipids (glycerophospholipids and sphingomyelin), (iii) 2 min to restore the initial conditions, and (iv) the last 7 min to re-equilibrate the column.

The flow rate is maintained at 0.5 mL/min.

The column is maintained at 40 °C.

The injection volume is 10 μL per sample.

The lipids extracted are weighed (100 mg) and dissolved in 500 μL of chloroform/methanol (88:12, vol/vol) and transferred into capped test tubes for HPLC analysis.

Each sample is injected three times.

4.2 HPLC/ELSD Calibration

The identification of glycerophospholipids and sphingomyelin is carried out by comparison with the retention time of pure standards.

To obtain a quantitative evaluation of the glycerophospholipids and sphingomyelin, five calibration curves are determined from the area values obtained by injecting 10 μL of chloroform/methanol (88:12, vol/vol) serially diluted solutions of PE, PC, PS, PI, and SM.

Each solution is prepared and injected in triplicate.

Calibration curves are calculated by applying the equations of the polynomial model to the area and concentration values.

The sum of glycerophospholipids (PE, PI, PS, and PC) and sphingomyelin concentration is regarded as total phospholipid concentration.

4.3 Results

Figure 1 shows the HPLC chromatograms recorded with the described protocol. The efficient separation of the different polar lipid classes permits their quantification.

Figure 2 shows the calibration curves obtained for the phospholipids (PC, PE, PI, PS) and sphingomyelin (SM).

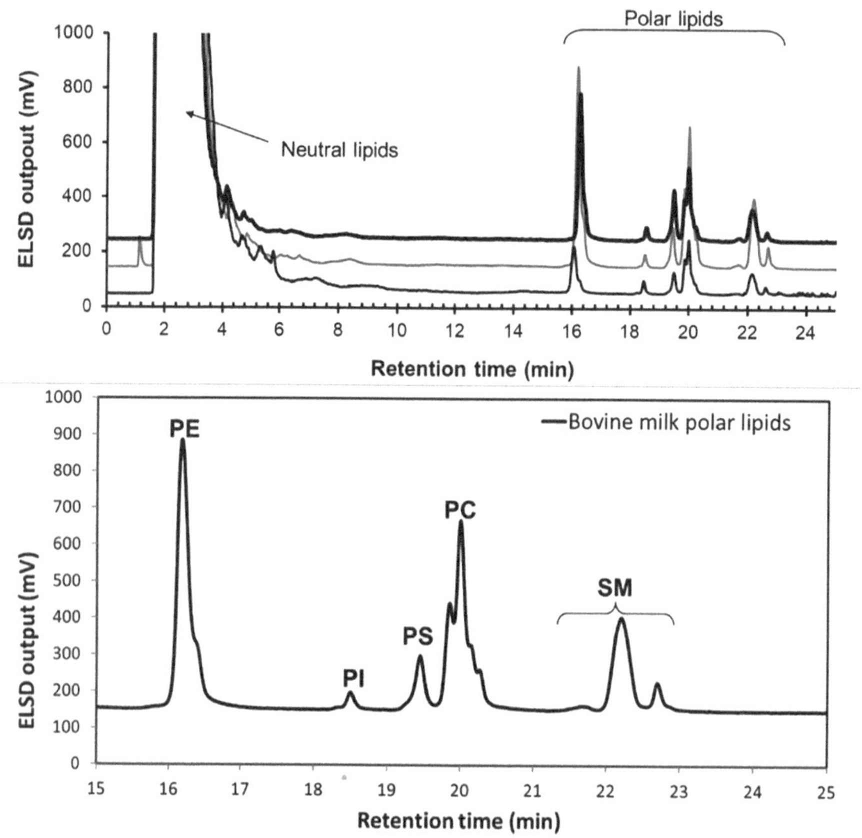

Fig. 1 Chromatograms obtained upon elution of total lipids, including non polar neutral lipids and polar lipids with the individual PE, PI, PS, PC, and SM

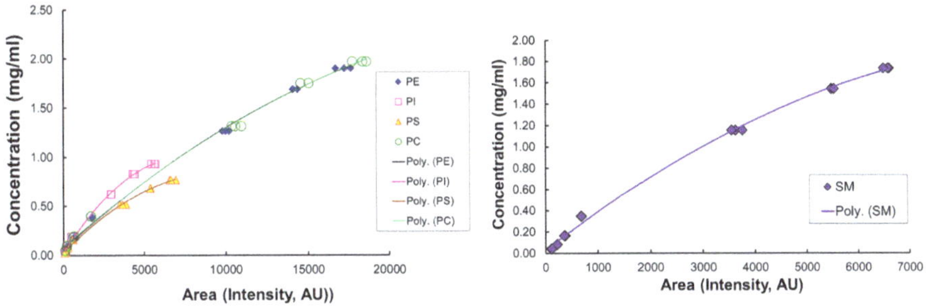

Fig. 2 Calibration curves

References

1. Folch J, Lees M, Stanley GHS (1957) A simple method for the isolation and purification of total lipides from animal tissues. J Biol Chem 226: 497–509
2. Lopez C, Briard-Bion V, Menard O, Rousseau F, Pradel P, Besle J-M (2008) Phospholipid, sphingolipid, and fatty acid compositions of the milk fat globule membrane are modified by diet. J Agric Food Chem 56:5226–5236. https://doi.org/10.1021/jf7036104
3. Rombaut R, Camp JV, Dewettinck K (2005) Analysis of phospho- and sphingolipids in dairy products by a new HPLC method. J Dairy Sci 88:482–488. https://doi.org/10.3168/jds.S0022-0302(05)72710-7

Chapter 10

Quantification of Tocopherols and Tocotrienols (Tocols) in Oils, Fats, and Food Products by HPLC Fluorescence

Lucie Ribourg-Birault and Anne Meynier

Abstract

Tocopherols and to a lesser extent tocotrienols were widely present in food from animal and vegetable origin. Some of these compounds exhibited a vitamin E activity and played a major action in the cell antioxidant defence system and as antioxidants per se in foods. Tocols were mainly present in the human diet. Thus, accurate quantification of tocols in fats, oils, and foods was required both for nutritional recommendations and food quality and shelf life in relation to their antioxidant properties. In this chapter, a step-by-step protocol was presented to quantify these compounds -after normal phase liquid chromatography separation and fluorescence detection of tocols in oils and some foods of different origins.

Key words Tocopherol, Tocotrienol , Tocos , Vitamin E , Antioxidant , Lipid oxidation , HPLC

1 Introduction

Vitamin E is a liposoluble vitamin, recognized as an antioxidant, thanks to its ability to inhibit lipid peroxidation through its reactivity with reactive oxygen species (ROS). The term "tocopherols or tocols or tocochromanols" [1] includes two groups of molecules: tocopherols and tocotrienols. Each family includes four substances: α, β, γ, and δ-tocopherol and α, β, γ, and δ-tocotrienol. All contain a chromacol ring and a hydrophobic side chain, namely, a phytyl chain in tocopherols and an isoprenyl chain (three double bounds) in tocotrienols. The term vitamin E denotes tocopherol and tocotrienol derivatives (Table 1) having biological activities identical to those of alpha-tocopherol, the most active biological form is alpha-tocopherol. Although playing essential roles in human and animal biology, tocopherols are synthesized only by organisms performing photosynthesis [2]. Besides the numerous health benefits, tocols are also used to increase the shelf life and stability of foods [3, 4].

The main dietary sources of vitamin E are vegetable oils. Vitamin E is also present in products of animal origin but in smaller

Christelle Lopez et al. (eds.), *Multidimensional Characterization of Dietary Lipids*, Methods and Protocols in Food Science,
https://doi.org/10.1007/978-1-0716-3758-6_10,

Table 1
Structure of tocopherol and tocotrienol molecules

Form	Structure	R^1	R^2	R^3
α-tocopherol	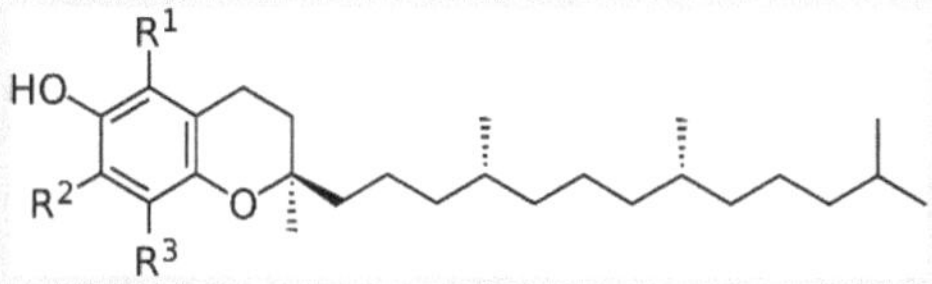	CH_3	CH_3	CH_3
β-tocopherol		CH_3	H	CH_3
γ-tocopherol		H	CH_3	CH_3
δ-tocopherol		H	H	CH_3
α-tocotrienol	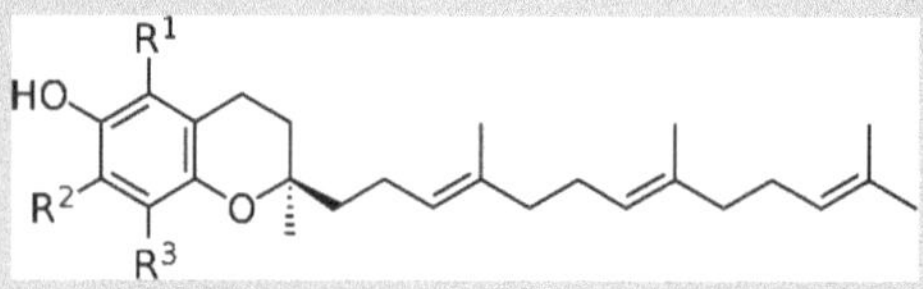	CH_3	CH_3	CH_3
β-tocotrienol		CH_3	H	CH_3
γ-tocotrienol		H	CH_3	CH_3
δ-tocotrienol		H	H	CH_3

quantities (Fig. 1a). There is no vitamin E deficiency observed in the population. It is therefore not necessary to consume food supplements containing vitamin E.

Nowadays, international agencies and nutritional recommendations encourage us to diversify our dietary oils to diversify and re-equilibrate our intake of fatty acids and to increase the portion of plant protein in our diet. Some of these new sources contain a noticeable amount of tocos (Fig. 1b).

Oils and foods consumed in Europe contain very little tocotrienols, but trends to diversify our diet and reduce our animal food intake could qualify this statement. For example, palm oil contains more tocotrienols than tocopherols [5].

Interested readers can refer to the following reviews related to the extraction and analysis of tocols [1, 6]. Sample preparation can include simple solubilization of the edible oils or fat in an apolar solvent such as n-hexane, liquid–liquid extraction, or solvent extraction after saponification [7]. Numerous chromatographic methods can separate and then quantify tocols. In this chapter, we will focus on a normal phase (NP)-HPLC method paired with a fluorimeter detector (Fig. 2) as this detector is one of the most sensitive for the targeted molecules [8].

2 Materials

2.1 Chemical Products

Prepare solutions using ultrapure water (conductivity = 18 MΩ. cm^{-1} at 25 °C) and/or HPLC-grade reagents (Table 2). Diligently follow safety instructions and waste disposal regulations when disposing waste materials.

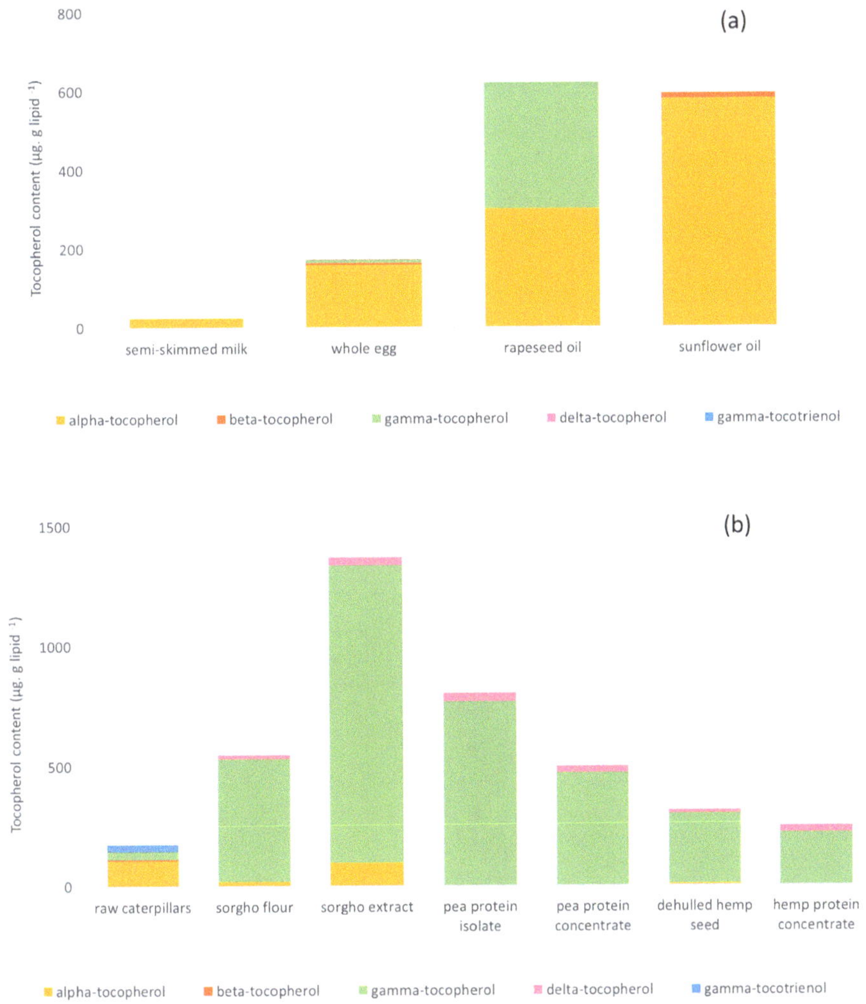

Fig. 1 (**a**) Example of tocols content and composition in various foods and edible oils and (**b**) example of tocols content and composition in emerging sources as alternatives to animal products. (Data unpublished to date obtained in our laboratory)

2.2 Mandatory Equipment

1. Analytical precision balance, sensitivity: 0.01 mg.
2. HPLC with fluorescence detector.
3. Silica HPLC column, 250 × 3 mm, particle size 3 μm (Dionex Polar advantage 2 for example).
4. UV–visible spectrophotometer; absorption semi-micro cells (optical path-length: 10 mm).
5. Positive displacement pipettes and disposable capillary piston tips (1000–50 μL).

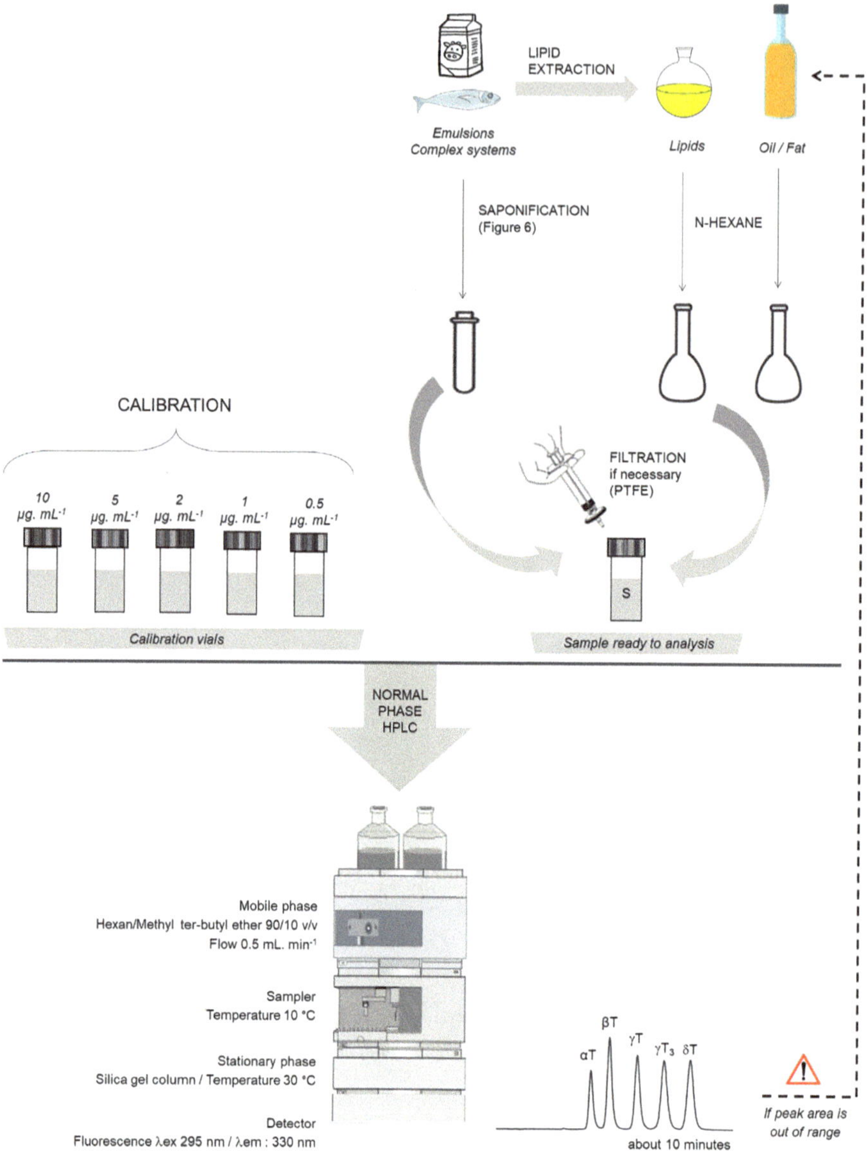

Fig. 2 Flow chart of the main steps involved in the extraction and analysis of tocopherols and tocotrienols in food products by HPLC fluorescence

6. Graduated cylinders: 500 mL, 50 mL.
7. 500 mL glass bottle.
8. HPLC glass vial.
9. Volumetric flasks.

Table 2
Chemical required for tocopherols and tocotrienols method

Chemical	Chemical formula	CAS number	Molecular weight (g.mol^{-1})	Quality/reference	Safety instructions
n-Hexane	C_6H_{14}	110-54-3	86.18	HPLC grade	
Methyl tert-butyl ether	$C_5H_{12}O$	1634-04-4	88.15	HPLC grade	
Tocopherol set (α, β, γ, δ-tocopherol)	α: $C_{29}H_{50}O_2$ β: $C_{28}H_{48}O_2$ γ: $C_{28}H_{48}O_2$ δ: $C_{27}H_{46}O_2$		430.71 416.68 416.68 402.65	Calbiochem item 613424	
γ-Tocotrienol	$C_{28}H_{42}O_2$	14101-61-02	410.63	Cayman chemical item 10008494	
Ethanol	C_2H_6O	64-17-5	46.07	For analysis	
Ascorbic acid	$C_6H_8O_6$	50-81-7	176.12		
Potassium hydroxide	KOH	1310-58-3	56.11		

3 Method

3.1 Calibration

3.1.1 Determination of the Concentration of Tocopherols and γ-Tocotrienol in Commercial Solutions

1. Depending on the cell quantity, disperse each individual commercial standard in the appropriate volume of hexane to obtain a final concentration of 10 mg mL^{-1}. Store these solutions in the frozen state (−80 °C if possible) in brown bottles closed under nitrogen.
2. Determine the actual concentration of tocopherol of these solutions; carry out dilutions in ethanol to obtain a concentration of approximately 10 μg mL^{-1}. Measure the absorbance with a UV–visible spectrophotometer at the wavelengths indicated in Table 3. As an indication, the absorbance is between 0.07 and 0.1 for solutions with a concentration of 10 μg. mL^{-1}.

   These values are taken from the AOCS [9]. For readers particularly interested in tocotrienols, the data are available in other articles [10, 11].

Table 3
Molar extinction coefficient (ε) and optimal absorption wavelength (λ_{max}) of each tocol isomer prepared in ethanol

	α-Toco	β-Toco	γ-Toco	δ-Toco	γ-Tocotrienol
λ_{max} (nm)	292	296	298	298	296
ε (mL.μg^{-1}.cm^{-1}) [9]	0.00756	0.00892	0.00912	0.00872	0.00904

Table 4
Concentrations of standards solutions

Solution	Concentration of each tocol
Commercial solutions	10 mg. mL^{-1}
Standard solution A	10 μg. mL^{-1}
Standard solution B	5 μg. mL^{-1}
Standard solution C	2 μg. mL^{-1}
Standard solution D	1 μg. mL^{-1}
Standard solution E	0.5 μg. mL^{-1}

3. Calculate the tocopherol or tocotrienol concentration (C, mg. mL^{-1}) of each commercial solution with the Beer–Lambert law taking into account the dilutions (*see* **Note 1**):

$$C = \frac{\text{Abs} \times \text{dilution}}{\varepsilon \times l \times 1000}$$

with ε = the molar extinction coefficient of tocopherol or tocotrienol (see Table 3), dilution = 1000 and l = optical path of the cell.

*3.1.2 Preparation of Calibration Solutions (Fig. 3, see **Note 2**)*

1. Dilute each commercial standard (10 mg. mL^{-1}) by 100 in n-hexane to obtain a concentration of 100 μg. mL^{-1}.
2. Prepare the calibration solution A: dilute by 10 each of the five individual isomers in n-hexane in the same vial (*see* **Note 3**).
3. Dilute sequentially this standard solution A by 2 (standard solution B), by 5 (standard solution C), by 10 (standard solution D), and by 20 (standard solution E) in n-hexane.

The concentrations of these standard solutions are indicated in Table 4.

As commercial standards are prepared individually, only the required ones (present in the targeted foods) can be analyzed.

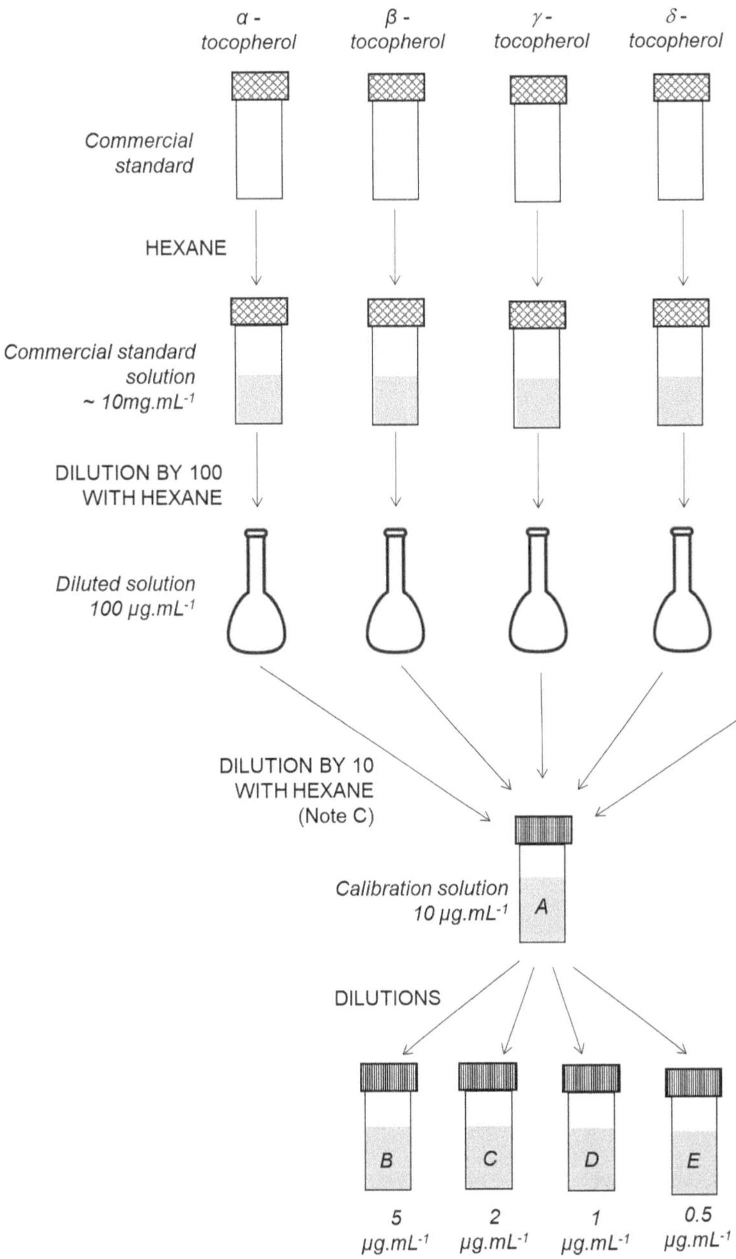

Fig. 3 Flow chart of the preparation of calibration solutions

3.1.3 HPLC Analysis

Forty microliters of these standard solutions are injected by normal phase HPLC equipped with an isocratic solvent delivery system and a fluorescence detector with excitation and emission wavelengths set at 295 nm and 330 nm, respectively. A silica gel column (25 cm × 3 mm ID, 3 μm) maintained at 30 degrees is used with the mobile phase of hexane/methyl tert-butyl ether (90/10 v/v) at a flowrate of 0.5 mL min^{-1}. The temperature of the sampler is set at 10 degrees.

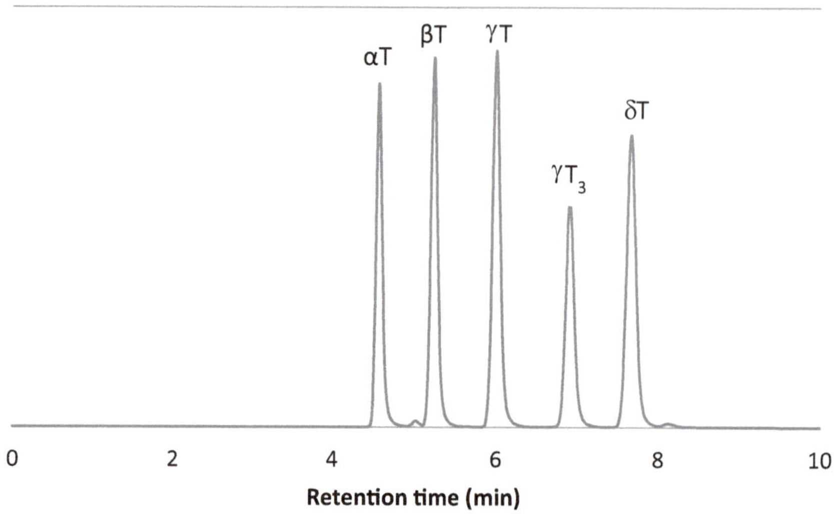

Fig. 4 Typical chromatogram of the standard solution A containing α-tocopherol (1), β-tocopherol (2), γ-tocopherol (3), γ-tocotrienol (4), and δ-tocopherol (5)

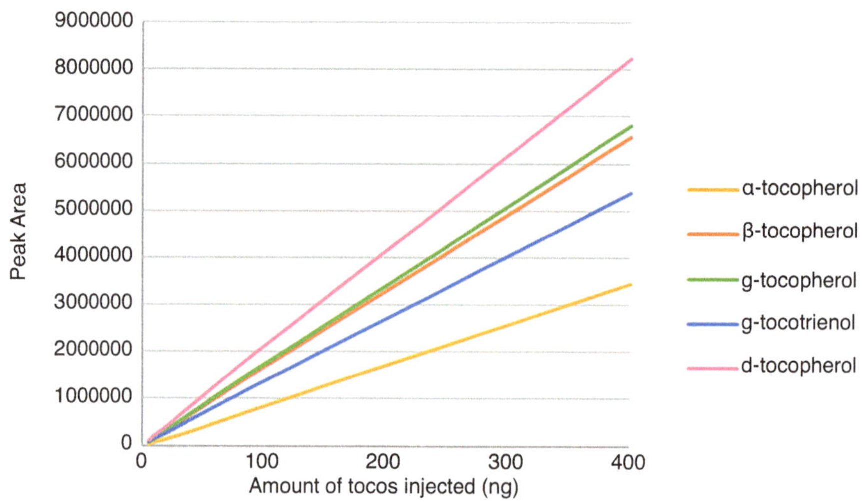

Fig. 5 Calibration curves produced from standard solutions containing α, β, γ, δ-tocopherol and γ-tocotrienol

The chromatogram shows five peaks (Fig. 4), in the following elution order, that is, one peak per isomer: α-tocopherol (αT), β-tocopherol (βT), γ-tocopherol (γT), γ-tocotrienol (γT_3), and δ-tocopherol (δT) identified by comparing chromatogram with those obtained by injecting the standard individually.

3.1.4 Calibration Curve

The five isomers exhibited different response factors (Fig. 5). If there is α, β, or δ-tocotrienol in the sample, the quantities must be expressed in ng of equivalent γ-tocotrienol (*see* **Note 5**).

Detector response was a linear function of the injected amounts. The calibration curves are $y = ax + b$ with y, peak area, and x, the quantity of tocopherol isomer in ng injected, obtained by multiplying the actual concentration of the standard in μg mL^{-1} by the volume injected in μL × 1000.

Linearity is validated between 10 and 400 ng injected (and down to 5 ng injected for α-tocopherol, β-tocopherol, and γ-tocopherol) (*see* **Note 4**).

3.2 Determination of Tocopherols and γ-Tocotrienol Content in the Sample

Oil or extracted lipids can be directly dissolved in hexane and then injected in HPLC. For complex samples such as milk, a saponification is first carried out and then the unsaponifiable fraction is analyzed by HPLC.

3.2.1 Sample Preparation

Edible Oils

A solution of lipids in n-hexane is prepared, and the chosen concentration depends on the presumed level of oxidation of the sample (the more oxidized, the less expected amount of tocols). For example, prepare a solution of concentration 10 mg mL^{-1} of lipids in n-hexane for a fresh and nonoxidized edible oil. For an oxidized or purified oil which has been stripped of most of its tocopherols, use a concentration of 200 mg mL^{-1}.

Weigh accurately (0.1 mg) a defined quantity of oil in a vial or flask and add a defined volume of hexane in the vial or complete the flask to the mark with hexane to obtain the targeted concentration. Vortex or shake with a magnetic bar.

Extracted Lipids

For the lipid extract of known concentration in a given solvent, introduce the required volume of the solution in a vial, dry with nitrogen, and add n-hexane to obtain the appropriate concentration, related to the expected range of tocols concentration, then vortex. If necessary, samples can be filtered through a 0.22 μm PTFE membrane before injection.

Others Sample with Lipids (Milk, Meat, Fish, Microalgae)

For this kind of complex sample, it is recommended to perform saponification before tocol analysis and then to analyze the unsaponifiable fraction which contains tocols. The saponification step allows the removal of lipids (TAG, PL, etc.) avoiding the overload of the HPLC column and maintaining a good resolution factor.

In terms of convenience and time saving, saponification can be preferred over the tedious lipid extraction step when there is no need to quantify lipids from a sample or to make additional characterization such as FA composition.

Figure 6 shows the steps of the saponification protocol for milk:

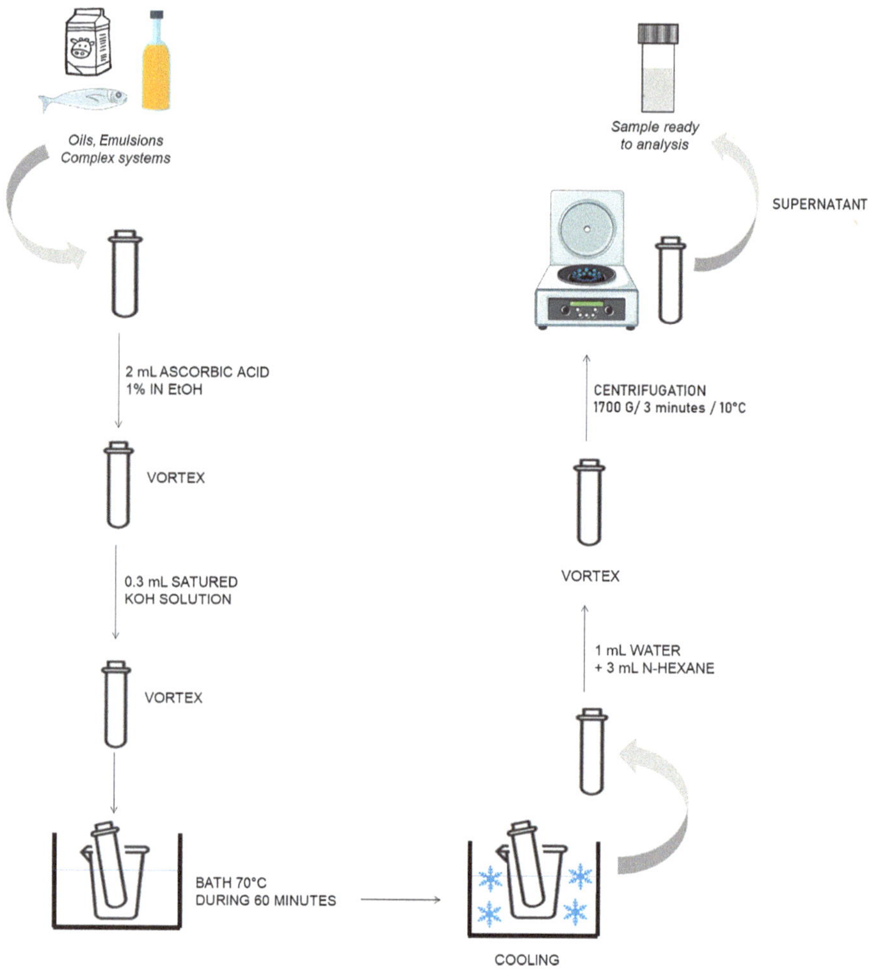

Fig. 6 Flow chart of the main steps involved in the saponification of sample

1. In a glass centrifuge tube, add 2 mL of milk then 2 mL of ascorbic acid and 1% in ethanol, vortex.
2. Add 0.3 mL of saturated solution of potassium hydroxide (about 500 g. L^{-1}) and vortex.
3. Immerse the tube in a bath at 70 degrees for 60 min then cool in an ice-water bath.
4. In the tube, add 1 mL of water and 3 mL of hexane, vortex.
5. Finally, centrifuge the tube at 1700 g for 3 min and take the supernatant corresponding to the hexane phase.
6. Inject directly in HPLC or concentrate before injection.

3.2.2 Determination of Tocopherols and γ-Tocotrienol Content

The peaks of the chromatogram are integrated and identified by their retention time (*see* **Note 5**).

With parameters of the calibration curve, calculate the amount of tocopherols and tocotrienols in the volume of the sample injected, then express the tocopherol content in micrograms per gram of sample or gram of lipids using the concentration in lipids of the sample.

The detection limit ranged from 1 to 2 ng injected depending on the isomers. Quantification limits were 5 ng injected for α-tocopherol, β-tocopherol, and γ-tocopherol and 10 ng for δ-tocopherol and γ-tocotrienol. For low tocol values, we recommend preparing a sample solution more concentrated in lipids or increasing the injected volume in the HPLC system.

4 Notes

1. Check the concentration of commercial solutions every 2 months before preparation of new calibration.
2. Prepare standard solutions every 2 months from checked commercial solutions.
3. For convenience, we recommend mixing the five standards, but it can be considered to prepare and inject each standard one by one.
4. For routine analysis, the calibration curve ranges from 10 ng to 400 ng, but for specific applications, lower calibration point can be added (5 ng).
5. In the case of matrices that are little known or studied, the use of a mass spectrometry detector may be relevant to validate the identifications of the peaks.

If the sample contains isomers of tocotrienols other than gamma-tocotrienol for which a standard is not available at the laboratory, their amount will be expressed in gamma-tocotrienol equivalent.

Acknowledgments

We would like to give special thanks to Thais-Christina Bennatti-Gallo, Elena Keuleyan, Perrine Gelebart, and Youssef Tahir for the work on sorghum, hemp, and pea protein ingredients that led to the data on tocols (Fig. 1b).

References

1. Saini RK, Keum YS (2016) Tocopherols and tocotrienols in plants and their products: a review on methods of extraction, chromatographic separation, and detection. Food Res Int 82:59–70. https://doi.org/10.1016/j.foodres.2016.01.025
2. Guilland JC, Lequeu B (2009) Les sources de vitamines. In: Encyclopédie des vitamines. Du nutriment au médicament. Tec&Doc, Paris, pp 849–1640

3. Kamal-Eldin A, Budilarto E (2015) Tocopherols and tocotrienols as antioxidants for food preservation. Elsevier Ltd
4. Seppanen CM, Song Q, Saari Csallany A (2010) The antioxidant functions of tocopherol and tocotrienol homologues in oils, fats, and food systems. J Am Oil Chem Soc 87:469–481. https://doi.org/10.1007/s11746-009-1526-9
5. Monde AA, Michel F, Carbonneau MA et al (2009) Comparative study of fatty acid composition, vitamin E and carotenoid contents of palm oils from four varieties of oil palm from Côte d'Ivoire. J Sci Food Agric 89:2535–2540. https://doi.org/10.1002/jsfa.3740
6. Moreau RA, Lampi AM (2012) Analysis methods for tocopherols and tocotrienols. Anal Antioxidant-Rich Phytochem 353–386. https://doi.org/10.1002/9781118229378.ch11
7. Rupérez FJ, Martín D, Herrera E, Barbas C (2001) Chromatographic analysis of α-tocopherol and related compounds in various matrices. J Chromatogr A 935:45–69. https://doi.org/10.1016/S0021-9673(01)01101-3
8. Cunha SC, Amaral JS, Fernandes JO, Oliveira MBPP (2006) Quantification of tocopherols and tocotrienols in Portuguese olive oils using HPLC with three different detection systems. J Agric Food Chem 54:3351–3356. https://doi.org/10.1021/jf053102n
9. Ce 8–89 De- Determination of tocopherols and tocotrienols in vegetable oils and fats by HPLC. In: AOCS Official Methods. AOCS
10. Podda M, Weber C, Traber MG, Packer L (1996) Simultaneous determination of tissue tocopherols, tocotrienols, ubiquinols, and ubiquinones. J Lipid Res 37:893–901. https://doi.org/10.1016/S0022-2275(20)37587-8
11. Kofler M, Sommer PF, Bolliger HR et al (1962) Physicochemical Properties and Assay of the Tocopherols. Vitam Horm 20:407–440. https://doi.org/10.1016/S0083-6729(08)60727-X

Chapter 11

Dietary Carotenoid: Identification and Quantification

Pascale Goupy and Béatrice Gleize

Abstract

In the diet, carotenoids, the lipophilic pigments, are mainly provided by fruits and vegetables. The chemical class of carotenoids includes approximately 750 molecules characterized in natural sources with different chemical structures (linear, double bond, ring, oxygen function, esterification, isomerization, etc.). Different analytical methods can be used depending on the plant matrices and target carotenoids for sample preparation, lipid extraction, and liquid chromatography method.

This chapter will present a general procedure to be implemented to identify and quantify unesterified and esterified carotenoids from fruits and vegetables from sample preparation to data processing.

Key words Carotene, Xanthophyll, Isomer, Esterified, Saponification

1 Introduction

Carotenoids, which are lipid-soluble micronutrients, exhibit a characteristic tetraterpene skeleton of 40 C-atoms involving conjugated double bonds (which serve as the light-absorbing chromophore) and are responsible for the yellow to red colors of many fruits and vegetables. The tetraterpene skeleton can undergo different modifications, such as cyclization at one end or both ends, hydrogenation, dehydrogenation, chain shortening or extension, double-bond migration, rearrangement, isomerization, introduction of oxygen functions, or a combination of certain processes, and therefore result in a multitude of compounds. By their structure, two subclasses can be defined: the carotenes (pure hydrocarbons) and the xanthophylls (oxygenated groups). The hydroxyl groups of xanthophylls can be esterified with fatty acids. The xanthophylls present in vegetables (green leaf) are mainly unesterified (free form), while those in ripe fruits are mainly esterified. Carotenoids in fruits and vegetables exist mainly in their most stable isomeric form, the all-*trans* (all-*E*), but *cis* (*Z*) isomers can also be found in the plant matrix or produced during food processing (Fig. 1) [1].

Christelle Lopez et al. (eds.), *Multidimensional Characterization of Dietary Lipids*, Methods and Protocols in Food Science,
https://doi.org/10.1007/978-1-0716-3758-6_11,

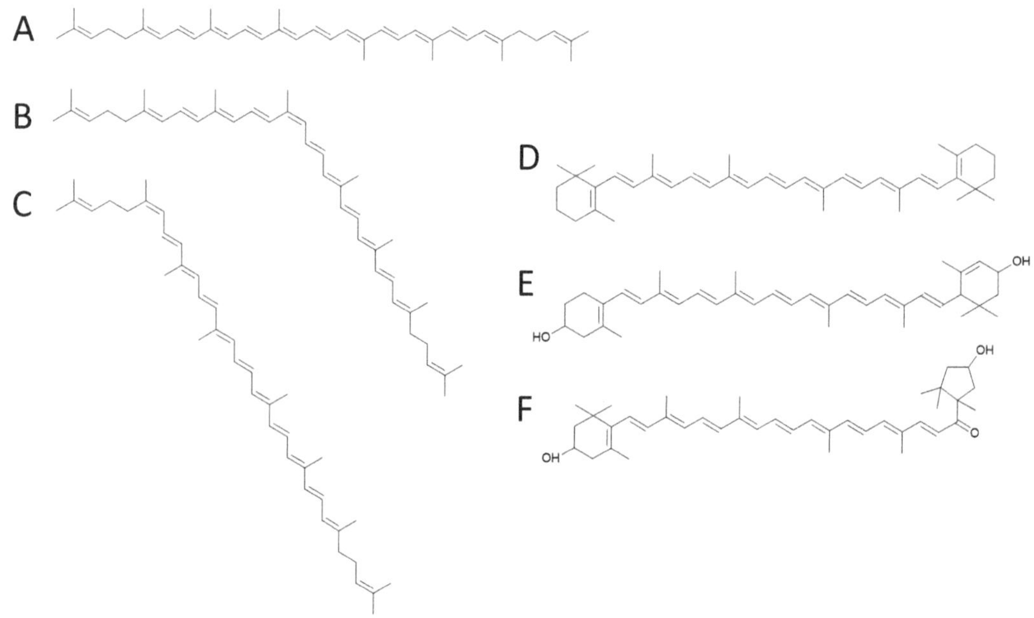

Fig. 1 Structure of main carotenoids of fruits and vegetables. (**a**) (all-*E*)-lycopene; (**b**) (13-*Z*)-lycopene; (**c**) (5-*Z*)-lycopene; (**d**) (all-*E*)-β-carotene; (**e**) (all-*E*)-lutein; (**f**) (all-*E*)-capsanthin

If carotenoids are relatively stable in the plant matrix, their chemical structure is sensitive to some reactions, such as oxidation and isomerization, especially due to light, heat, acids, and oxygen exposure when these molecules are in solution [2, 3].

There are no standard methods for carotenoid analysis; they were adapted according to food matrices and target carotenoids. However, most analysis methods follow a common path involving the release of components from the plant matrix by disrupting tissue and liquid–solid or liquid–liquid extraction, followed by analysis in liquid chromatography.

The general procedure to identify and quantify unesterified and esterified carotenoids will be described step-by-step, and two UPLC DAD MS_n methods are presented (Fig. 2).

2 Materials

Prepare all solutions using ultrapure water (sensitivity of 18 MΩ.cm at 25 °C). The chemical products required are displayed in Table 1. Prewash carefully all glasses with dichloromethane before use.

2.1 Sample Preparation

1. Homogenizer Silent Crusher M with the dispersion tool 18F (Heidolph, Schwabach, Germany).
2. Analytical mill A11 basic (IKA, Staufen, Germany).
3. Liquid nitrogen (*see* **Note 1**).

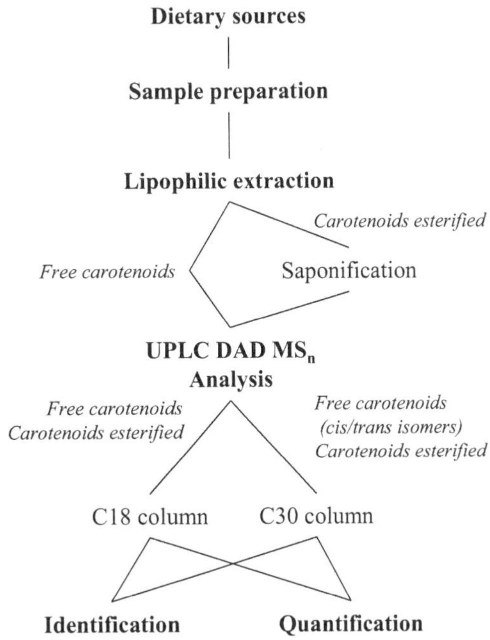

Fig. 2 Decision tree for carotenoids analysis

2.2 Lipophilic Extraction

1. Homogenizer SPEX 1600 MiniG® (SPEX SamplePrep, Delta-Labo, France).
2. 0.1 mm zirconia/silica beads (BioSpec Products, Bartlesville, USA).
3. 2 mL screw cap tube.
4. Saturated aqueous NaCl solution: Weigh 36 g of NaCl in a bottle and add 100 mL of ultrapure water. Shake using a magnetic stirrer for 15 min at room temperature.
5. n-Hexane.
6. Dichloromethane.
7. Ethyl acetate.

2.3 Liquid Chromatography Analysis

1. Carotenoid analysis in liquid chromatography can be done with different chromatographic systems (HPLC or UPLC) coupled with a diode array detector and/or a mass spectrometry detector. The methods presented here were carried out by UPLC using an ACQUITY UPLC® system (Waters Corp., Milford, MA, USA) coupled with a diode array detector (UPLC DAD, Waters Corp., Milford, MA, USA) and an ion trap MS

Table 1
Chemical products

Chemical	Chemical formula	CAS number	Molecular weight (g/mol)	Quality	Safety instructions
Formic acid	CH_2O_2	64-18-6	46.03	LC-MS grade	
Hydrochloric acid	HCl	7674-01-0	36.46	37%	
Methanol	CH_4O	67-56-1	32.04	LC-MS grade	
Ammonium formate	$CHO_2{\bullet}NH_4$	540-69-2	63.06	99%	
Dichloromethane	CH_2Cl_2	75-09-2	84.93	Analytical grade	
Ethyl acetate	$C_4H_8O_2$	141-78-6	88.12	Analytical grade	
Hexane	C_6H_{14}	110-54-3	86.17	Analytical grade	
Methyl-*tert*-butyl-ether (MTBE)	$C_5H_{12}O$	75-09-2	88.15	Analytical grade	

equipped with an atmospheric pressure chemical ionization source (HCT Ultra Ion Trap, Bruker Daltonics, Bremen, Germany).

2. Carotenoid analysis is commonly performed using the C18 or C30 stationary phase column. Two different columns are used in the presented methods:
   (a) ACQUITY UPLC® HSS T3 C18 column (150 mm × 2.1 mm i.d., 1.8 μm; Waters Corp., Milford, MA, USA).
   (b) YMC Carotenoid C30 column (150 mm × 4.6 i.d., 3 μm; YMC America Inc., Allentown, PA, USA).
3. Carotenoid analysis is carried out with different elution and injection solvents depending on the stationary phase of the columns.

   For the presented method using the C18 column, the elution and injection solutions are as follows:
   (a) Methanol/water (80/20, v/v) with 10 mM ammonium formate and 0.05% formic acid (*see* **Note 2**): Weigh 630 mg of ammonium formate in a 200 mL volumetric flask. Add 500 μL of formic acid and complete to 200 mL with ultrapure water. Transfer the solution to a 1 L bottle and add 800 mL of methanol. Stock the solution at room temperature in a ventilated cabinet.
   (b) Ethyl acetate.
   (c) Methanol.

   For the presented method using the C30 column, the elution and injection solutions are as follows:
   (a) Methanol with 0.05% formic acid: Add 500 μL of formic acid to 1 L of methanol. Stock the solution at room temperature in a ventilated cabinet.
   (b) Methyl *tert*-butyl ether.
   (c) Methanol/dichloromethane (65/35, v/v): Mix 65 mL of methanol and 35 mL of dichloromethane in a 100 mL bottle. Stock the mix at room temperature in a ventilated cabinet.

2.4 Saponification of Esterified Carotenoids

1. 5% KOH methanolic solution: Weigh 5 g of KOH in a 100 mL volumetric flask and complete with methanol. Shake the solution until KOH dissolution occurs (the solution is cloudy). The solution must be prepared on the day of handling.
2. Glass tube with a screw cap.
3. 37% HCl.
4. n-Hexane.

3 Methods

Carry out all procedures under dim light and at room temperature. Carry out each handling of solvents using an extractor hood.

3.1 Sample Preparation

Direct analysis of fruits and vegetables is not possible because first they must be transformed into a homogeneous sample. Sampling and homogenization can be done directly on fresh or frozen samples. The freeze-drying of the samples is not recommended for carotenoid analysis (*see* **Note 3**).

3.1.1 Fresh Samples

1. If necessary, cut the sample of fruits and vegetables into little pieces.
2. Weigh 200 to 500 g of sample in a beaker.
3. Homogenize to obtain a puree using the homogenizer set at 16,000 rpm for 3 to 5 min.
4. Use the homogeneous sample immediately for extraction or store it at −80 °C under argon.

3.1.2 Frozen Samples

1. Cool the blades of the mill with liquid nitrogen (*see* **Note 1**).
2. Put the frozen pieces of the sample in the mill bowl.
3. Cool the mill bowl and the sample with liquid nitrogen.
4. Remove excess nitrogen when everything is cool.
5. Screw the bowl onto the mill and then grind.
6. Use the sample powder immediately for extraction or store it at −80 °C under argon.

3.2 Lipophilic Extraction

1. Weigh 500 mg of sample in a 2-mL screw cap tube containing 80–100 mg of zirconia/silica beads.
2. Add 100 μL of saturated aqueous NaCl solution.
3. Add 100 μL of n-hexane and shake the solution for 45 s at full speed and high energy with the homogenizer SPEX 1600 MiniG®.
4. Centrifuge the solution at 13,000 g at 5 °C for 5 min.
5. Add 200 μL of dichloromethane and shake as in **step 3**.
6. Centrifuge as in **step 4**.
7. Add 800 μL of ethyl acetate and shake as in **step 3**.
8. After the last centrifugation (as in **step 4**), recover the supernatant (upper phase) and transfer it to a tube.
9. Re-extract the pellet three times (as in **steps 3–8**) until discoloration of the supernatant occurs.

10. Pool the supernatants and evaporate the solvent under argon at room temperature.
11. Use the dry extract immediately for UPLC analysis or store it at −20 °C under argon for a maximum of three days before UPLC analysis.

3.3 Liquid Chromatography Analysis

The principles of using an HPLC/UPLC will not be detailed here.

3.3.1 Liquid Chromatography Method

Two different methods are presented in this section depending on the stationary phase of the columns. They allow the separation of *E/Z* isomers of carotenoids and carotenoids esterified with fatty acids with more or less resolution depending on the solvents used and target carotenoids.

However, the setting of the sample manager and DAD and MS detectors is common to both methods:

1. The sample manager is set at 10 °C.
2. The wavelength range of DAD is set from 190 to 800 nm and working wavelengths of 440 nm for detection of xanthophylls and 470 nm for carotenes (*see* **Note 4**).
3. The ion trap mass spectrometry detector is equipped with an atmospheric pressure chemical ionization source in positive mode. The ionization source parameters are set as follows: dry temperature of 300 °C, nebulizer pressure of 10 psi, dry gas flow of 4 L min^{-1}, vaporizer temperature of 350 °C, capillary voltage of +2.0 kV, and corona current of 3200 nA.

Method Using the C18 Column

1. Place methanol/water with 10 mM ammonium formate and 0.05% formic acid solution as eluent A and ethyl acetate as eluent B.
2. Program the elution gradient as follows: A, 100%, 0–2 min; A, 100–30%, 2–20 min; A, 30%, 20–22 min; A, 30–20%, 22–23 min and return to the initial conditions (eluent A, 100%) for 5 min.
3. The flow is set at 0.4 mL min^{-1}.
4. Set the temperature of the C18 column to 50 °C.
5. Switch on the sample manager set at 10 °C, the DAD and the acquisition of the MS.
6. Equilibrate the system with 100% eluent A for 30 min.
7. Dissolve the dry extract of the sample in 1 mL of methanol.
8. Inject 2 μL of the extract (*see* **Note 5**).
9. After analysis, rinse the column with 100% eluent A for 15 min (*see* **Note 6**).

Method Using the C30 Column

1. Place methanol with 0.05% formic acid as eluent A and MTBE as eluent B.
2. Program the elution gradient as follows: A, 100%, 0–1 min; A, 100–20%, 1–15 min; A, 20–0%, 15–20 min and returned to the initial conditions (eluent A, 100%) for 5 min.
3. The flow is set at 0.8 mL min^{-1}.
4. Set the temperature of the C30 column to 30 °C.
5. Dissolve the dry extract of the sample in 1 mL of methanol/ dichloromethane solution (65/35, v/v).
6. Inject 5 μL of the extract (*see* **Note 5**).
7. After analysis, rinse the column with 50% eluent A and 50% eluent B for 15 min and then 100% eluent A for 15 min (*see* **Notes 6** and 7).

3.3.2 Identification of Carotenoids

Identification of carotenoids is established on the basis of both spectroscopic characteristic of carotenoids (*see* **Note 8**) and mass spectrometric data (*see* **Note 9**) by comparison with pure standards (when they are available) and/or with the literature.

In practice, on the device software:

1. Select the chromatogram at the maximum wavelength of carotenoid (usually at 460 nm or in a range between 440 and 470 nm).
2. Select the Total Ion current Chromatogram, the Base Peak Chromatogram, or the Extraction Ion Chromatogram for specific research of *m/z*.
3. Select one peak at the peak maximum and note the retention time (RT).
4. Analyze the UV-visible spectrum and note the characteristics of spectral fine structures (*see* **Note 8**).
5. Analyze the MSn spectra and note the *m/z* of the most abundant ion and of different obtained fragments. If necessary, scan *m/z* around the peak maximum (or all the peaks) and analyze the average *m/z* spectrum to highlight the signal.
6. Compare the retention time, the UV-visible spectral characteristics, and mass spectrometric data with the literature or pure standards analysis to identify the carotenoid.

To illustrate the identification of carotenoids, two examples are presented:

- Identification of free carotenoids in a spinach extract (using the C18 method).
- Identification of esterified carotenoids in a red pepper extract (using the C30 method).

Case of Free Carotenoids

The typical UPLC DAD chromatogram of the spinach extract at 470 nm revealed the presence of 5 main peaks and 6 minor peaks with UV-visible spectral characteristics of carotenoids (*see* **Note 8** and Fig. 3), and 8 peaks with characteristic UV-visible spectra of chlorophylls (Fig. 4). The UPLC DAD retention time, UV-visible spectral characteristics and mass spectrometric data of the main peaks are given in Table 2. For example, the analysis of peaks 7, 8, and 9 is detailed below:

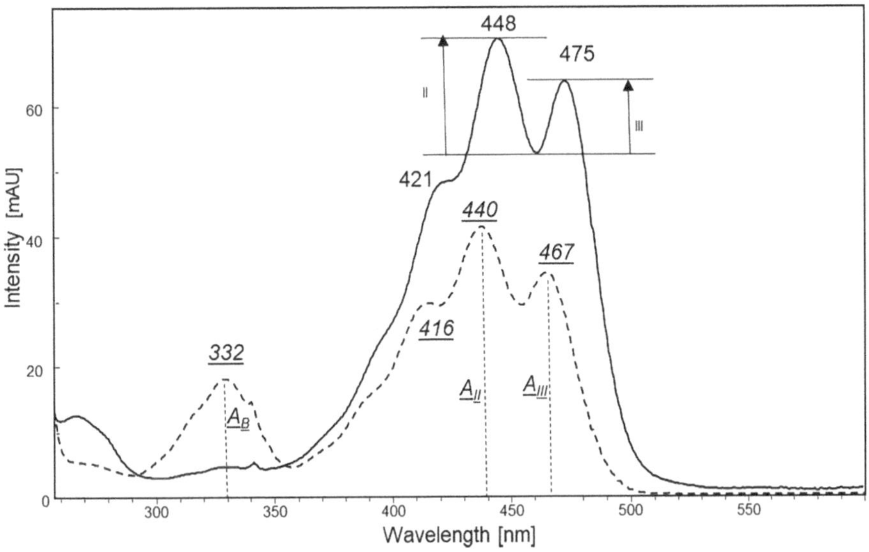

Fig. 3 UV-visible spectra of (all-*E*)-lutein (full line) and (13-Z)-lutein (dotted line). A_B: cis-peak; A_{II}: absorbance of the middle absorption peak II; A_{III}: absorbance of the longest absorption peak III; and 0: the minimum between the two peaks II and III

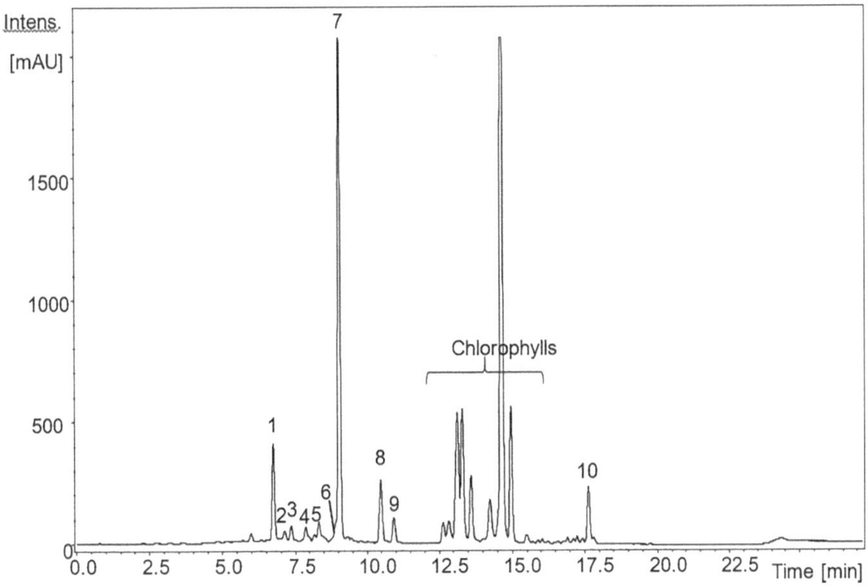

Fig. 4 UPLC chromatogram of the spinach extract at 470 nm using a C18 column. For UPLC conditions, *see* Subheading "Method Using the C18 Column". *See* Table 2 for identifications

Table 2
Identification of carotenoids from the spinach extract

Peak	RT (min)	λ max (nm)	MS ions *m/z* [b]	MS2 ions *m/z* [b]	Cis peak	%III/II	Identification
1	6.8	328/415/439/469	601.3 $[M+H]^+$	583.3/ 565.3		86	(All-*E*)-Neoxanthin
2	7.1	400/423/449	601.3 $[M+H]^+$	583.3		98	(All-*E*)-Luteoxanthin
3	7.4	399/423/449	601.3 $[M+H]^+$	583.3		98	(All-*E*)-Mutaxanthin
4	7.7	(423)[a]/447/474	585.3	565.3		nd	nd
5	7.9	(422)[a]/445/474	585.4	565.3		nd	nd
6	8.9	(425)[a]/451/478	569.34 $[M+H]^+$ 551.35 $[M+H-18]^+$	551.34 $[M+H-18]^+$ 533.40 $[M+H-36]^+$		20	(All-*E*)-Zeaxanthin
7	9.0	333/423/446/474	551.27 $[M+H-18]^+$ 533.30 $[M+H-36]^+$	533.29 $[M+H-36]^+$		61	(All-*E*)-Lutein
8	10.4	330/418/441/470	551.32 $[M+H]^+$		0.024	64	(9*Z*)-Lutein
9	10.9	332/416/440/467	551.32 $[M+H]^+$		0.437	35	(13*Z*)-Lutein or (13′*Z*)-Lutein
10	17.5	(428)[a]/452/480	537.4 $[M+H]^+$			10	(All-*E*)-β-carotene

UPLC DAD MS_n analysis was performed using a C18 column, *see* Subheading "Method Using the C18 Column"
[a]Parentheses indicate a shoulder
[b]*m/z* underlined indicate the maximum intensity ion; and *nd* Not determined

1. Peak 7 had a retention time at RT 9.0 min and displayed three absorption maxima at 421, 446, and 475 nm, a most abundant ion at *m/z* 551 (Fig. 5a) corresponding to the loss of water fragment $[M+H-18]^+$, no absorption around 332 nm and a % III/II of 60. The loss of water fragment $[M+H-18]^+$ is due to the presence of the hydroxyl group in an allylic position (a hydroxyl group located in the ε-ring) instead of the corresponding protonated molecule at *m/z* 569. This fragmentation allows to distinguish lutein from zeaxanthin (with a most abundant ion at *m/z* 569, Fig. 5b), which differs in the location of the double bond in one of the end rings and gives to lutein three chiral centers, whereas zeaxanthin has two [4]. The

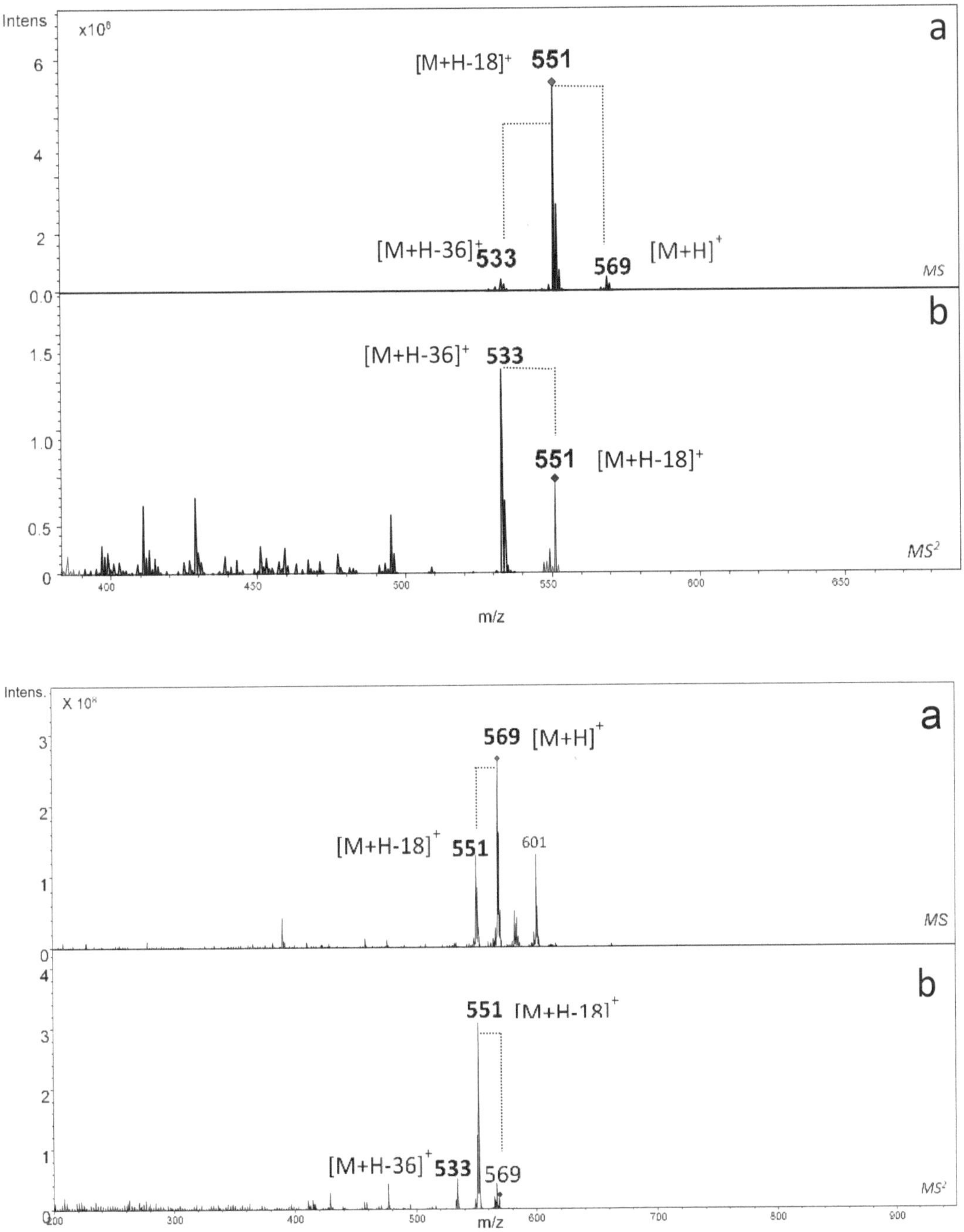

Fig. 5 Mass spectrum fingerprints of (all-*E*)-lutein (peak 7, **a**) and (all-*E*)-zeaxanthin (peak 6, **b**) from the spinach extract

injection of pure standard of (all-*E*)-lutein shows the same RT and the same fragmentation pattern, which confirms the identification. If UPLC APCI MS in positive ion mode on the C18 column allows us to separate and identify the main carotenoid from spinach, only the high-selective C30 column allows a good separation of the stereoisomers, lutein and zeaxanthin, which differ only in the position of one double bond in one ionone ring system because of the characteristic loss of water from the protonated lutein molecule.

2. Peaks 8 and 9 with retention time, respectively, at 10.4 and 10.9 min and displaying a 2–6 nm hypsochromic shift of their absorption maxima compared to the (all-*E*)-lutein and a most abundant ion at *m/z* 551 were identified as *Z*-isomers of lutein.
   (a) Peak 8 that displayed three absorption maxima at 418, 441, and 470 nm, no absorption around 332 nm, and a %III/II of 64 and a small ratio $\%A_B/A_{II}$ (2.4%) was identified as (9-*Z*)-lutein (*see* **Note 8**).
   (b) Peak 9 that displayed three absorption maxima at 416, 440, and 467 nm, an important *cis*-peak at 332 nm (Fig. 3), and a %III/II of 35 and a high ratio $\%A_B/A_{II}$ (43.8%) was identified as (13-*Z*)-lutein.

Case of Esterified Carotenoids

The UPLC DAD chromatogram of the red pepper extract at 460 nm revealed the presence of many peaks (Fig. 6). The identification of carotenoids was carried out by the analysis of the retention times, UV-visible spectral characteristics, and the mass spectra of the compounds, as presented in Table 3. For example, the analysis of peaks 3, 6, and 11 of the red pepper extract is detailed below:

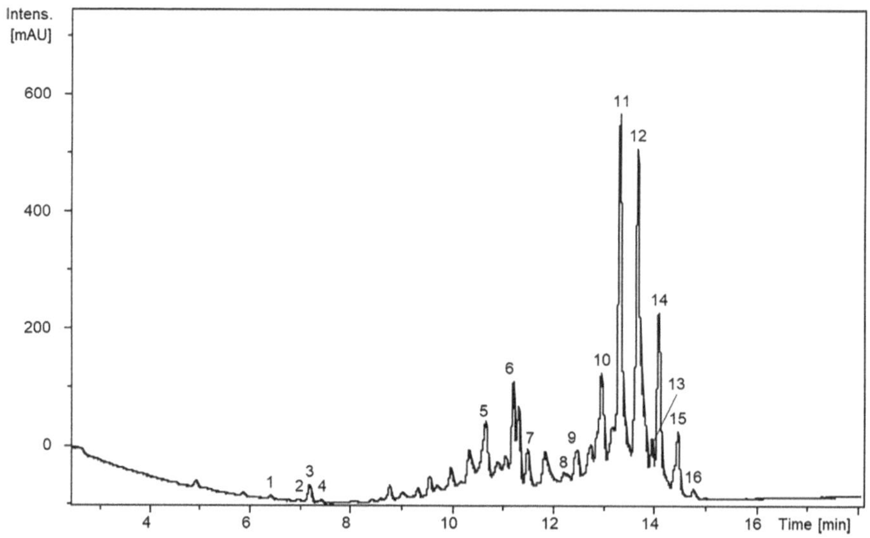

Fig. 6 UPLC chromatogram of the red pepper extract at 460 nm using a C30 column. For UPLC conditions, *see* Subheading "Method Using the C30 Column". *See* Table 3 for identifications

Table 3
Identification of carotenoids from the red pepper extract

Peak	RT (min)	λ max (nm)	MS ions *m/z*	MS^2 ions *m/z*	Identification
1	6.5	473	585.09 $[M+H]^+$	567.09 $[M+H-18]^+$	(*Z*)-Capsanthin
2	7.0	487	601.07 $[M+H]^+$	583.08 $[M+H-18]^+$	Capsorubin
3	7.2	473	585.10 $[M+H]^+$	567.10 $[M+H-18]^+$	(all-*E*)-Capsanthin
4	7.5	423/450/478	569.12 $[M+H]^+$	551.13 $[M+H-18]^+$	Zeaxanthin
5	10.7	422/446/473	795.28 $[M+H]^+$	567.02 $[M+H-228]^+$	Antheraxanthin-myristate (C14:0)
6	11.2	473	795.32 $[M+H]^+$	777.33 $[M+H-18]^+$ 567.13 $[M+H-228]^+$	Capsanthin-myristate (C14:0)
7	11.9	422/445/473	823.36 [M+H]+	805.33 $[M+H-18]^+$ 567.09 $[M+H-256]^+$	Antheraxanthin-palmitate (C16:0)
8	12.2		807.32 $[M+H]^+$	789.36 $[M+H-18]^+$ 551.12 $[M+H-256]^+$	Zeaxanthin-palmitate (C16:0)
9	12.5	423/450/473	735.26 $[M+H]^+$	535.09 $[M+H-200]^+$	β-Cryptoxanthin-laurate (C12:0)
			763.31 $[M+H]^+$	535.08 $[M+H-228]^+$	β-Cryptoxanthin-myristate (C14:0)
10	13	473	949.43 $[M+H]^+$	749.30 $[M+H-200]^+$ 549.12 $[M+H-400]^+$	Capsanthin-di-laurate (C12:0/C12:0)
11	13.3	473	977.47 $[M+H]^+$	749.30 $[M+H-228]^+$ 549.12 $[M+H-428]^+$	Capsanthin-laurate-myristate (C12:0/C14:0)
12	13.7	473	1005.49 $[M+H]^+$	777.31 $[M+H-228]^+$ 549.09 $[M+H-456]^+$	Capsanthin-di-myristate (C14:0/C14:0)
13	14	422/450/456	989.52 $[M+H]^+$	761.29 $[M+H-228]^+$	Zeaxanthin-di-myristate (C14:0/C14:0)

(continued)

Table 3
(continued)

Peak	RT (min)	λ max (nm)	MS ions *m/z*	MS^2 ions *m/z*	Identification
				533.0 $[M+H-456]^+$	
14	14.1	473	1033.56 $[M+H]^+$	805.35 $[M+H-228]^+$ 777.31 $[M+H-256]^+$ 549.32 $[M+H-484]^+$	Capsanthin-myristate-palmitate (C14:0/C16:0)
15	14.5	473	1061.56 $[M+H]^+$	805.35 $[M+H-256]^+$ 549.09 $[M+H-512]^+$	Capsanthin-di-palmitate (C16:0/C16:0)
16	14.8		1045.59 $[M+H]^+$	789.32 $[M+H-256]^+$ 533.30 $[M+H-512]^+$	Zeaxanthin-di-palmitate (C16:0/C16:0)

UPLC DAD MS_n analysis was performed using an YMC C30 column, *see* Subheading "Method Using the C30 Column"

1. Peak 3 has a retention time at 7.2 min displaying one absorption maxima at 473 nm (*see* **Note 10**) and a most abundant ion at *m/z* 585 corresponding to the protonated molecule $[M + H]^+$ (Fig. 7a). The MS^2 spectrum shows a major fragment ion at *m/z* 567, which represents a loss of one molecule of water ($[M+H-18]^+$). Comparison of the results obtained with the literature allows identifying peak 3 as (all-*E*)-capsanthin. The injection of pure standard of (all-*E*)-capsanthin shows the same RT and the same fragmentation pattern which confirms the identification.
2. Peak 6 has a retention time at 11.2 min displaying one absorption maxima at 473 nm and a most abundant ion at *m/z* 795 corresponding to the protonated molecule $[M+H]^+$ (Fig. 7b). The MS^2 spectrum shows two fragment ions at *m/z* 777 and *m/z* 567, which represent a loss of one molecule of water ($[M+H-18]^+$) and of one molecule of myristic acid ($[M+H-228]^+$), respectively. This peak is identified as a monoester of capsanthin, capsanthin myristate.
3. Peak 11 has a retention time at 13.3 min displaying one absorption maxima at 473 nm and a most abundant ion at *m/z* 977 corresponding to the protonated molecule $[M+H]^+$ (Fig. 7c). The MS^2/MS^3 spectra show two fragment ions at *m/z* 749 and *m/z* 549, which represent a loss of one molecule

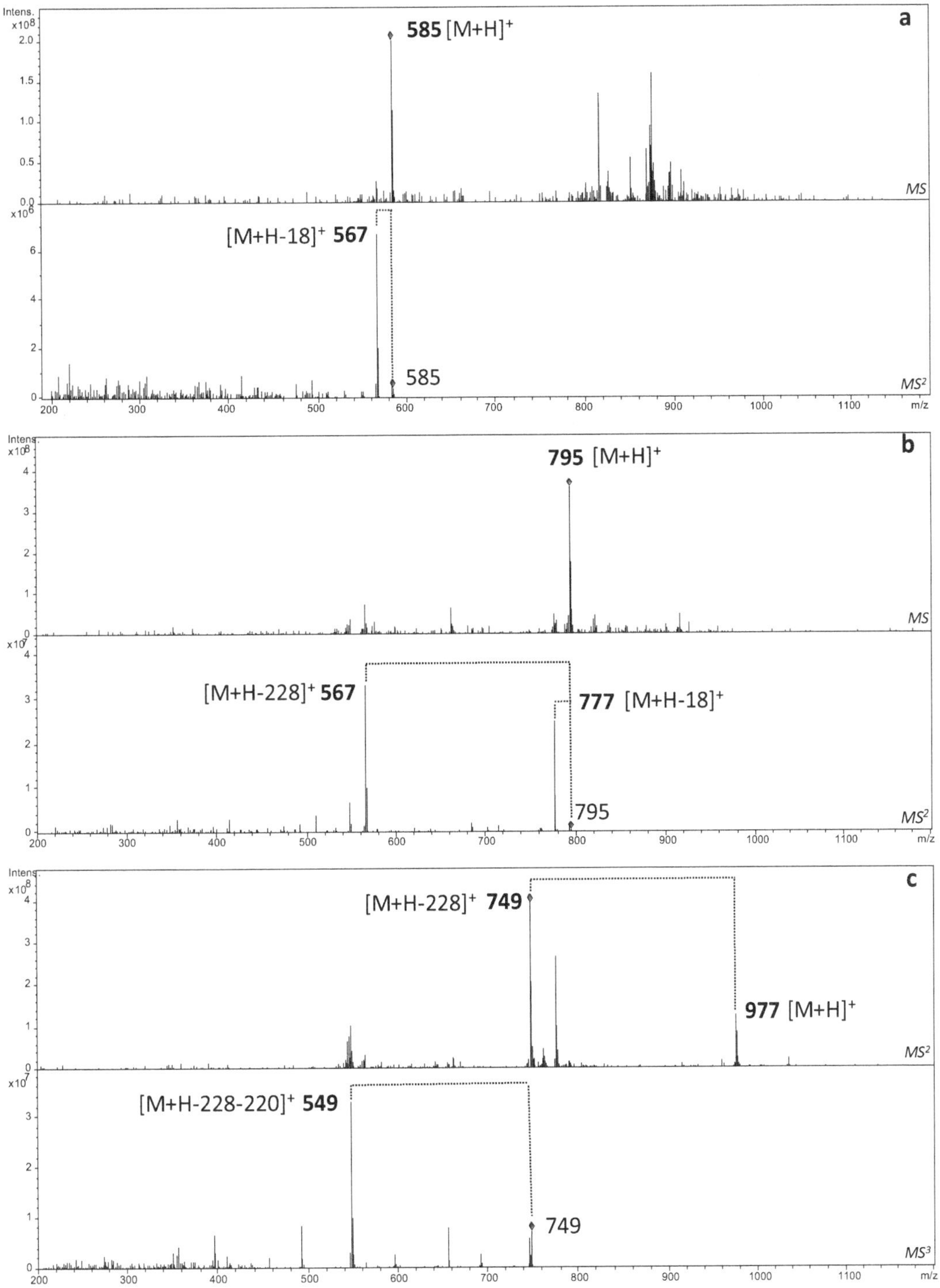

Fig. 7 Mass spectrum fingerprints of (all-*E*)-capsanthin (peak 3, **a**), capsanthin-myristate (peak 6, **b**), and capsanthin-laurate-myristate (peak 11, **c**) from the red pepper extract

of myristic acid ($[M+H\text{-}228]^+$) and of one molecule of lauric acid ($[M+H\text{–}228\text{-}200]^+$), respectively. This peak is identified as a diester of capsanthin, capsanthin laurate myristate.

The UPLC DAD MS_n analysis has identified the carotenoids present in the red pepper extract (capsanthin, capsorubin, antheraxanthin, β-cryptoxanthin, and β-carotene) in free form or as mono- or di-ester of fatty acid.

3.3.3 Quantification of Carotenoids

The quantification of carotenoids present in a sample is calculated from calibration curves of pure standards under the same chromatographic conditions. The carotenoid standards are not always available, especially the standards of esterified carotenoids. In the case of free carotenoids, the quantification is expressed as the equivalent of other carotenoid for which the standard is available. In the case of esterified carotenoids, a saponification step is necessary to release the free forms of carotenoids (*see* protocol in Subheading 3.4) and to allow their quantification as equivalent of free form (*see* **Note 11**).

The use of an internal standard is necessary to follow the losses during the extraction step. As the carotenoid contents are very different depending on the matrix but also in the same matrix, an external calibration is recommended. The choice of the internal standard must be made according to the matrix: it must be absent from the analyzed matrix; moreover, it should not co-eluate with another compound (*see* **Note 12**). To perform the quantification, add the internal standard at the extraction step (*see* **Note 13**) and inject the internal standard using the same liquid chromatography method at the same concentration as the sample.

Calibration Curve of the Carotenoid Standard

1. Prepare the stock solution of the carotenoid standard by dissolving the dry standard in adequate solvent (2 mg mL^{-1}).
2. Determine the absorbance of the stock solution using a UV–V is spectrophotometer.
3. Calculate the exact concentration of the stock solution using the Beer-Lambert law and the molar absorption coefficient ($A^{1\%/1cm}$, Table 4) as follows:

$$\text{Carotenoid } (\mu g/mL) = \frac{\text{Absorbance sample at } \lambda max}{\frac{A1\%carotenoid}{10000}}$$

4. Prepare daughter solutions at a 0.5 mg mL^{-1} final concentration.
5. Prepare seven dilutions from 0.1 to 20 μg mL^{-1} in methanol (each dilution in triplicate, *see* **Note 14**).

Table 4
UV-visible absorption data and absorption coefficients ($A^{1\%/1cm}$) of common food carotenoids

Carotenoid	Solvent	λ max (nm)[a]	%III/II[b]	$A^{1\%/1cm}$
Antheraxanthin	Ethanol	422/<u>446</u>/472	55	2350
Astaxanthin	Ethanol	<u>478</u>		2100
Capsanthin	Ethanol	<u>474</u>		2076
α-Carotene	Hexane	422/<u>445</u>/473	55	2710
β-Carotene	Hexane	(425)/<u>450</u>/477	25	2590
β-Cryptoxanthin	Hexane	(428)/<u>450</u>/478	27	2460
Lutein	Ethanol	422/<u>445</u>/474	60	2550
Lycopene	Petroleum ether	444/<u>470</u>/502	65	3450
Neoxanthin	Ethanol	415/<u>439</u>/467	80	2243
Zeaxanthin	Ethanol	(428)/<u>450</u>/478	26	2540

Data from [1, 6]

[a] Parentheses indicate a shoulder. Underlined wavelengths are used for quantification

[b] Percentage of the ratio of the height of the longest wavelength absorption peak (III) to that of the middle absorption peak (II) taking the minimum between the two peaks as the baseline (*see* Fig. 3)

6. Inject each solution using the same liquid chromatography method as the sample (*see* Subheading 3.3.1 Liquid Chromatography).
7. Construct a calibration curve by plotting the concentration of carotenoid against area as a linear regression:

$$y = ax + b$$

where x is the concentration of the carotenoid and y is the peak area of the corresponding standard.

8. For the validation of the method, *see* **Note 14**. Calculate the means of the slopes (S) and the standard deviation of the y-intercepts (σ) (*see* **Note 15**).

Calculation of Carotenoid Content in the Sample

1. Calculate the recovery of the internal standard as follows:

$$\text{Recovery of intenal standard} = \frac{\text{Area of internal standard in sample}}{\text{Area of internal standard alone}}$$

2. Calculate the concentration of a specific carotenoid in the extract from the calibration curve obtained for this carotenoid (or the reference carotenoid) as follows:

$$\text{Carotenoid } (\mu g/mL) = \frac{\frac{\text{Area of carotenoid peak}}{\text{Recovery of internal standard}} - b}{a}$$

where a is the slope of the calibration curve and b is the y-intercept.

3. Calculate the content of a specific carotenoid in the fruit and vegetable sample as follows:

$$\text{Carotenoid } (\mu g/100g) = \frac{\text{Carotenoid } (\mu g/mL) \times 100}{\text{Sample weight (g)}}$$

3.4 Saponification of Esterified Carotenoids

1. Dissolve the dry extract of the sample in 1 mL of 5% KOH methanolic solution.
2. Flush the tube under argon.
3. Incubate at room temperature under shaking (120 rpm) for at least 5 h.
4. Add 10 μL of 37% HCl.
5. Vortex for 10 s.
6. Add 2 mL of n-hexane and 1 mL of ultrapure water.
7. Vortex for 10 s.
8. Centrifuge for 5 min at 500 *g* and at 10 °C.
9. Recover the supernatant (upper phase) and transfer it to a tube.
10. Wash the supernatant by adding 0.5 mL of methanol and 0.5 mL of ultrapure water.
11. Vortex for 10 s.
12. Centrifuge for 5 min at 500 *g* and at 10 °C.
13. Recover the supernatant and transfer it to a tube.
14. Evaporate the solvent under argon at 30 °C.
15. Use the dry extract immediately for UPLC analysis or store it for maximum of 3 days at −20 °C under argon before UPLC analysis.

4 Notes

1. The use of liquid nitrogen requires working in a ventilated room to avoid the accumulation of nitrogen, which can cause discomfort. To avoid the risk of burns, wear a long-sleeved lab coat, safety glasses, safety shoes, and cryogloves.
2. Ammonium formate improves the recovery of carotenoids from the column during the separation. Formic acid facilitates ionization of compounds and improves sensitivity.

3. Freeze-drying induces an increase in oxygen exposure and therefore a rapid loss and degradation of carotenoids [2].
4. The majority of carotenoids exhibit absorption in the visible region of the spectrum, between 400 and 500 nm, or in some cases, UV region.
5. It is recommended to inject firstly a blank sample (i.e., methanol or a standard as β-apo-8′-carotenal) to allow optimal equilibration of the system.
6. Acidic conditions decrease the lifespan of UPLC or HPLC columns, so clean the column with organic solvent without formic acid if you don't use the column for several hours.
7. The MTBE attacks the piston seals, so the tubing must be put under methanol if the UPLC system stops for several hours.
8. Carotenoids possess spectral fine structures leading to characteristic UV-visible spectra with three absorption maxima between 400 and 500 which can also be expressed as a percentage of the ratio between the peak heights III/II (Fig. 3). The absorption maximum of the (*Z*)-isomers is hypsochromically shifted a few nm compared to the all-*E* configuration, usually 2–6 nm for mono-(*Z*)-isomers and 8–12 nm for di-(*Z*)-isomers. A new peak observed in the 330–340 nm region or about 142 nm below the peak III, named the *cis*-peak, proves the presence of (*Z*)-isomers. The relative intensity of the percentage of the ratio between the absorbance of the *cis*-peak (A_B) and the absorbance of the middle and main absorption band II (A_{II}) allows the determination of the position of the *Z*-double bond(s): the greater is the ratio (% A_B/A_{II}), the closer the (*Z*)-double bond is to the center of the molecule [5].
9. Commonly, carotenoids are ionized by APCI in the positive mode and show the protonated molecule $[M+H]^+$.
10. Some carotenoids, such as capsanthin, canthaxanthin, and echinenone, have in their structure a carbonyl group conjugated to the conjugated double bond. This causes a bathochromic shift and a loss of fine spectral structure (the three maxima spectrum).
11. Quantification of esterified carotenoids is partial because it does not take account the weight represented by fatty acids.
12. The most commonly used internal standard is the β-apo-8′-carotenal except for green vegetables (co-elution with chlorophyll b) for which echinenone should be used.
13. Internal standard solution at 0.5 mg/mL. For example for β-apo-8′-carotenal solution, weigh 10 mg of β-apo-8′-carotenal in a 20 mL volumetric flask and dissolve with some droplets of dichloromethane. Complete to 20 mL with ethyl acetate. Add 10 μL of this solution at **step 3** of extraction (*see* Subheading 3.2 Lipophilic Extraction).

14. The range of concentration of calibration solution need to be adapted according to the amount in the sample.
15. With the parameter of the three calibration curves, calculate the means of the slopes (S) and the standard deviation of the y-intercepts (σ). Determine the limit of detection (LOD = 3.3 (σ/S)) and the limit of quantification (LOQ = 10 (σ/S)) of the method. Verify the linearity of the method by five repeated injections of the same standard solution.

Acknowledgement

This chapter is in memory of Dr. Catherine Caris-Veyrat who devoted her career to study the carotenoids.

References

1. Rodriguez-Amaya DB (2001) A guide to carotenoid analysis in foods. ILSI Press, Washington, DC
2. Craft NE, Wise SA, Soares JH (1993) Individual carotenoid content of SRM 1548 total diet and influence of storage temperature, lyophilization, and irradiation on dietary carotenoids. J Agric Food Chem 41:208–213. https://doi.org/10.1021/jf00026a012
3. Meléndez-Martínez AJ, Mandić AI, Bantis F, Böhm V, Borge GIA, Brnčić M, Bysted A, Cano MP, Dias MG, Elgersma A, Fikselová M, García-Alonso J, Giuffrida D, Gonçalves VSS, Hornero-Méndez D, Kljak K, Lavelli V, Manganaris GA, Mapelli-Brahm P, Marounek M, Olmedilla-Alonso B, Periago-Castón MJ, Pintea A, Sheehan JJ, Tumbas Šaponjac V, Valšíková-Frey M, Meulebroek LV, O'Brien N (2021) A comprehensive review on carotenoids in foods and feeds: *status quo*, applications, patents, and research needs. Crit Rev Food Sci Nutr 1–51. https://doi.org/10.1080/10408398.2020.1867959
4. Rivera SM, Christou P, Canela-Garayoa R (2014) Identification of carotenoids using mass spectrometry. Mass Spectrom Rev 33:353–372. https://doi.org/10.1002/mas.21390
5. Britton G, Liaaen-Jensen S, Pfander H (1995) Carotenoids. Volume 1B: Spectroscopy. Birkhäuser Verlag, Basel, Boston, Berlin
6. Britton G, Liaaen-Jensen S, Pfander H (2004) Carotenoids. Handbook. Birkhäuser Verlag, Basel, Boston, Berlin

Part III

Evaluation of the Oxidation of Dietary Lipids

Chapter 12

Quantification of Conjugated Dienes in Oxidizing Oils and Emulsions

Lucie Ribourg-Birault, Claire C. Berton-Carabin, and Claude Genot

Abstract

UV spectroscopy is a simple method to detect and quantify hydroperoxides containing conjugated diene structures obtained from the oxidation of lipids containing linoleic and linolenic fatty chains. The protocol detailed below can be applied to the extracted dietary lipids, oil-in-water emulsions, and biological matrices stabilized by proteins. The UV spectra of the lipids solubilized in 2-propanol are recorded, and the absorbance at 233 nm is converted into molar concentration of conjugated diene hydroperoxides using their molar extinction coefficient. In addition to this classical approach, we recommend correcting the data using a nonoxidized blank sample selected by checking the second derivative of the UV-visible spectra.

Key words Lipid, Oxidation, Conjugated dienes, Hydroperoxides, Dietary fats and oils, UV spectra

1 Introduction

Measurement of conjugated dienes (CD) by UV spectroscopy is a simple and nondestructive method to detect and quantify conjugated hydroperoxides obtained from the oxidation of dietary lipids containing linoleic and linolenic fatty chains. The assay requires a small amount of sample (less than 1 mg) and is a sensitive method to follow the early stage of oxidation. It is however not suitable for oils containing low levels of polyunsaturated fatty acids (PUFA), such as olive and high oleic oils, and when the lipids are subjected to conditions that decompose hydroperoxides (e.g., heating) [1].

The first step of oxidation of PUFA with nonconjugated *cis-cis* double bonds is the abstraction of *bis*-allylic hydrogen. Then, migration of the double bond generates conjugated dienes. O_2 fixation and further rearrangements lead finally to the formation of different isomers of conjugated hydroperoxides (*cis-trans*, *trans-cis*, and *trans-trans*) in proportions varying with oxidation conditions [2]. These compounds absorb in the ultraviolet region, with a

Christelle Lopez et al. (eds.), *Multidimensional Characterization of Dietary Lipids*, Methods and Protocols in Food Science,
https://doi.org/10.1007/978-1-0716-3758-6_12,

peak maximum around 231–235 nm, which is the basis for the present assay.

The measurement of lipid absorbance around 233 nm (A233) was proposed as a simple test to evaluate lipid oxidation long ago [3]. Authors thus often only report changes in A233 of a given lipid material to highlight the formation of CD and therefore primary oxidation products. This is also the basis for normalized methods [4, 5]. Then, specific extinction coefficients (E1%, 1 cm) are calculated. The sensitivity of the assay is reported to be about 0.2 mEq/kg [6]. To increase sensitivity and reproducibility, some authors used the ratio between A333 and the absorbance at the isobestic point (around 212–214 nm) [7–9]. This normalization procedure eliminates the dispersion of data due to pipetting and dilution steps. For mass balance, A233 is converted into equivalent linoleate hydroperoxides, by using an average extinction coefficient. The real coefficient depends on the formed isomers and the solvent [10, 11]. Standard conjugated hydroperoxides can also be used when available.

During the initial steps of oxidation of lipids containing high levels of linoleic and linolenic acids, the absorbance at 233 nm (A233) generally follows the same trends with similar order range of oxygen consumption and hydroperoxide formation [12, 13]. For oils rich in monounsaturated fatty acids (e.g., oleic acid), the assay under-evaluates the formation of primary products of oxidation because oleic hydroperoxides do not possess the conjugated structure [14]. In the case of lipids containing high levels of long-chain PUFA, such as EPA and DHA, the detection of CD formation could also be limited due to the instability of the formed hydroperoxides (rapid cyclization of monohydroperoxides into epidioxides and formation of di- and tri-hydroperoxides lacking the conjugated structure, or rapid decomposition into secondary oxidation products). This explains why the method is more suited to assess the oxidation of lipids rich in 18:2 n-6 and 18:3 n-3 than of lipids containing n-3 long-chain PUFAs, or majorly composed of monounsaturated fatty acids [12]. It also implies that the changes in the UV spectrum and the calculated amounts of CD for a given lipid material should be viewed as a relative measurement of oxidation.

As the changes in UV spectra vary in quality and magnitude depending on the lipid fatty acid composition and the oxidation conditions, CD values obtained for different lipids should not be directly compared in terms of oxidation degree [6, 15]. Moreover, it should be kept in mind that the CD peak is already present in nonoxidized lipids containing naturally conjugated fatty acids such as milk fat or processing-induced conjugated fatty acids (hydrogenated fats). It also overlaps with the broad peak exhibited by lipid structures (maximum around 215 nm for carboxylic acids and esters and around 206 nm for *cis* double bonds) and with the absorption peak of nonlipid materials such as phenolic antioxidants [16]. For low levels of oxidation, CD absorption is thus only viewed as a shoulder in the broad peak (210–250 nm) due to lipid structures [17]. It is thus recommended to consider a blank made of nonoxidized lipids and to subtract the blank from the spectra of oxidized samples [11, 18].

Calculating second derivatives of the spectra is a way to increase the sensitivity and the information provided by the assay [8, 19–23]. The second derivative spectra present several peaks, the negative peaks around 233 and 242 nm being attributed to *trans-trans* and *cis-trans* isomers, respectively. As the formation of *cis-trans* isomers is favored in the presence of H-donors, the ratio between these peaks gives information about the oxidation mechanisms and the redox state of the medium [24]. In the protocol that we propose herein, we suggest selecting a nonoxidized sample as a blank sample and identifying it as such, thanks to its second derivative spectrum (thus showing no peak).

The present protocol schematically represented in Fig. 1 is applied in our laboratory for various dietary lipids and oil-in-water emulsions.

It can be very easily implemented in any laboratory. The method comprises first the determination of the UV spectra of the lipids solubilized in 2-propanol. The solvent has the advantage for complex samples such as emulsions or biological samples of simultaneously precipitating proteins and dissolving lipids, thus avoiding cumbersome and restrictive steps of lipid extraction. Then, data treatment is done by subtracting the blank corresponding to a non-oxidised sample and converting results into molar concentration by using the extinction coefficient (27,000 M^{-1} cm^{-1}) [25].

2 Materials

2.1 Chemical Products

The optical quality of 2-propanol (Table 1) should be tested before using: it should not absorb in the UV region (210 nm and above). This solvent can be replaced by any solvent of lipids with similar optical characteristics.

2.2 Mandatory Equipment

1. Analytical precision balance, sensitivity: 0.01 mg.
2. UV-visible spectrophotometer (in the following the method is described for a double-beam spectrophotometer) and semi-micro optical quartz cells (optical path length: 10 mm).
3. Positive displacement pipettes and disposable capillary piston tips.

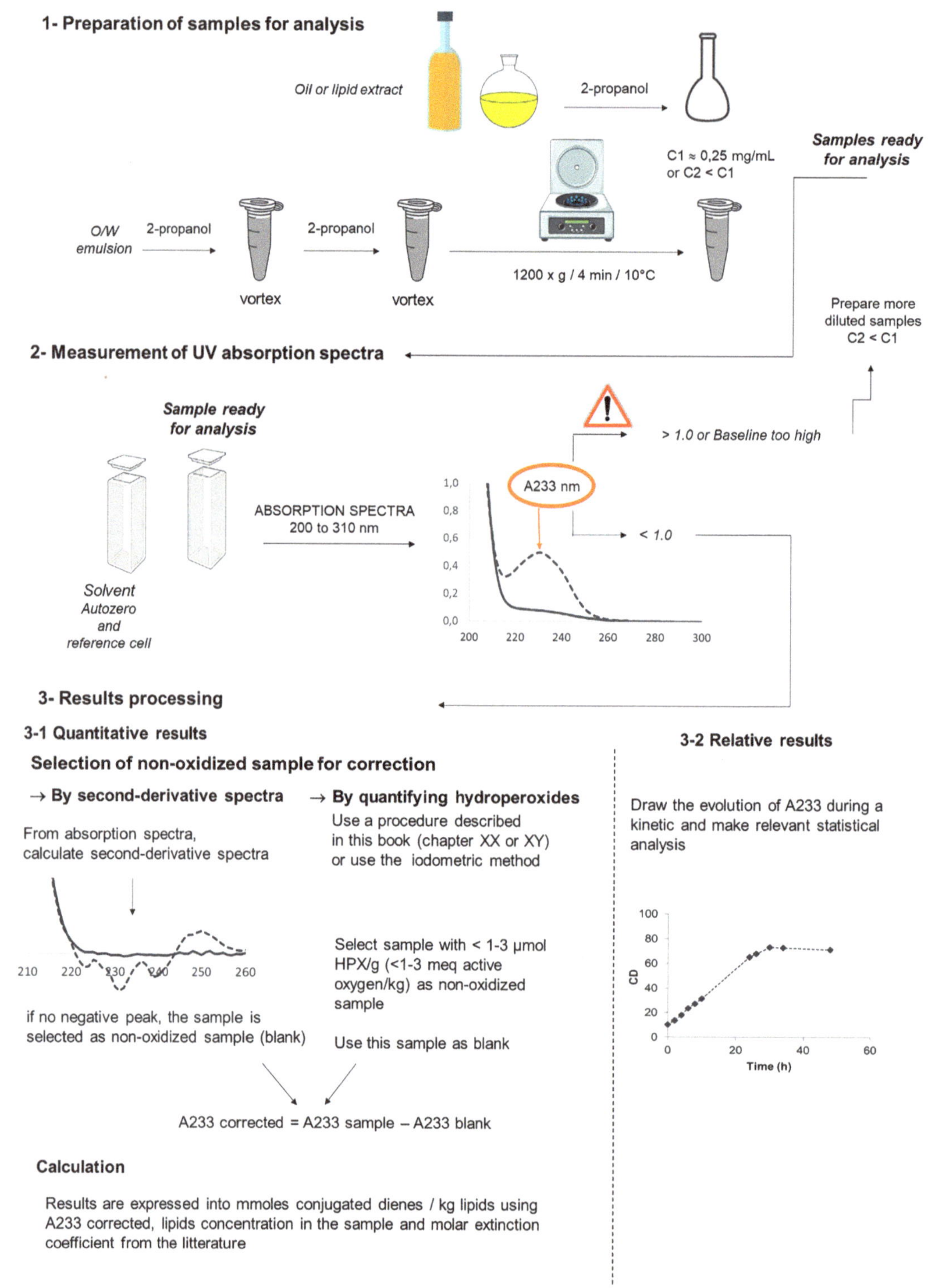

Fig. 1 Flowchart of the method

Table 1
Chemical required to analyze conjugated dienes

Chemical	Chemical formula	CAS number	Molecular weight ($g.mol^{-1}$)	Quality/ reference	Safety instructions
2-Propanol propan-2-ol	$(CH_3)_2CHOH$	67–63-0	60.10	RPE for analysis	

4. Microtubes (1.5 mL).
5. Vortex agitator.
6. Centrifuge for microtubes (in the case of emulsion analysis).

3 Method

3.1 Preparation of Samples for Analysis

3.1.1 Edible Oils

Weigh accurately 25 mg of oil (m, mg) in a microtube and add 975 μL of 2-propanol to obtain a concentration of around 25 mg of lipids per milliliter in this first tube. Carry out 2 successive ten-fold dilutions with 2-propanol to obtain a concentration of around 0.25 mg of lipids per milliliter in the final tube. This concentration (C_{oil}, $mg.mL^{-1}$) is recommended for fresh oils or poorly oxidized lipids and may be lowered if the lipids are highly oxidized. Vortex the solution thoroughly ("sample ready for analysis").

Calculate the final concentration C_{lipids} ($mg.mL^{-1}$) $= m/(V_{total} \times \text{dilution})$

3.1.2 Oil-in-Water Emulsion

Introduce a known volume of emulsion (V_{em}, μL) into a microtube and make up to 1 mL with 2-propanol to obtain a concentration of around 25 mg of lipids per milliliter in this first tube. Carry out 2 successive ten-fold dilutions with 2-propanol to obtain a concentration of around 0.25 mg of lipids per milliliter. This concentration (C_{lipids}, $mg.mL^{-1}$) is recommended for fresh or poorly oxidized lipids and may be lowered if the lipids are highly oxidized. For example, 90 μL are required for a 30% oil-in-water emulsion (w/w) in a tube with 910 μL of 2-propanol followed by 2 successive ten-fold dilutions. The concentration in this case is 0.27 mg of lipids per milliliter in the final tube ("sample ready for analysis").

Centrifuge the mixture for 4 min at 1200 × g. Collect the supernatant.

Calculate the lipid concentration (C_{lipids}, $mg.mL^{-1}$) in the supernatant ("sample ready for analysis" in Fig. 1) taking into account the volume of the emulsion taken (V_{em}, μL) and the mass fraction of lipids in the emulsion (ϕ_m, wt %) and considering that the density of the emulsion is equal to 1 (if the actual density is known, take this into account).

$$C_{\text{lipids}} = (\phi_m \times V_{\text{em}} \times 1000)/(V_{\text{total}} \times \text{dilution})$$

with C_{lipids} in mg.mL^{-1}, ϕ_m in wt %, and V_{em} and V_{total} in mL.

Calculate also the 2-propanol/water ratio in the "sample ready for analysis."

3.2 Measurement of UV Absorption Spectra

3.2.1 Edible Oils

Introduce 2-propanol in the reference and sample quartz microcells and record the autozero of the spectrophotometer. Then, introduce the sample into the sample microcell and record the absorbance spectrum of the sample between 200 and 310 nm, step 1 nm (*see* **Notes 1** and **2**).

3.2.2 Oil-in-Water Emulsion

In the case of the analysis of an emulsion, the autozero is carried out on 2-propanol containing water in the same proportion as in the final mixture.

For example, for 90 μL of emulsion 30% (m/m) (consider as 90 mg of emulsion) in 1 mL then 2 dilutions (ten-fold each): in a 10-mL volumetric flask, weigh 6.3 mg of ultrapure water and make the volume up to 10 mL with 2-propanol.

Then, introduce the sample into the sample microcell and record the absorbance spectrum between 200 and 310 nm (step 1 nm or below) (Fig. 2).

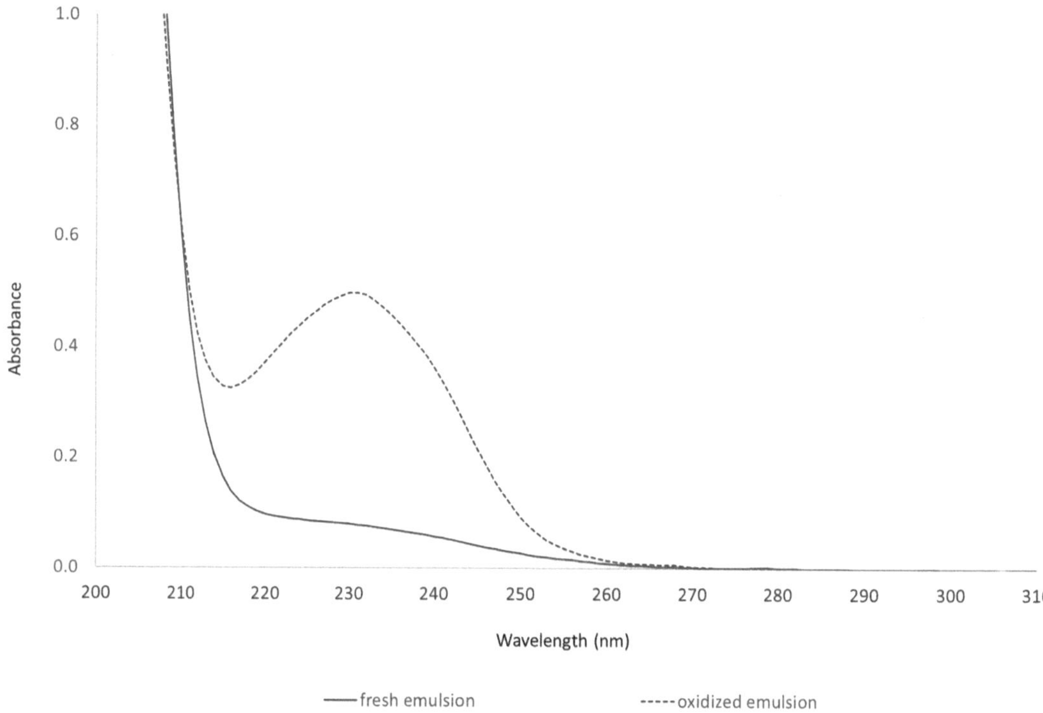

Fig. 2 Example of UV spectra obtained from fresh and oxidized rapeseed oil-in-water emulsions. Preparation of samples and recording conditions are described in [27]

If the absorbance at 233 nm is greater than 1 (A233 > 1), prepare less concentrated "sample for analysis" from the original sample (Fig. 1) and record again the UV spectrum to get A233 < 1.

Check for the absorbance at 310 nm (A310 nm) and the shape of the spectrum. If it is far from zero, with no return to the baseline, adjust the experimental conditions by additional dilution and/or stronger centrifugation step to reduce the background noise as much as possible. At least, subtract the absorbance at 310 nm from that at 233 nm (A233–A310). Check also for the presence of other chromophores absorbing in this spectral region that may interfere with the measurement.

3.3 Results Processing

To express the amounts of conjugated hydroperoxides in a sample as mmol kg lipids^{-1}, A233 is converted, after blank subtraction, according to the Beer–Lambert's law by using the molar extinction coefficient of linoleic hydroperoxides [25]. As explained in the introduction part, the blank subtraction allows to take into account and eliminate the contribution at 233 nm of nonoxidized lipid structures and possible nonlipid material and/or baseline shift. This contribution varies with the composition of the lipid extract and differs depending on the analyzed oil or fat.

The "blank sample" is a nonoxidized (or very weakly oxidized) sample of similar origin and lipid composition and concentration to the analyzed samples.

To select the blank sample, calculate the second derivative spectra from the UV spectra recorded previously. The second derivative spectra can be calculated with a spreadsheet in two steps as shown in Fig. 3 or by using Savitzky–Golay algorithm [26] or data treatment software.

$$\frac{A_{202} - A_{200}}{202 - 200}$$

Wavelength (nm)	Absorbance	Slope = First derivative	Slope = Second derivative
200	0.9003		
201	1.2552	0.3058	
202	1.5118	0.2433	-0.1491
203	1.7417	0.0075	0.0012

$$\frac{\text{Slope (203 nm)} - \text{Slope (201 nm)}}{203 - 201}$$

Fig. 3 Example of 2-step calculation of the second derivative spectrum with a spreadsheet

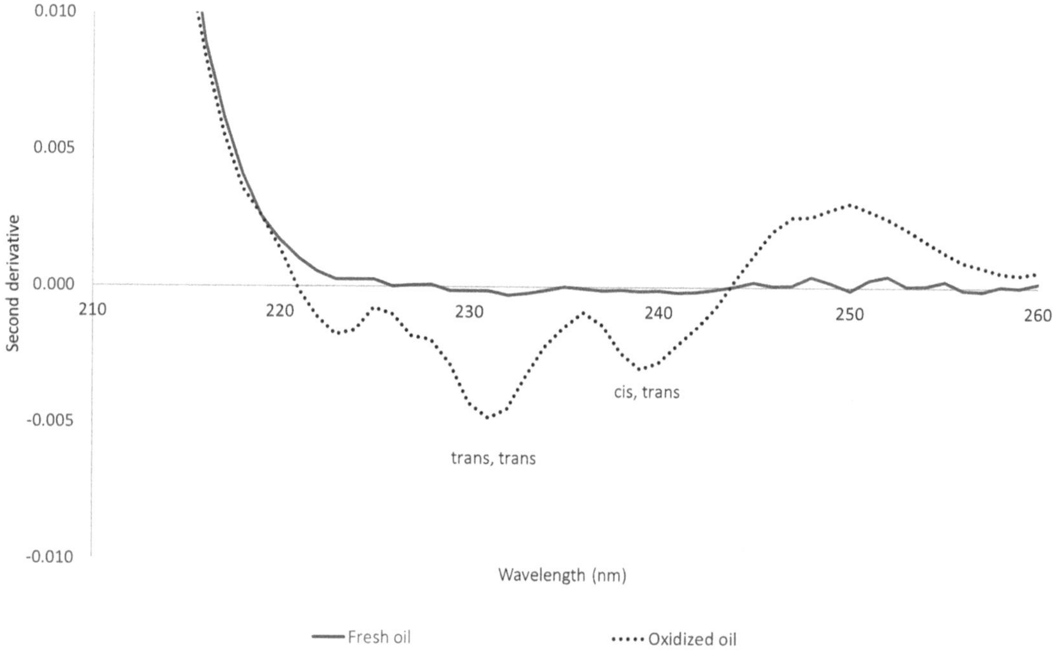

Fig. 4 Example of second derivative spectra of conjugated dienes obtained from fresh and oxidized rapeseed oil

Plot the derivative spectra and then choose a sample showing no negative peak as a blank sample (Fig. 4).

Another way is to select a sample whose hydroperoxide content, determined, for instance, by the method described in chapter 13 and 14 or by the iodometric method, is as low as possible (<1 to 3 μmol.g^{-1} depending on the lipids). Finally, subtract the absorbance at 233 nm of the nonoxidized sample from the absorbance of the analyzed samples:

$$A_{233\ \text{corrected}} = A_{233}\text{sample} - A_{233}\text{blank}.$$

3.4 Calculation of the Amount of Conjugated Dienes in the Sample

Calculate the amount of CD using the following formula:

$$\text{CD} = \frac{A233\,\text{corrected}}{C_{\text{lipids}} \times 10^{-3} \times \varepsilon \times l} \times 1000$$

where CD is the concentration of conjugated dienes in the oil in mmol kg oil^{-1}; $A_{233\ \text{corrected}}$ corresponds to A_{233} sample—A_{233} blank; C_{lipids} is the concentration of lipids in "sample ready for analysis" in mg mL^{-1} (CF Subheading 3.1); ε is the extinction molar coefficient: 27000 L mol^{-1} cm^{-1} for C18:2 hydroperoxides; and l is the optical path length: 1 cm.

4 Notes

1. The scan speed must be adapted to the procedure chosen to process the results. Choose a low speed if you have to trace the spectrum of the second derivatives or a higher speed if you do screening.
2. In case of insufficient quality of spectra (i.e., high background), it may be necessary to choose a lower scan speed.

References

1. Frankel EN (2005) Methods to determine extent of oxidation. Lipid oxidation. The Oily Press PJ Barnes & associates, Bridgewater, pp 99–127
2. Frankel EN, Hydroperoxide formation. (2005) Lipid oxidation. The Oily Press PJ Barnes & Associates, Bridgewater, pp 25–50
3. Farmer EH, Koch HP, Sutton DA (1943) The course of autoxidation reactions in polyisoprenes and allied compounds. Part VII. Rearrangement of Double Bonds during Autoxidation. J Chem Soc:541–547
4. ISO 3656. Animal and vegetable fats and oils - Determination of ultraviolet absorbance expressed as specific UV extinction 2011. p. 8
5. AOCS Official Method Th 1a-64 Diene value of drying oils. 2017
6. Dobarganes MC, Velasco J (2002) Analysis of lipid hydroperoxides. Eur J Lipid Sci Technol 104(7):420–428
7. Klein RA (1970) Detection of oxidation in liposome preparations. Biochim Biophys Acta 210(3):486–489
8. Kansci G, Genot C, Meynier A, Gandemer G (1997) The antioxidant activity of carnosine and its consequences on the volatile profiles of liposomes during iron/ascorbate induced phospholipid oxidation. Food Chem 60(2): 165–175
9. Genot C, Metro B, Viau M, Bouchet B (1999) Characterisation and stability during storage of liposomes made of muscle phospholipids. Food Sci Technol 32(3):167–174
10. Chan HWS, Levett G (1977) Autoxidation of methyl linolenate—Analysis of methyl hydroxylinolenate isomers by high-performance-liquid-chromatography. Lipids 12(10):837–840
11. Recknagel RO, Glende EA (1984) Spectrophotometric detection of lipid conjugated dienes. Methods Enzymol 105:331–337
12. Viau M, Genot C, Ribourg L, Meynier A (2016) Amounts of the reactive aldehydes, malonaldehyde, 4-hydroxy-2-hexenal, and 4-hydroxy-2-nonenal in fresh and oxidized edible oils do not necessary reflect their peroxide and anisidine values. Eur J Lipid Sci Technol 118(3):435–444
13. Kulas E, Ackman RG (2001) Different tocopherols and the relationship between two methods for determination of primary oxidation products in fish oil. J Agric Food Chem 49(4):1724–1729
14. Marmesat S, Morales A, Velasco J, Ruiz-Mendez MV, Dobarganes MC (2009) Relationship between changes in peroxide value and conjugated dienes during oxidation of sunflower oils with different degree of unsaturation. Grasas Aceites 60(2):155–160
15. Gray JI (1978) Measurement of lipid oxidation - Review. J Am Oil Chem Soc 55(6):539–546
16. Gruszka J, Pawlak A, Kruk J (2008) Tocochromanols, plastoquinol, and other biological prenyllipids as singlet oxygen quenchers-determination of singlet oxygen quenching rate constants and oxidation products. Free Radic Biol Med 45(6):920–928
17. Lezerovich A (1986) Derivative UV spectra of lipid conjugated dienes. J Am Oil Chem Soc 63(7):883–888
18. Esterbauer H, Striegl G, Puhl H, Rotheneder M (1989) Continuous monitoring of in vitro oxidation of human low-density lipoprotein. Free Radic Res Commun 6(1):67–75
19. Recknagel RO, Ghoshal AK (1966) Lipoperoxidation of rat liver microsomal lipids induced by carbon tetrachloride. Nature. 210(5041): 1162
20. Corongiu FP, Milia A (1983) An improved and simple method for determining diene conjugation in autoxidized poly-unsaturated fatty-acids. Chem Biol Interact 44(3):289–297
21. Corongiu FP, Poli G, Dianzani MU, Cheeseman KH, Slater TF (1986) Lipid peroxidation and molecular damage to polyunsaturated

fatty-acids in rat liver - Recognition of 2 classes of hydroperoxides formed under conditions in vivo. Chem Biol Interact 59(2):147–155

22. Sergent O, Morel I, Cogrel P, Chevanne M, Pasdeloup N, Brissot P et al (1993) Simultaneous measurements of conjugated dienes and free malonaldehyde, used as a micromethod for the evaluation of lipid peroxidation in rat hepatocytes cultures. Chem Phys Lipids 65(2): 133–139
23. Baron CP, Bro R, Skibsted LH, Andersen HJ (1997) Direct measurement of lipid peroxidation in oil-in-water emulsions using multiwavelength derivative UV-spectroscopy. J Agric Food Chem 45(5):1741–1745
24. Porter NA, Weber BA, Weenen H, Khan JA (1980) Autoxidation of poly-unsaturated lipids—Factors controlling the stereochemistry of product hydroperoxides. J Am Chem Soc 102(17):5597–5601
25. Pryor WA, Castle L (1984) Chemical methods for the detection of lipid hydroperoxides. Methods Enzymol 105:293–299
26. Savitzky A, Golay MJE (1964) Smoothing and differentiation of data by simplified least squares procedures. Anal Chem 36(8): 1627–1639
27. Berton C, Ropers MH, Viau M, Genot C (2011) Contribution of the interfacial layer to the protection of emulsified lipids against oxidation. J Agric Food Chem 59(9):5052–5061

Chapter 13

Quantification of Hydroperoxides in Oils and Fats, Oil-in-Water Emulsions, and Food Products by Ferrous Oxidation–Xylenol Orange Method

Lucie Ribourg-Birault and Claude Genot

Abstract

Lipid hydroperoxides (HPX) are primary products of lipid oxidation and key markers of chemical alteration of dietary oils and fats. The ferrous ion–xylenol orange (FOX2) assay allows for HPX quantification in oils and lipid-containing food products. It is based on the oxidation of ferrous ions in acid and alcoholic medium by HPX, followed by the formation of a colored xylenol orange–ferric ion complex. The complex is finally detected and quantified by spectrophotometry at 560 nm. This method is rapid, is very sensitive (<0.5 mmol HPX/kg lipids), and applies to very small amounts of sample (about 10 mg lipids). Here, we detail the procedure to quantify HPX in dietary oils or lipid extracts, in oil-in-water (O/W) emulsions, or in complex food matrices. The steps for the optimization of HPX extraction from a new matrix and the factors to be carefully monitored throughout the execution of the method are also specified.

Key words Dietary lipids, Emulsion, Food, Hydroperoxides, Lipid oxidation, Xylenol orange–FeIII complex, FOX2 assay

1 Introduction

The formation of hydroperoxides (HPX; also sometimes abbreviated as ROOH), the primary lipid oxidation products, is an early and critical step of lipid oxidation. The hydroperoxide content, often expressed as peroxide value (PV, meq active oxygen/kg), is accordingly considered a very useful criterion to follow lipid oxidation and evaluate the chemical alteration and shelf life of foods. Hydroperoxides are odorless and tasteless, but they are reactive and further decompose into secondary products, including odorant volatile compounds that are generally detrimental to product quality (i.e., development of rancid flavors) and biologically active and potentially toxic reaction products [1–4]. This is why it is highly recommended to measure HPX amounts in dietary oils and fats and in foods after processing and storage in conjunction

Christelle Lopez et al. (eds.), *Multidimensional Characterization of Dietary Lipids*, Methods and Protocols in Food Science,
https://doi.org/10.1007/978-1-0716-3758-6_13,

with secondary products of oxidation [5–7]. Among the methods usually used to measure HPX, the ferrous oxidation-xylenol orange (FOX) method, for which we detail an operational protocol below, allows for quantification of HPX in dietary lipids, in O/W emulsions, and in various foods [8]. The method has a greater sensitivity, requires less sample, and is less sensitive to oxygen and light than the classical iodometric procedure, initially developed by Lea in 1946 [9], which also involves subjective endpoint detection. Moreover, when the iodometric method applies only to extracted oils or fats, requires an oil sample mass of 5 g (for PV > 1 meq active oxygen per kg) to 10 g (for PV $\leq$ 1), and fails to quantify low levels of HPX (PV < 30) [10], the FOX method allows for quantification of low amounts of HPX (0.1–0.5 mmol/kg lipids) [11] and, with appropriate dilutions, high amounts (up to 250 mmol/kg) in small amounts of oils or complex matrices (about 10 mg lipids).

The first step of the FOX method is the extraction of HPX from the matrix. Thus, it does not require a separate lipid extraction step ab initio, which is an advantage since such an extraction can be a major bias in the determination of lipid oxidation products. The FOX method is therefore useful to measure HPX in valuable and expensive products such as model foods formulated for research applications.

The FOX method was initially developed by Gupta in 1973 to quantify hydrogen peroxide in irradiated solutions [12]. It was then adapted to various matrices, including oils, fats, and foods to quantify lipid or protein HPX [13–18]. Two main versions of the FOX method have been described [15]. The first version (FOX1: ferrous oxidation-xylenol orange version 1) was set up to measure submicromolar concentrations of H_2O_2 in aqueous media. With minor modifications, it can be applied for the quantification of HPX in different biological samples and several aqueous food matrices. The reagent is prepared in an aqueous medium and the oxidation of the ferrous ion is amplified by sorbitol.

The second version of the FOX method, often referred to as ferrous oxidation-xylenol orange version 2 or FOX2 assay, was adopted to measure hydroperoxides in edible oils [15, 19] and in various food products [20–23]. Another method for oils and lipid extracts using chloroform/methanol as solvent instead of methanol, sometimes referred to as the mFOX method, was also proposed [24, 25].

The method is based on the ability of hydroperoxides to oxidize ferrous ions into ferric ions in acidic medium (Eqs. 1 and 2):

$$\mathrm{ROOH} + \mathrm{H}^+ + \mathrm{Fe}^{2+} \rightarrow \mathrm{Fe}^{3+} + \mathrm{RO}^\circ + \mathrm{H_2O} \quad (1)$$

$$\mathrm{RO}^\circ + \mathrm{Fe}^{2+} + \mathrm{H}^+ \rightarrow \mathrm{ROH} + \mathrm{Fe}^{3+} \quad (2)$$

A blue–purple complex is then formed between ferric ions and xylenol orange [*o*-cresolsulfonphthalein-3,3-*bis*(methyl-iminodiacetic acid sodium salt; XO] (Eq. 3):

$$Fe^{3+} + XO \rightarrow Fe\text{-}XO \tag{3}$$

The absorption maximum of the XO–Fe^{3+} complex is generally located around 550–600 nm. Its stoichiometry is 1:1, whereas the global stoichiometry of the reactions is not clear because the number of oxidized Fe^{2+} ions varies with the reaction conditions and the nature of measured HPX [18, 20, 26]. Briefly, between 2 and 3 moles of Fe^{3+} are formed per mol of lipid or protein HPX [14]. In certain conditions (addition of sorbitol, absence of butyl hydroxyl-toluene (BHT), etc.), an amplification of the reaction occurs [17]. The assay allows for the detection of monohydroperoxides but also mono-, bi-cyclic, and serial peroxides that can be present in samples containing long-chain polyunsaturated fatty chains [27].

The quantification of HPX is achieved by means of an external calibration curve (i.e., t-butyl-hydroperoxide, cumene hydroperoxide, or H_2O_2, depending on the authors). A fair agreement between the results obtained from this assay and the iodometric method was reported [24, 28, 29]. Data obtained on fish oils and lipids extracted from fish oil-enriched milk drink and mayonnaise also showed fairly good linear correlations between the FOX2 and iodometric titration methods and between the FOX2 and diphenyl-1-pyrenylphosphine and thiocyanate methods [30–32], whereas Nielsen et al. noticed that "*the methods employed gave very different PVs for the same sample and there was no consistency in the levels of PV determined by different methods*" [30]. Several authors noticed, however, higher values with the FOX method than with the iodometric method [19, 24, 29], whereas lower PV values than those obtained with the iodometric titration, by Near Infra-Red (NIR) determination or with the peroxy safe kit were recorded for soybean oils and PUFA-rich oils [31, 33].

The present protocol of the FOX-2 method, schematically represented in Fig. 1, is routinely applied in our laboratory. It has been successfully applied to various vegetable and fish dietary oils [28], model O/W emulsions [34, 35], during simulated digestion of emulsions [36], and in several foods such as fatty fish minces [37], milk and dairy products, and dried and smoked edible insects. We obtained very similar and correlated values for HPX measured by the present FOX method and the iodometric method (PV) in oxidizing edible oils [28]. Importantly, as highlighted in [21] and because several parameters and factors may interfere with the color development, the conditions to extract HPX from a new matrix should be tested and optimized as suggested below (appropriate solvent, shaking time, etc.). To have an overview of these

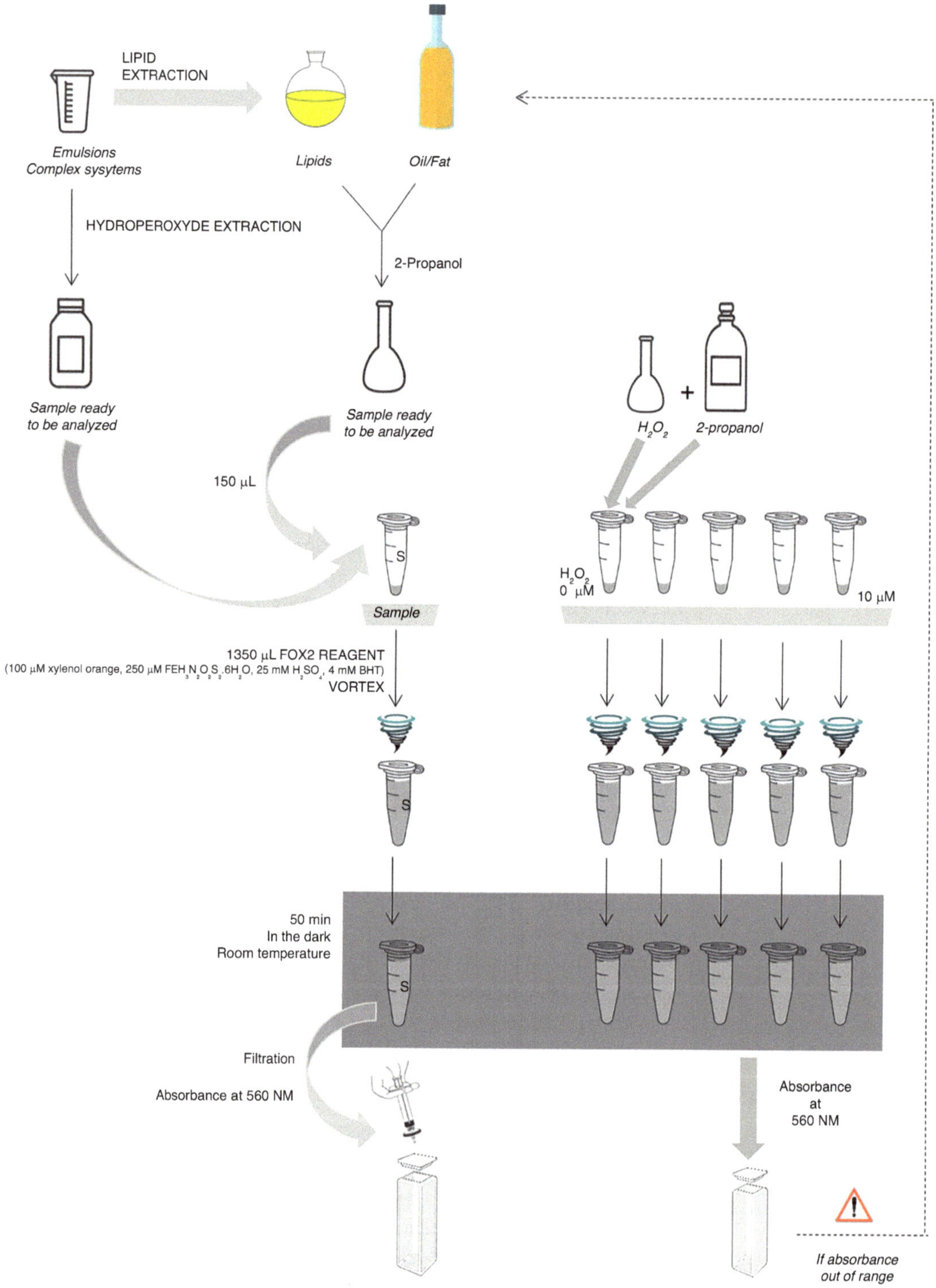

Fig. 1 Flowchart of the main steps involved in the quantification of hydroperoxides in oils, fats, emulsions, and complex food products by the FOX2 method

parameters and factors possibly affecting the method, we recommend the review paper of Bou et al. [20]. Hydrogen peroxide is proposed here as the standard compound. Amounts of hydroperoxides are finally expressed as μmole H_2O_2 equivalent per gram of lipids. The method allows for the quantification of HPX from 0.5 to 250 μmol eq H_2O_2/g lipids.

2 Materials

2.1 Chemicals (Table 1)

Prepare solutions using ultrapure water (conductivity = 18 MΩ. cm^{-1} at 25 °C) and/or HPLC-grade reagents. Check the spectral quality of the solvents: they should not absorb between 200 and 650 nm. Diligently follow safety instructions and waste disposal regulations when disposing waste materials.

2.2 Mandatory Equipment

1. Analytical precision balance, sensitivity: 0.01 mg.
2. UV-visible spectrophotometer (in the following, the method is described for a double-beam spectrophotometer) and semi-micro optical quartz cells (optical path length: 10 mm) with Teflon caps or a multiplate reader with solvent compatible plates and films (UV-star® microplates, Greiner, reference 675801 or 655801, viewseal film, Greiner reference 676070).
3. Positive displacement pipettes and disposable capillary piston tips (1000 μL, 250 μL, and 50 μL).
4. Magnetic stirrer and bar magnets.
5. 1- or 2-mL graduated glass pipette.
6. Volumetric glass flasks: 10 mL, 25 mL, 50 mL, and 500 mL.
7. 500-mL glass bottle.
8. Weighing shoes.
9. 2- or 1.5-mL polypropylene microtubes (ideally 2 mL).
10. Syringes (preferably 5 mL), needles, and cellulose regenerated syringe filters (diameter: 25 mm and pore size: 0.45 μm).
11. Bottle-top dispenser (ideally total volume: 50 mL) enabling to distribute a volume of 1.35 mL (or 2 positive displacement pipettes and capillary piston tips of 1000 μL).

2.3 Preparation of the FOX2 Reagent

*2.3.1 Sulfuric Acid Solution: 500 mM (see **Note 1**)*

Introduce about 12 mL water into a 50-mL volumetric flask and add 1.4 mL sulfuric acid 96% with a graduated glass pipette (*see* **Note 1**). Make up to 50 mL with water.

Table 1
Chemicals required for FOX2 method

Chemicals	Chemical formula	CAS number	Molecular weight (g. mol^{-1})	Quality	Safety instructions
Ferrous ammonium sulfate hexahydrate	$H_8FeN_2O_8S_2$, $6H_2O$ Fe $(NH_4)_2SO_4)_2$ 6 H_2O	7783-85-9	392.14		 H315 H319 H335
Sulfuric acid 96%	H_2S0_4	7664-93-9	98.078	96 wt% 1.84 g/cm^3 (20 °C)	 H290 H314
Xylenol orange (XO) o-cresolsulfonephthalein 3,3′-bis(N,N-bis (carboxymethyl) aminomethyl tetrasodium salt	$C_{31}H_{28}N_2Na_40_{13}S$ 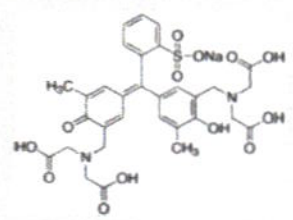	3618-43-7	760.59		
Butylhydroxytoluene (BHT) 2,6-di-tert-butyl-4-methylphenol	$(CH_3)_3C]_2$ $C_6H_2(CH_3)0H$	128-37-0	220.36		 H410
Methanol[a]	CH_30H	67-56-1	32.042	RPE for analysis	 H225  H301 H311 H331  H370

2-Propanol Propan-2-ol Isopropyl alcohol[a]	$(CH_3)_2CHOH$	67-63-0	60.10	RPE for analysis	 H225  H319 H336
Hydrogen peroxide	H_2O_2	7722-84-1	34.01	30 wt%	 H314 H335 H302 H332  H272  stored at 4 °C in darkness.

[a]Prerequisites check of the spectral quality of the solvents: record the solvent spectrum between 200 and 650 nm against air after having recorded the baseline with empty reference and sample cells. The solvents should not absorb over this wavelength range

2.3.2 Ferrous Ammonium Sulfate Solution: 5 mM $FeH_8N_2O_2S_2$

Weigh 100 mg of ferrous ammonium sulfate in a 50-mL volumetric flask. Add sulfuric acid solution. Dissolve thoroughly and make up to 50 mL with sulfuric acid solution.

*2.3.3 FOX2 Reagent: 100 μM Xylenol Orange, 250 μM $FeH_8N_2O_2S_2.6H_2O$, 25 mM H_2SO_4, 4 mM BHT in MeOH/H_2O (95/5; v/v) (see **Note 2**)*

Weigh 38 mg xylenol orange and 430 mg of BHT in separate weighing shoes. Introduce the chemicals into a 500 mL volumetric flask; rinse abundantly the weighing shoes with methanol. When methanol is around up to 300 mL, introduce 25 mL of ferrous ammonium sulfate solution. Dissolve thoroughly and make up to 500 mL with methanol. Store in a brown bottle (*see* **Note 2**) at 4 °C in darkness.

3 Method

3.1 Calibration

*3.1.1 Preparation of H_2O_2 Standard Solutions (see **Notes 3** and **4**)*

Prepare the H_2O_2 standard solutions every day just before use. The H_2O_2 calibration solutions are usually prepared in 2-propanol or in the solvent chosen during the optimization step of HPX extraction (*see* Subheading "Other Samples of Lipid-Containing Material").

H_2O_2 standard solution A: 150 mM

Weigh 150 μL of hydrogen peroxide in a 10 mL volumetric flask and make up to 10 mL with 2-propanol. Mix with a vortex agitator.

H_2O_2 standard solution B: 3 mM

Introduce 200 μL of solution A into a 10-mL volumetric flask and make up to the mark with 2-propanol. Then, vortex them.

H_2O_2 diluted solutions 1 (300 μM) and 2 (60 μM)

Introduce 1 mL and 200 μL of solution B into 10 mL volumetric flasks and make up to the mark with 2-propanol. Then, vortex them.

*3.1.2 Determination of the Concentration of H_2O_2 Standard Solution (see **Note 5**)*

Introduce 2-propanol into the reference and sample quartz microcells and record the autozero of the spectrophotometer. Dilute with 10-, 20-, and 50-fold factors the standard solution A with 2-propanol and measure the absorbance at 240 nm (A240nm) of the three diluted solutions.

Calculate the H_2O_2 concentration ([H_2O_2], mM) of the solutions using the following formula (Eq. 4):

$$[H_2O_2] = A240\,nm \times dilution \times 1000/\varepsilon \quad (4)$$

where ε = Molar extinction coefficient of H_2O_2 at 240 nm in 2-propanol = 49.5 $M^{-1}.cm^{-1}$. Finally, calculate the mean of the three values to obtain the concentration of the standard solution.

3.1.3 Preparation of Tubes for the Calibration Curve

Depending on the level of oxidation of the sample and the amounts of hydroperoxides it contains, we recommend different ranges of H_2O_2 calibration concentrations in the reacting medium to build the calibration curve:

1. 0.5 to 6 μmol.L^{-1} for low lipid HPX levels <50 μmol/g; it is obtained with H_2O_2 diluted solution 2.
2. 3 to 10 μmol.L^{-1} for HPX level > 20 μmol/g lipids; the calibration curve is built from H_2O_2 diluted solution 1.

For sample with unknown oxidation level, we suggest to prepare an enlarged calibration curve (0.5 to 10 μmol.L^{-1}) with standard diluted solutions 1 and 2.

Prepare microtubes with 0 to 150 μL of diluted solution 1 or 2 and make up the total volume to 150 μL with 2-propanol or the required solvent according to Table 2. Use positive displacement pipettes.

3.1.4 Reacting Conditions and Absorbance Measurement (see ***Notes 6*** *and* ***7****)*

Check the absorbance at 560 nm (A560 nm) of the FOX2 reagent as follows: Introduce 2-propanol into the reference and sample microcells and record the autozero of the spectrophotometer. Keep 2-propanol in the reference microcell and introduce FOX2 reagent into the sample microcell and record the spectrum between 300 and 650 nm. The FOX2 reagent is considered to be of good quality when A560 nm is less than 0.4.

To each microtube, add 1350 μL of FOX2 reagent with a dispenser (*see* **Note 5**) and vortex the mixture thoroughly. Then let the microtubes stand for 50 min at laboratory temperature (18–25 °C) in darkness (*see* **Note 6**) and vortex them again. Measure the absorbance at 560 nm of the reacted calibration solutions against a reference cell filled with 2-propanol. Finally draw the calibration curve (A560 nm = f([H_2O_2])) and fit to the linear model (Eq. 5):

Table 2
Preparation of tubes for the H_2O_2 calibration curve

	Calibration tube												
	Concentration of H_2O_2 (μmol.L^{-1})												
Volume of reactant (μL)	**0**	**0.5**	**1.0**	**2.0**	**3.0**	**3.0**	**4.0**	**5.0**	**5.0**	**6.0**	**7.0**	**8.0**	**10.0**
Diluted solution 2 (60 μmol.L^{-1})	0	13	25	50	75		100	125		150			
Diluted solution 1 (300 μmol.L^{-1})	0					15			25		35	40	50
2-propanol	150	138	125	100	75	135	50	25	125	0	115	110	100

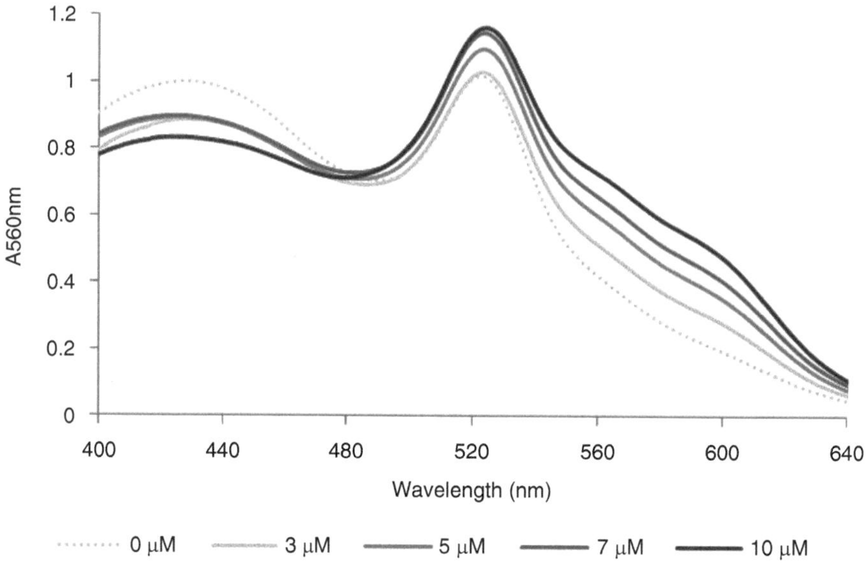

Fig. 2 Usual spectra obtained with H_2O_2 calibration solutions after reaction with the FOX reagent

$$A560\ \mathrm{nm} = a \times [\mathrm{H_2O_2}] + b \tag{5}$$

Check for linearity ($r^2 > 0.99$). The usual slope (a) is around 0.03 M^{-1} and the intercept ranges between 0.3 and 0.5.

We advise to also record the absorbance spectra between 350 or 400 and 650 nm for several samples and check if they conform to usual spectra (Fig. 2).

3.2 Determination of the HPX Content in Samples

3.2.1 Sample Preparation (see **Note 3**)

The present protocol enables to determine the HPX content of oils, fats, or previously extracted lipids, provided the extracts do not contain pigments absorbing in the visible region (*see* Subheadings "Edible Oils" and "Extracted Lipids"). Indeed, the lipids should have been extracted with an appropriate method avoiding elevated temperature and limiting contact with oxygen and light as detailed in Chapter XX. As the lipid extraction step may result in the formation or loss of HPX (i.e., during the drying step or during the storage of the extract), we recommend, when possible, to extract HPX directly from the matrix (*see* Fig. 1 and Subheading "Oil-in-Water (O/W) Emulsion"). In the following, 2-propanol is the extracting solvent. This has been validated in our laboratory for different food matrices. However, other matrices may require optimization of the extraction conditions and adaptation of the conditions of the assay to the matrix. Tests should be also performed to ensure the absence of interferences. The testing and optimization step is described in Subheading 3.3.

Edible Oils

Weigh about 10 mg (± 0.01 mg) of oil in a polypropylene microtube and add 1000 μL of 2-propanol or the chosen solvent for obtaining a concentration of around 10 mg of lipids per milliliter of fresh, nonoxidized oils (*see* **Note 3**). This concentration (C_{oil}, mg. mL^{-1}) can be lowered if the lipids are highly oxidized.

Extracted Lipids

For lipid extracts of known lipid concentration in a solvent such as pentane, hexane, or chloroform, introduce the required volume of the solution into a microtube, dry with nitrogen flow, and add 2-propanol or the chosen solvent to obtain about 10 mg lipids per milliliter (*see* **Note 3**). Vortex. Record the exact lipid concentration (C_{lipids}, mg.mL^{-1}) in the solution.

Oil-in-Water (O/W) Emulsion

Calculate the emulsion volume to sample (V_e, μL) containing about 10 mg (m, mg) of oil from the oil volume fraction (Φ) and oil density (d) (Eq. 6):

$$V_e = m/d.\Phi \tag{6}$$

Introduce into a microtube this volume of emulsion.

Add 2-propanol (usually 1000 μL) to obtain a lipid concentration of around 10 mg oil per milliliter and vortex. Calculate and record the exact lipid concentration in the mixture (C_{lipids}, mg.mL^{-1}).

Other Samples of Lipid-Containing Material

Calculate the product mass (m_s, mg) containing about 10 mg of lipids from its total lipid content ($L\%$, g/100 g). Weigh this given mass (m_s) in a tube and add the required solvent volume (usually 1000 μL). Calculate the exact lipid amount present in the weighed product and the theoretical lipid concentration in the mixture (C_{lipids}, mg mL^{-1}). Vortex or grind depending on the physical state of the sample, stir with a magnetic bar or a rotary agitator, and centrifuge the mixture. Collect the supernatant and, if necessary (for instance, in case of turbidity), filter the supernatant through a syringe filter.

3.2.2 Preparation of Reaction Tubes

Into each microtube, introduce 150 μL (or the required volume) of each sample solution with a positive displacement pipette.

*3.2.3 Reaction Conditions and Absorbance Measurement (see **Notes 5** and **6**)*

The procedure is similar to the one described in Subheading 3.1.4, with an additional filtration step.

To each microtube, add 1350 μL of FOX2 reagent with a dispenser (*see* **Note 5**), vortex, and let the microtubes stand at laboratory temperature (18–25 °C) in the dark. After 50 min, vortex the microtube again and collect the reacting medium with a syringe equipped with a needle. Filter the reacting medium through a regenerated cellulose membrane and collect the filtrate in the optical microcells directly. Measure the absorbance at 560 nm against the required solvent. You may use the same syringe, needle, and filter for triplicates if you rinse with the solvent after each use.

If the sample absorbance is outside the range of the calibration curve ($0.3 < A560 < 0.8$) corresponding to 0.5 to 10 μM eq.

H_2O_2), dilute the sample solution to be analyzed or prepare more concentrated sample solution from the initial sample (oil, lipid extract, emulsion, milk, etc.). The color of the medium after the reaction should appear orange or brown. If it is blue or purple, then the amount of hydroperoxides in the sample is too high.

3.2.4 Determination of the HPX Content (see Notes 8 and 9)

From A560 nm measured on the sample reaction medium and parameters of the calibration curve (*see* Subheading 3.1.2), calculate the concentration of HPX ([H_2O_2], µmoles eq H_2O_2 L^{-1}) in the reaction medium (Eq. 7):

$$[H_2O_2] = (A560\,\mathrm{nm} - b)/a \quad (7)$$

Then, calculate the theoretical concentration of lipids in the reaction medium (C, g/L) (Eq. 8):

$$C(\mathrm{g/L}) = (C' \times V)/V' \quad (8)$$

where $C' = C_{\mathrm{oil}}$ or C_{lipids} (mg mL^{-1}), i.e., the concentration of lipids in the sample solution (*see* Subheading 3.2.1); V (µL), the volume of the sample, usually 150 µL; and V', the total volume of the reaction medium, 1500 µL.

Lastly, calculate the HPX concentration (HPX, µmol H_2O_2 eq. per gram of lipids or mmol H_2O_2 eq. per kg) in the analyzed product (Eq. 9):

$$\mathrm{HPX}(\mu\mathrm{mol/g}) = [H_2O_2]/C \quad (9)$$

It can also be relevant to express the result in µmol H_2O_2 eq. per gram of the initial wet or dry matter.

3.3 Optimization of HPX Extraction and Reaction Conditions for a New Matrix

HPX are more polar than nonoxidized lipids they come from. Thus, the first step of the sample preparation for FOX2 method consists of extracting and solubilizing HPX in an appropriate solvent. Alcoholic solvents such as methanol or, as chosen in the present protocol, 2-propanol, allow for HPX extraction from most oils and matrices. However, when solubility and/or turbidity problems occur, methanol/water or polar/apolar solvent mixtures (i.e., pentane (or hexane)/2-propanol, chloroform (or dichloromethane)/methanol/water) can be used, provided the final solvent composition of the reacting medium makes it monophasic, compatible with the formation of the Fe^{3+}–XO complex and with A560 reading. The conditions selected by different authors for extracting hydroperoxides from different food products before using the FOX2 method are reported in Table 3.

Interestingly, the chloroform/methanol 2/1 (v/v) mixture was noticed to extract from cheese more hydroperoxides with lipids than using extraction according to the Bureau of Dairy Industry and that chloroform/methanol may be substituted by either

Table 3
Solvents selected for extracting and measuring hydroperoxides by the FOX2 assay in various oils and food and biological products and solvent composition of the reacting medium

Sample	Extracting solvent	Sample concentration	Solvent composition of the reacting medium (v/v)	Reference
Vegetable and fish oils, lipid extracts, O/W emulsions, simulated digestion media, dairy products, animal foods	2-Propanol	10 mg/mL	Methanol/2-propanol/water: 95/11.1/5 to 95/10.1/6 (v/v/v) depending on sample water content	Our work [1, 28, 34, 35, 37]
Vegetable oils, fresh and oxidized soybean oils Fish and oxidized tuna and rapeseed oils, mayonnaise, milk drink	2-Propanol	10 ml/mL 5–40 nmol eq. ROOH	Methanol/2-propanol/water: 90/11.1/10 (v/v/v)	[19, 33] [30]
PUFA-rich fish oils	2-Propanol	0.7–80 mg/mL	Methanol/2-propanol/water: 90/11.1/10 (v/v/v)	[31]
Animal tissues (liver, heat, brain, muscle)	Cold (5 °C) methanol	0.1 to 0.2 g/mL	Water/methanol: 99/1 tp 90/10 (v/v)	[17]
Chicken meat	Cold (−20 °C) methanol	0.2 g/mL	Methanol/water: 75/25 (v/v)	[22]
Fatty fish muscle (horse mackerel)	Cold (4 °C) methanol	0.1 g/mL 1:10 w/v	Methanol/water (90/10)	[21]
Invertebrate tissues (crabs and marine worms)	Methanol	1:9 1 g/9 mL?	Methanol/water: 90/2 to 90/12 (v/v)	[48]
Plant tissues	Ethanol/water or methanol/water (80/20; v/v)	20–110 mg/mL	Methanol/water (90/10)	[29]
Egg yolk phosphatidyl-choline liposomes	Buffered water	0.76 mg/mL	Methanol/water: 60/40 → peak at 610 nm or methanol/water: 90/10 → peak at 560 nm	[45, 49]
Marine diatoms	Tris-HCl buffer	$37.5\ 10^4$ to $37.5\ 10^5$ cells/mL	Methanol /water: 90/35 (v/v)	[42]
Soybean and fish oils, butter, lipid extracts from beef and chicken	Chloroform/methanol: 7/3 (v/v); dried	1–30 mg/mL	Chloroform/methanol/water: 7/3/0.1 (v/v/v)	[24]

(continued)

Table 3
(continued)

Sample	Extracting solvent	Sample concentration	Solvent composition of the reacting medium (v/v)	Reference
Vegetable oils, fish oil, butter	Chloroform/methanol (7/3; v/v)	20 mg/mL	Chloroform/methanol/water: 7/3/0.15 (v/v/v)	[11, 25]
Fish hydrolysates Fish by-products	Chloroform/methanol 2/1 (v/v)	50–200 mg/1000 μL	Methanol/chloroform/water: 90/6.8/4.6 (v/v/v)	[50]
Lipid extract from colored microalgae	Chloroform/methanol 7/3 v/v	0.1 mg lipid/mL	Chloroform/methanol/water: 7/2.8/0.1 (v/v/v)	[51]
Sea cucumber lipids pigmented by carotenoids	Dichloromethane/methanol (3/2 v/v)	Up to 125 mg/mL	Dichloromethane/ethanol/methanol/water: 9/6/4/1. then 1 vol water to extract the iron–XO complex	[52]
Edible oils Fried snacks	Dichloromethane/ethanol (3/2); v/v)	0.1 to 25 mg/mL	Dichloromethane/ethanol/methanol/water: 9/6/4/1 (v/v/v/v)	[23, 32]

methanol/decanol/hexane (3:2:1, v/v/v) or hexane/2-propanol/methanol (5:7:2, v/v/v) for HPX determination [38].

As detailed in Fig. 3, for certain matrices the above conditions for the assay need to be adapted.

In these cases, the optimization of the extraction conditions (appropriate solvent, sample/solvent ratio, shaking time, etc.) should be performed [17, 21, 22].

The suggested optimization process for testing and adjusting the method to a new material is schematized in the decision tree given in Fig. 4.

The extract is usually prepared at an equivalent lipid concentration of about 10 mg/mL, but this concentration can be lowered or increased depending on the HPX content of the sample to keep the final absorbance at 560 nm of the reacting medium between 0.3 and 0.8. We recommend preparing all samples in triplicate.

During the optimization step, we advise to record the absorbance spectrum of the reacting medium at least between 400 and 650 nm and not only at 560 nm. The spectra should be compared to the spectra obtained with the calibration solutions and with usual spectra (Fig. 2) to evidence, for example, problems due to turbidity or the presence of pigments. Afterward, on routine, the absorbance

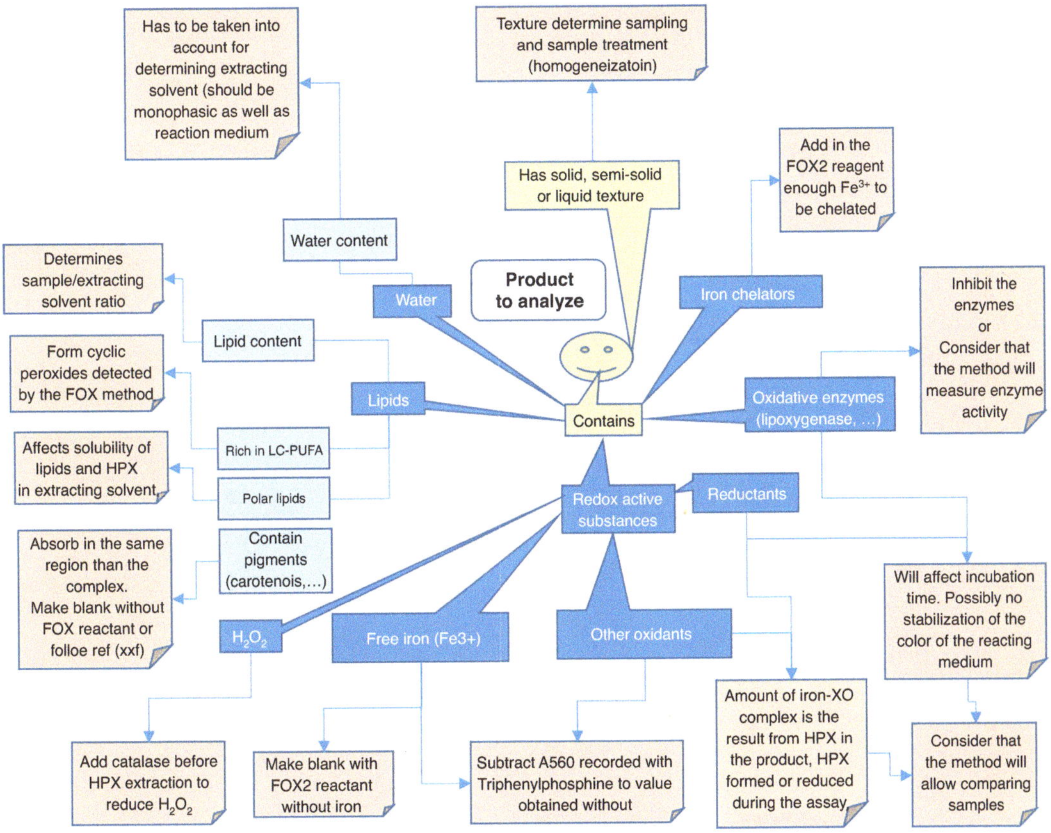

Fig. 3 Characteristics of the analyzed food product to be taken into account for conducting the FOX2 assay. (See the review of Bou et al., 2008 [20] for information about possible interferences)

only at 560 nm can be recorded. The stability of the developed color and the incubation time of the reaction medium should be also checked.

4 Notes

1. Add acid to water for safety reasons. Let the acid mixture cool down to ambient temperature while stirring and then make it up to 50 mL with water.
2. The brown bottle can be replaced by a transparent glass bottle covered with aluminum foil. The FOX2 reagent can be stored for 1 month at 4 °C. After the storage period, check if the calibration curve fits previous curves with a linear increase of A560 nm with H_2O_2 concentration. If it is not the case, prepare a fresh reagent.
3. As it is more hydrophobic than methanol, 2-propanol facilitates the solubilization of oils, phospholipids, especially those

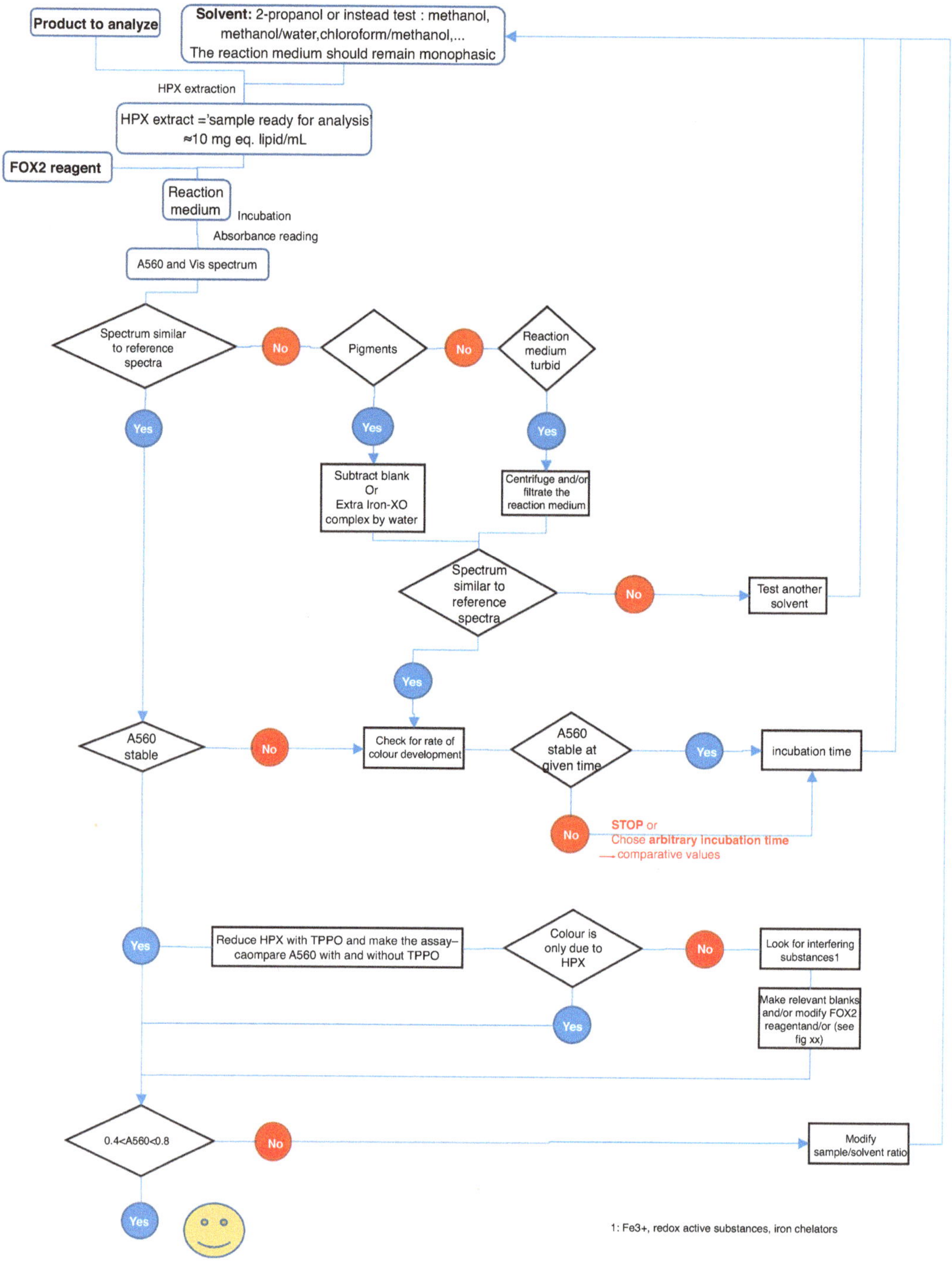

Fig. 4 Decision tree for optimization of hydroperoxide extraction from food complex matrices and reaction conditions

containing omega-3 long-chain PUFA, and their hydroperoxides, close to 10 mg/mL [23, 31]. However, it is prone to form HPX [20, 23]. Thus, only use recently opened bottles and preferentially BHT-stabilized solvent. Moreover, during the experiment, keep 2-propanol in a stoppered bottle and close the bottle after each sampling, unless peroxides can be produced.

4. Several authors built their calibration curve with iron (II) chloride [11, 24]. However, as the final XO-Fe stoichiometry and the intensity of the color (A560 nm) vary with the composition of the reaction medium [13, 18, 20, 39], we recommend to perform calibration with authentic peroxides (i.e., hydrogen peroxide or cumene hydroperoxide). Cumene hydroperoxide (CAS number: 80–15-9) was formerly used as the calibration standard in our laboratory. As it is highly toxic (https://pubchem.ncbi.nlm.nih.gov/compound/Cumene-hydroperoxide, `retrieved 21-01-2021`), we now use H_2O_2 as the calibration standard (after having confirmed similar results with both standards).
5. The molar extinction coefficient of H_2O_2 at 240 nm in 2-propanol (49.5 $M^{-1}.cm^{-1}$) was determined using the molar extinction coefficient of H_2O_2 at 240 nm in water (43.6 $M^{-1}.cm^{-1}$; Lei B, Adachi N, Arai T. Measurement of the extracellular H_2O_2 in the brain by microdialysis. Brain Res Brain Res Protoc. 1998;3 (1):33–36) after comparison of absorbance at 240 nm of similar concentrations of H_2O_2 in 2-propanol and water.
6. We suggest to put FOX2 reagent in a bottle-top dispenser just before use. It is more convenient and repeatable than using two positive displacement pipettes (one for 1000 μL and another for 350 μL) to add 1350 μL of FOX2 reagent.
7. In the case of complex samples, it may be necessary to extend the incubation time depending on the time needed to reach the plateau.
8. The specific absorbance of the iron–XO complex depends on the solvent used in the assay and the source of xylenol orange [18].
9. Checking and eliminating interferences.

Fe^{3+} ions in the analyzed sample may participate in the formation of the iron–xylenol complex in the reaction medium, leading to overestimation of HPX amounts. The interference due to Fe^{3+} ions can be solved by preparing a FOX reactant in which Fe^{2+} ions are omitted.

More generally, to confirm that the formation of the iron–XO complex is only due to HPX present in the analyzed sample, a sample blank containing 2 mM of the HPX reducing agent triphenyl-phosphine (TPP), can be prepared and incubated for 30 min to allow the complete reduction of any present hydroperoxides, including cyclic and serial hydroperoxides. The assay is finally performed as previously described. As the alcohol analogs formed by TPP reduction do not respond to the FOX assay, A560 nm of this reaction medium (blank) is subtracted from the value measured with the original FOX reagent and the result is attributed to HPX [11, 29, 40, 41]. If the TPP reaction medium becomes opalescent, the reduction of HPX can be achieved with 50 mM tris-(2-carboxyethyl) phosphine incubated for 1 h [42].

The possible contribution of H_2O_2 in the sample to the color development can be eliminated by preliminary treatment of the samples by catalase (500–1000 units for 15–30 min) [13, 29].

Ascorbic acid can react with oxygen and generate H_2O_2 through superoxide ion as intermediate. This contribution can be eliminated by performing the assay in inert (nitrogen) atmosphere [43].

More generally, reducing or oxidizing substances such as ascorbate and ascorbic acid, glutathione and cysteine, and compounds that bind ferric ions such as EDTA may also interfere with the FOX assay [15, 17]. In contrast, glucose, fructose, galactose, cholesterol, triglycerides, linoleic acid, reduced gluthathione, vitamin E, TCA, thiols, urate, and salts were found not to influence the development of the colored complex [15, 44].

A. Appendix: Complementary Experiments

We detail in this appendix several tests carried out to validate several parameters mentioned above.

Appendix I/BHT Content of FOX2 Reagent

The original FOX2 reagent contains 4 mM BHT to avoid the formation of lipid hydroperoxides from nonoxidized lipids and protect the sample from oxidation during incubation of the reacting medium [13–15]. According to Wolff (1992), "*in the presence of a high background of nonperoxidized lipid, alkoxyl radicals generated in the ferrous oxidation step react rapidly with native lipid, generating further hydroperoxide in a chain reaction.... Inclusion of the chain-breaking antioxidant butylated hydroxytoluene (BHT) overcomes this problem. BHT presumably repairs alkyl radicals produced by the reaction of alkoxyl radicals with unsaturated lipids. Experimentally, BHT at a concentration of 4 mM was found to provide a firm endpoint when measuring phosphatidylcholine and low-density lipoprotein peroxide content.*"

Indeed, this concentration is by far higher than the BHT concentration currently used to protect lipid from oxidation during extraction and storage (0.01% solvent; about 1 mg.g-1 lipids). It

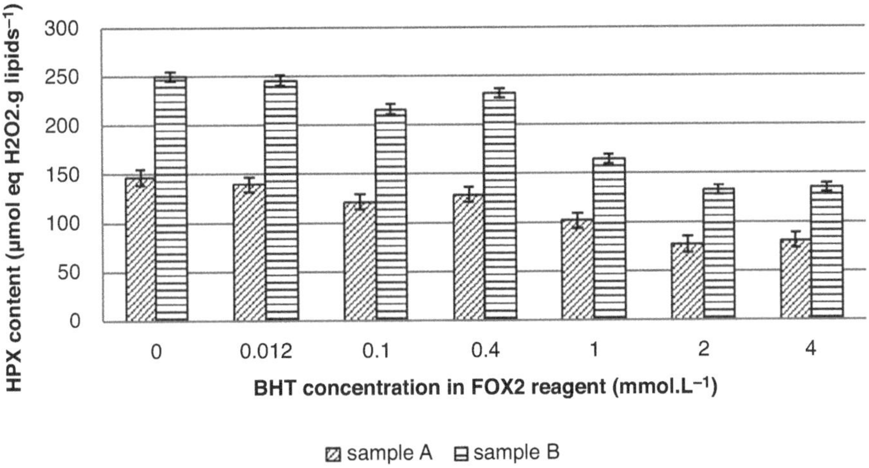

Fig. 5 HPX content depending on BHT concentration in FOX2 reagent (the mean of triplicate analysis; error bars = standard deviation)

was thus criticized by authors who either reduced this concentration or removed it from the reagent [17, 21, 22, 24, 30, 45]. In fact, when performed without BHT, in the presence of oxygen-sensitive compounds, the FOX assay would work as an induced method giving information on the oxidizability of the material rather than on its actual oxidation level at a given time [22, 26].

To determine the BHT concentration needed to avoid the formation of peroxidized species during the analysis, we tested the effect of 0 to 4 mM BHT in the reagent on the formation of XO–Fe3+ complex (A560 nm) with two samples of sunflower oil characterized by two levels of oxidation.

As shown in Fig. 5, for both oil samples, the measured HPX amount was maximum for the lowest BHT concentrations (0 or 0.012 mM), then the absorbance decreased until a plateau observed for 2 mM BHT was higher. The plateau corresponds to the reaction of the HPX initially present in the oil samples with Fe^{2+} ions that oxidize into Fe^{3+} and the formation of the colored Fe^{3+}–XO complex. At BHT concentrations lower than 2 mM, additional HPX were formed upon the assay procedure, and so did the complex amounts. This result confirms the need for at least 2 mM BHT in the FOX2 reagent to avoid the formation of hydroperoxides during the reaction. It validates the use of FOX2 reagent containing 4 mM BHT as recommended by the previous authors [14, 15, 19].

Appendix II/pH of the Reaction Mixture

The optimum for the formation of the ferric ion–xylenol orange complex in H_2SO_4 was suggested at pH 1.7–1.8 by several authors [40, 46]. [40] suggested that a better sensitivity of the test was attained when the pH was adjusted to 1.7–1.8 with Na_2HPO_4, whereas [47] observed a better sensitivity and reproducibility of the assay when perchloric acid was used instead of sulfuric acid.

We measured the pH of the FOX2 reagent (100 μM xylenol orange, 250 μM $FeH_8N_20_2S_2.6H_20$, 25 mM H_2SO_4, and 4 mM BHT) and found a value of 0.9. The pH of the reaction medium for the H_2O_2 calibration solution was 0.6, which is far away from the optimum pH given in the literature. To test the effect of pH on the formation of the complex, the FOX2 reagent (100 μM xylenol orange, 250 μM $FeH_8N_2O_2S_2.6H_2O$, and 4 mM BHT) was prepared at pH 1.8 by replacing 25 mM H_2SO_4 by 3 mM H_2SO_4. Under this condition, the color of the reagent changed continuously over time, as well as the absorbance of the reaction medium obtained for the H_2O_2 calibration points. We concluded that it is not suitable to use a FOX2 reagent prepared at this pH.

Appendix III/Effect of Incubation Time

Incubation time may vary with the tested matrix. For instance, Delong et al. reported that, for plant tissues, maximal color development varied between 120 and 180 min or even longer [29]. They chose to measure A560 exactly 10 min following sample introduction in the reaction mixture, a period during which the absorbance developed linearly in all systems tested. Earlier, Shantha and Decker pointed out the instability of the reagent upon storage [24].

To test the stability of the color of the FOX2 reaction medium and determine the required incubation time, we used two H_2O_2 solutions taken in the calibration points (3 and 10 $\mu mol.L^{-1}$) and oxidized sunflower oil. The reaction medium was incubated at ambient temperature in the dark. As shown in Fig. 6, A560 nm

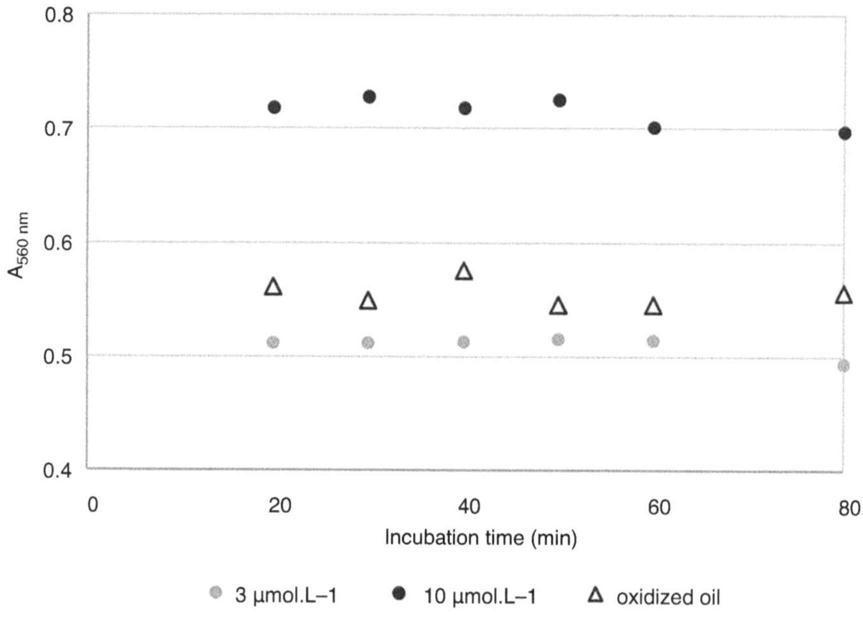

Fig. 6 Evolution with the incubation time of the absorbance at 560 nm of FOX2 reaction medium for two H_2O_2 solutions (3 and 10 $\mu mol\ L^{-1}$) and for a sample of oxidized vegetable oil

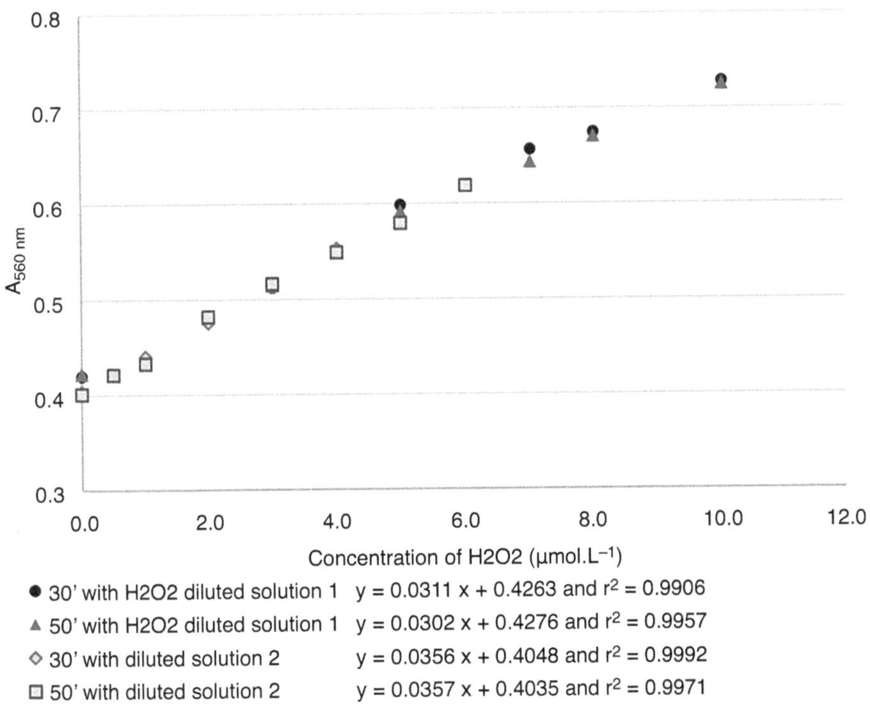

Fig. 7 H_2O_2 calibration curves (A560 nm of the reacting medium) depending on incubation time: 30 vs 50 min

remained stable between 20 and 50 min of incubation and then the absorbance tended to decrease slightly.

Finally, we compared calibration curves obtained with the two H_2O_2 standard solutions after 30 or 50 min of incubation. The calibration parameters (slope and intercept) are similar for both incubation times (Fig. 7). Finally, in the protocol developed above, we chose a 50 min incubation time for practical reasons, but we could have also chosen 30 min.

References

1. Awada M, Soulage CO, Meynier A, Debard C, Plaisancie P, Benoit B et al (2012) Dietary oxidized n-3 PUFA induce oxidative stress and inflammation: role of intestinal absorption of 4-HHE and reactivity in intestinal cells. J Lipid Res 53(10):2069–2080. https://doi.org/10.1194/jlr.M026179
2. Kubow S (1993) Lipid oxidation products in foods and atherogenesis. Nutr Rev 51(2): 33–40. https://doi.org/10.1111/j.1753-4887.1993.tb03064.x
3. Vieira S, Zhang GD, Decker EA (2017) Biological implications of lipid oxidation products. J Am Oil Chem Soc 94(3):339–351. https://doi.org/10.1007/s11746-017-2958-2
4. Grootveld M, Percival BC, Leenders J, Wilson PB (2020) Potential adverse public health effects afforded by the ingestion of dietary lipid oxidation product toxins: significance of fried food sources. Nutrients 12(4). https://doi.org/10.3390/nu12040974
5. Frankel EN, Meyer AS (2000) The problems of using one-dimensional methods to evaluate multifunctional food and biological antioxidants. J Sci Food Agric 80(13):1925–1941. https://doi.org/10.1002/1097-0010

(200010)80:13%3C1925::AID-JSFA714%3E3.0.CO;2-4
6. Schaich KM (2013) Challenges in analyzing lipid oxidation: are one product and one sample concentration enough? In: Logan A, Nienaber U, Pan XS (eds) Lipid oxidation. Challenges in food systems. AOCS Press, Urbana. https://doi.org/10.1016/B978-0-9830791-6-3.50005-9
7. Schaich KM (2016) Analysis of lipid and protein oxidation in fats, oils, foods. In: Hu M, Jacobsen C (eds) Oxidative stability and shelf-life of foods containing oils and fats. AOCS Press, Urbana. https://doi.org/10.1016/B978-1-63067-056-6.00001-X
8. Dobarganes MC, Velasco J (2002) Analysis of lipid hydroperoxides. Eur J Lipid Sci Technol 104(7):420–428. https://doi.org/10.1002/1438-9312(200207)104:7%3C420::AID-EJLT420%3E3.0.CO;2-N
9. Lea CH (1946) The determination of the peroxide values of edible fats and oils: The iodimetric method. J Soc Chem Ind October:286–91
10. 3960 I. ISO 3960 (2017) Animal and vegetable fats and oils - Determination of peroxide value - Iodometric (visual) endpoint determination. ISO Online browsing platform [Internet]. Available from: https://www.iso.org/obp/ui/#iso:std:iso:3960:ed-5:v1:en
11. Dermis S, Can S, Dogru B (2012) Determination of peroxide values of some fixed oils by using the mFOX method. Spect Lett 45(5):359–363
12. Gupta BL (1973) Microdetermination techniques for H_2O_2 in irradiated solutions. Microbiochem J:363–374. https://doi.org/10.1080/00387010.2012.666702
13. Jiang ZY, Hunt JV, Wolff SP (1992) Ferrous ion oxidation in the presence of xylenol orange for detection of lipid hydroperoxide in low--density-lipoprotein. Anal Biochem 202(2):384–389. https://doi.org/10.1016/0003-2697(92)90122-n
14. Jiang ZY, Woollard ACS, Wolff SP (1991) Lipid hydroperoxide measurement by oxidation of Fe^{2+} in the presence of xylenol orange—Comparison with the TBA assay and an iodometric method. Lipids 26(10):853–856. https://doi.org/10.1007/BF02536169
15. Wolf SP (1994) Ferrous ion oxidation in presence of ferric ion indicator xylenol orange for measurement of hydroperoxides. Methods Enzymol 233:182–189. https://doi.org/10.1016/S0076-6879(94)33021-2
16. Nourooz-Zadeh J (1999) Ferrous ion oxidation in presence of xylenol orange for detection of lipid hydroperoxides in plasma. Oxidants and antioxidants. Method Enzymol 300:58–62. https://doi.org/10.1016/s0076-6879(99)00113-5
17. Hermes-Lima M, Willmore WG, Storey KB (1995) Quantification of lipid-peroxidation in tissue-extracts based on Fe(III)xylenol orange complex-formation. Free Rad Biol Med 19(3):271–280. https://doi.org/10.1016/0891-5849(95)00020-x
18. Gay C, Collins J, Gebicki JM (1999) Hydroperoxide assay with the ferric-xylenol orange complex. Anal Biochem 273(2):149–155. https://doi.org/10.1006/abio.1999.4208
19. Nourooz-Zadeh J, Tajaddini-Sarmadi J, Birlouez-Aragon I, Wolff SP (1995) Measurement of hydroperoxides in edible oil using the ferrous oxidation in xylenol orange assay. J Agric Food Chem 43(1):17–21. https://doi.org/10.1021/jf00049a005
20. Bou R, Codony R, Tres A, Decker EA, Guardicila F (2008) Determination of hydroperoxides in foods and biological samples by the ferrous oxidation-xylenol orange method: a review of the factors that influence the method's performance. Anal Biochem 377(1):1–15. https://doi.org/10.1016/j.ab.2008.02.029
21. Eymard S, Genot C (2003) A modified xylenol orange method to evaluate formation of lipid hydroperoxides during storage and processing of small pelagic fish. Eur J Lipid Sci Technol 105(9):497–501. https://doi.org/10.1002/ejlt.200300768
22. Grau A, Codony R, Rafecas M, Barroeta AC, Guardiola F (2000) Lipid hydroperoxide determination in dark chicken meat through a ferrous oxidation-xylenol orange method. J Agric Food Chem 48(9):4136–4143. https://doi.org/10.1021/jf991054z
23. Navas JA, Tres A, Codony R, Boatella J, Bou R, Guardiola F (2004) Modified ferrous oxidation-xylenol orange method to determine lipid hydroperoxides in fried snacks. Eur J Lipid Sci Technol 6(10):688–696. https://doi.org/10.1002/ejlt.200400987
24. Shantha NC, Decker EA (1994) Rapid, sensitive, iron-based spectrophotometric methods for deptermination of peroxide values of food lipids. J AOAC Int 77(2):421–424. https://doi.org/10.1093/jaoac/77.2.421
25. Burat KM, Bozkurt O (1996) Improvement of calibration curve for determining peroxide values of food lipids by the modified ferrous oxidation-xylenol orange method. J AOAC Int 79(4):995–997. https://doi.org/10.1093/jaoac/79.4.995

26. Tres A, Nuchi CD, Bou R, Codony R, Guardiola F (2009) Assessing rabbit and chicken tissue susceptibility to oxidation through the ferrous oxidation-xylenol orange method. Eur J Lipid Sci Technol 111(6):563–573. https://doi.org/10.1002/ejlt.200800230
27. Yin HY, Porter NA (2005) New insights regarding the autoxidation of polyunsaturated fatty acids. Antiox Redox Signal 7(1–2): 170–184. https://doi.org/10.1089/ars.2005.7.170
28. Viau M, Genot C, Ribourg L, Meynier A (2016) Amounts of the reactive aldehydes, malonaldehyde, 4-hydroxy-2-hexenal, and 4-hydroxy-2-nonenal in fresh and oxidized edible oils do not necessary reflect their peroxide and anisidine values. Eur J Lipid Sci Technol 18(3):435–444. https://doi.org/10.1002/ejlt.201500103
29. DeLong JM, Prange RK, Hodges DM, Forney CF, Bishop MC, Quilliam M (2002) Using a modified ferrous oxidation-xylenol orange (FOX) assay for detection of lipid hydroperoxides in plant tissue. J Agric Food Chem 50(2): 248–254. https://doi.org/10.1089/ars.2005.7.170
30. Nielsen NS, Timm-Heinrich M, Jacobsen C (2003) Comparison of wet-chemical methods for determination of lipid hydroperoxides. J Food Lipids 10(1):35–50. https://doi.org/10.1111/j.1745-4522.2003.tb00004.x
31. Ugalde-Benitez V, Rojo-Dominguez A, Jaramillo-Flores ME, Guerrero-Legarreta I (2014) Effect of sample solubilized aliquot volume and concentration on hydroperoxide detection in PUFA-rich fish oil. Int J Food Properties 17(3):650–660. https://doi.org/10.1080/10942912.2012.660719
32. Santas J, Guzman YJ, Guardiola F, Rafecas M, Bou R (2014) High-throughput analysis of lipid hydroperoxides in edible oils and fats using the fluorescent reagent diphenyl-1-pyrenylphosphine. Food Chem 162:235–241. https://doi.org/10.1016/j.foodchem.2014.04.059
33. Yildiz G, Wehling RL, Cuppett SL (2003) Comparison of four analytical methods for the determination of peroxide value in oxidized soybean oils. J Am Oil Chem Soc 80(2): 103–107. https://doi.org/10.1007/s11746-003-0659-3
34. Rampon V, Lethuaut L, Mouhous-Riou N, Genot C (2001) Interface characterization and aging of bovine serum albumin stabilized oil-in-water emulsions as revealed by front-surface fluorescence. J Agric Food Chem 49(8):4046–4051. https://doi.org/10.1021/jf001170y
35. Kabri T-H (2013) Conception raisonnée d'émulsions submicroniques vectrices d'acides gras polyinsaturés n-3. Impact de la structure moléculaire et supramoléculaire des lipides sur l'oxydation. Thèse de doctorat. Université deNantes-Oniris
36. Kenmogne-Domguia HB, Moisan S, Viau M, Genot C, Meynier A (2014) The initial characteristics of marine oil emulsions and the composition of the media inflect lipid oxidation during in vitro gastrointestinal digestion. Food Chem 52:146–154. https://doi.org/10.1016/j.foodchem.2013.11.096
37. Eymard S, Carcouët E, Rochet M-J, Dumay J, Chopin C, Genot C (2005) Development of lipid oxidation during manufacturing of horse mackerel surimi. J Sci Food Agric 85:1750–1756. https://doi.org/10.1001/jsfa.2145
38. Mortensen G, Sorensen J, Stapelfeldt H (2002) Comparison of peroxide value methods used for semihard cheeses. J Agric Food Chem 50(18):5007–5011. https://doi.org/10.1021/jf0200220
39. Wei C, Hua J, Bin C, Klassen K (2010) Impact of lipid emulsion containing fish oil on outcomes of surgical patients: systematic review of randomized controlled trials from Europe and Asia. Nutrition 26(5):474–481. https://doi.org/10.1016/j.nut.2009.09.011
40. Banerjee D, Madhusoodanan UK, Sharanabasappa M, Ghosh S, Jacob J (2003) Measurement of plasma hydroperoxide concentration by FOX-1 assay in conjunction with triphenylphosphine. Clin Chim Acta 337(1–2):147–152. https://doi.org/10.1016/j.cccn.2003.08.004
41. Sodergren E, Nourooz-Zadeh J, Berglund L, Vessby B (1998) Re-evaluation of the ferrous oxidation in xylenol orange assay for the measurement of plasma lipid hydroperoxides. J Biochem Biophys Methods 37(3):137–146. https://doi.org/10.1016/s0165-022x(98)00025-6
42. Orefice I, Gerecht A, d'Ippolito G, Fontana A, Ianora A, Romano G (2015) Determination of lipid hydroperoxides in marine diatoms by the FOX2 assay. Mar Drugs 13(9):5767–5783. https://doi.org/10.3390/md13095767
43. Bleau G, Giasson C, Brunette I (1998) Measurement of hydrogen peroxide in biological samples containing high levels of ascorbic acid. Anal Biochem 263(1):13–17. https://doi.org/10.1006/abio.1998.2801
44. Erel O (2005) A new automated colorimetric method for measuring total oxidant status. Clin Biochem 38(12):1103–1111. https://doi.org/10.1016/j.clinbiochem.2005.08.008

45. Fukuzawa K, Fujisaki A, Akai K, Tokumura A, Terao J, Gebicki JM (2006) Measurement of phosphatidylcholine hydroperoxides in solution and in intact membranes by the ferric-xylenol orange assay. Anal Biochem 359(1): 18–25. https://doi.org/10.1016/j.ab.2006.09.011
46. Gay C, Collins J, Gebicki JM (1999) Determination of iron in solutions with the ferric-xylenol orange complex. Anal Biochem 273(2):143–148. https://doi.org/10.1006/abio.1999.4207
47. Gay CA, Gebicki JM (2002) Perchloric acid enhances sensitivity and reproducibility of the ferric-xylenol orange peroxide assay. Anal Biochem 304(1):42–46. https://doi.org/10.1006/abio.2001.5566
48. Monserrat JM, Geracitano LA, Pinho GLL, Vinagre TM, Faleiros M, Alciati JC et al (2003) Determination of lipid peroxides in invertebrates tissues using the Fe(III) xylenol orange complex formation. Arch Environ Contam Toxicol 45(2):177–183. https://doi.org/10.1007/s00244-003-0073-x
49. Fukuzawa K, Shibata A, Okamura C, Fujiwara Y, Akai K, Tsuchiya K et al (2009) Measurement of lipid hydroperoxides by the ferric-xylenol orange method (1) Characteristics of the ferric-xylenol orange/membrane phosphatidylcholine complex. J Nutr Sci Vitaminol 55(1):9–14. https://doi.org/10.3177/jnsv.55.9
50. Oroy C, Ribourg-Birault L, Genot C, Leroy G, Lepoudere A, Le Grand F et al (2021) Development and adaptation of methods to quantify lipid oxidation in complex fish matrices. 18th Euro-Fed Lipid Congress and Expo, On-line, 18–21 October 2021
51. Gheysen L, Dejonghe C, Bernaerts T, Van Loey A, De Cooman L, Foubert I (2019) Measuring primary lipid oxidation in food products enriched with colored microalgae. Food Anal Methods 12(10):2150–2160. https://doi.org/10.1007/s12161-019-01561-0
52. Abuzaytoun R, Budge S, Hansen LT, MacKinnon S (2020) Modification of the ferrous oxidation-xylenol orange method for determination of peroxide value in highly pigmented sea cucumber viscera lipid. J Am Oil Chem Soc 97(5):509–516. https://doi.org/10.1002/aocs.12344

Chapter 14

Measurement of Peroxide Values (PV) in Oils by Triphenylphosphine/Triphenylphosphine Oxide (TPP/TPPO) Assay Coupled with ATR-FTIR Spectroscopy

Erwann Durand, Claire Bourlieu-Lacanal, Françoise Michel Salaun, and Pierre Villeneuve

Abstract

One of the first indicators of oxidation in oils and fats is the presence of hydroperoxides (LOOH); therefore, it is important to find a reliable method to quantify these oxidation products. Herein, we propose a procedure to measure oxidation levels of oils using Fourier transform infrared (FTIR) spectroscopy with an attenuated total reflectance (ATR) mode based on the stoichiometric conversion of triphenylphosphine (TPP) into triphenylphosphine oxide (TPPO) by LOOH. ATR-FTIR spectroscopy allows for simple and accurate detection of the TPPO formed by selective reaction with LOOH, with measurement of its specific adsorption band at 542 cm^{-1}. Calibration was made with TPPO solubilized in butan-2-one and covers a range of peroxide values (PV) from ~2 to ~800 mEq/kg of fat. This method is simple and fast while limiting the use of toxic solvents and is suitable for routine analysis of oils and fats.

Key words Lipid oxidation, Peroxide value, Hydroperoxide, Triphenylphosphine (TPP), Triphenylphosphine oxide (TPPO), ATR-FTIR

1 Introduction

Lipid oxidation is probably the most important nonmicrobial spoilage mechanism in foods, causing vast economic, organoleptic, and nutritional loss. Although the pathway of lipid degradation is multifaceted, the lipid hydroperoxides (LOOH) have been recognized as the first stable products that are released, and therefore are evaluated to probe the level of oxidation. Although several methods are available [1, 2], the way to assess LOOH remains controversial due to problems and variation of results. Typically, the concentration of LOOH is expressed as peroxide value (PV), which measures the amount of peroxide (in milliequivalents of active oxygen, mEq) per kilogram of sample (e.g., oil).

Christelle Lopez et al. (eds.), *Multidimensional Characterization of Dietary Lipids*, Methods and Protocols in Food Science,
https://doi.org/10.1007/978-1-0716-3758-6_14,

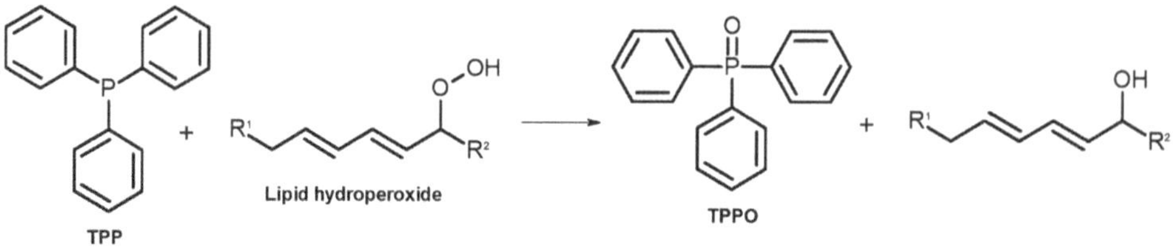

Fig. 1 Conversion of TPP to TPPO with lipid hydroperoxide

Fourier transform infrared (FTIR) spectroscopy is a rapid and nondestructive analytical tool that has been already used to detect and evaluate lipid oxidation [3–7]. The PV estimation of oil in the near-infrared (NIR) spectroscopy region (4000–12000 cm^{-1}) is very helpful because the samples can be directly detected without any additional treatment; yet, it requires a multivariate approach for a single product with mathematical and statistical treatment for calibration, making it difficult to implement for large-scale applications and miscellaneous lipid-based formulations. Absorption bands in the mid-infrared region (MIR, 400–4000 cm^{-1}) allow for a finer and more definite signal with less overlapping, therefore allowing an easier qualitative and quantitative assessment. The direct measurement of LOOH by FTIR spectroscopy in MIR is possible [8] but still difficult due to the low sensitivity and necessity of calibration model and deconvolution development to discriminate LOOH signals from the sample background. An interesting way to provide a simple means of determining PV by FTIR spectroscopy is to form and analyze triphenylphosphine oxide (TPPO) that is formed after selective reaction between LOOH and triphenylphosphine (TPP) (Fig. 1). This stoichiometric reaction has been known for a very long time [9], and this strategy has been already used to estimate PV by measuring the TPPO concentration by HPLC-UV [10], EI-MS [11], and infrared spectroscopy [12]. Although efficient, the procedure for analyzing TPPO remains somewhat fastidious (long and/or difficult sample preparation), is time-consuming, and involves a large volume of organic and toxic solvents. With the development of Attenuated Total Reflection (ATR) equipment, the FTIR signal acquisition has become more efficient and rapid, giving more accurate results with low signal-to-noise ratio. Moreover, only small sample volumes either solid or liquid (μL or mg) are necessary.

Herein, we describe a method to measure the PV in oils by ATR-FTIR spectroscopy based on the stoichiometric conversion of triphenylphosphine (TPP) by hydroperoxides into triphenylphosphine oxide (TPPO) [13]. This practical method allows the assessment of the oxidative status of lipids within a short time period, while minimizing waste, reagent disposal issues, and expenses and eliminating the key drawbacks associated with other analyses (e.g., titration assay). This approach perfectly matches with large-scale

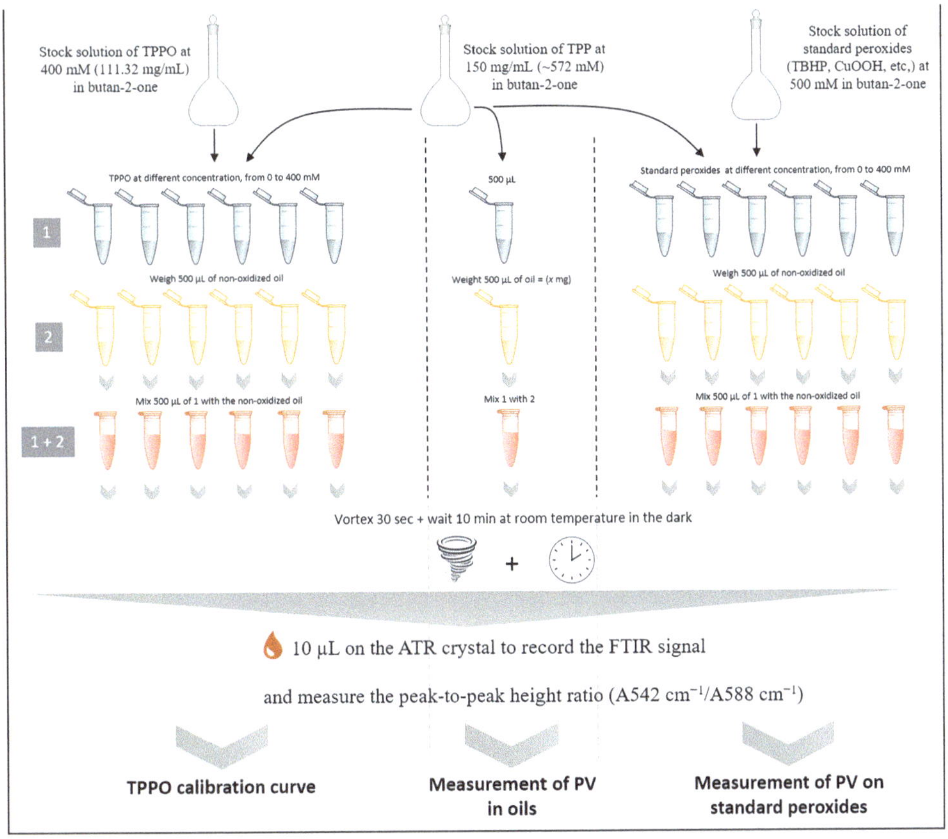

Fig. 2 Overview of the experimental procedure

analysis owing to its minimal sample preparation and fast and easy clean-up by simply removing the sample and cleaning the surface of the crystal. Analysis of samples can also be done in their natural states and is particularly adapted for viscous and strongly absorbing samples. The conditions for peroxide analysis were adjusted to allow a large and accurate range of PV estimation from ~2 to ~800 mEq/kg, that is, the range of concentrations mostly encountered in the literature when dealing with oxidation in real-time and accelerated aging assays (Fig. 2).

2 Chemicals, Materials, and Equipment

Reagent-grade triphenylphosphine (TPP >99%), triphenylphosphine oxide (TPPO ≥98%), tert-butyl hydroperoxide (TBHP), cumene hydroperoxide (CuOOH), butan-2-one, oil (e.g., sunflower oil purchased in local retail outlets has been taken as an example in this chapter) (Table 1).

Table 1
Chemicals required for the TPP/TPPO assay

Chemical	Chemical formula	CAS number	Molecular weight (g. mol^{-1})	Quality	Safety instructions
Triphenylphosphine (TPP)	$(C_6H_5)_3P$	603-35-0	262.29	Reagent grade >99%	GHS05, GHS07, GHS08
Triphenylphosphine oxide (TPPO)	$(C_6H_5)_3PO$	791-28-6	278.28	≥98% (HPLC)	GHS07
Tert-butyl hydroperoxide (TBHP)	$C_4H_{10}O_2$	75-91-2	90.12	5.0–6.0 M in decane	GHS02, GHS05, GHS06, GHS08, GHS09
Cumene hydroperoxide (CuOOH)	$C_9H_{12}O_2$	80-15-9	152.19	Technical grade, 80%	
Butan-2-one	C_4H_8O	78-93-3	72.10	≥99.5% (GC)	GHS02, GHS07

Tubes >3 mL with caps, volumetric flasks, mechanical pipettes <1 mL, spatula, FTIR equipped with an ATR accessory, vortex, and electronic balance.

3 Methods

3.1 Preparation of TPPO at Different Concentrations for ATR-FTIR Measurements

1. Prepare a stock solution of TPP at 150 mg/mL (~572 mM) in butan-2-one.
2. Prepare a stock solution of TPPO at 400 mM (111.31 mg/mL) in butan-2-one (*see* **Note 1**).
3. Dilute the TPPO stock solution with the TPP solution in butan-2-one to obtain different TPPO concentrations from 1 to 400 mM.
4. To separate glass tubes, add 500 μL of each TPPO concentration (from 1 to 400 mM) and 500 μL of the nonoxidized oil (e.g., sunflower; *see* **Note 2**) to obtain TPPO quantities in the mixture ranging from 0.5 to 200 μmol (from 0.5 to 200 mM). Prepare a control with 500 μL of the TPP solution in butan-2-one (150 mg/mL, ~572 mM) and 500 μL of the nonoxidized oil (e.g., sunflower).
5. Cap the tubes and vortex the samples for 30 s.
6. Allow to stand for 10 min at room temperature in the dark.

7. Collect and drop (10 μL) from each sample on the ATR crystal to record the FTIR signal (see Subheading 3.5 for signal acquisition).

3.2 TPPO Calibration Curve

1. Convert the FTIR transmission data (from Subheading 3.1) of TPPO spectra (Fig. 3) to absorption.
2. Calculate the peak-to-peak height ratio between the characteristic vibration bands of TPPO at 542 cm^{-1} and butan-2-one at 588 cm^{-1} (*see* **Note 3**).
3. Draw a linear plot of the peak-to-peak height intensities of the spectrum bands as a function of the TPPO quantity that ranges from 0.5 to 200 μmol (Beer's law is obeyed, *see* **Notes 4** and **5**) (Fig. 4).

3.3 Measurement of PV on Standard Peroxides (TBHP and CuOOH) with the TPP/TPPO-FTIR Method

Standard peroxides having different chemical structures and reactivities (namely tert-butyl hydroperoxide and cumene hydroperoxide) may be used for calibration:

1. Prepare a stock solution of TPP at 150 mg/mL (~572 mM) in butan-2-one.
2. Prepare stock solutions of TBHP and CuOOH at 500 mM in butan-2-one.
3. Dilute stock solutions of standard peroxides with the TPP solution in butan-2-one (150 mg/mL, ~572 mM) to obtain

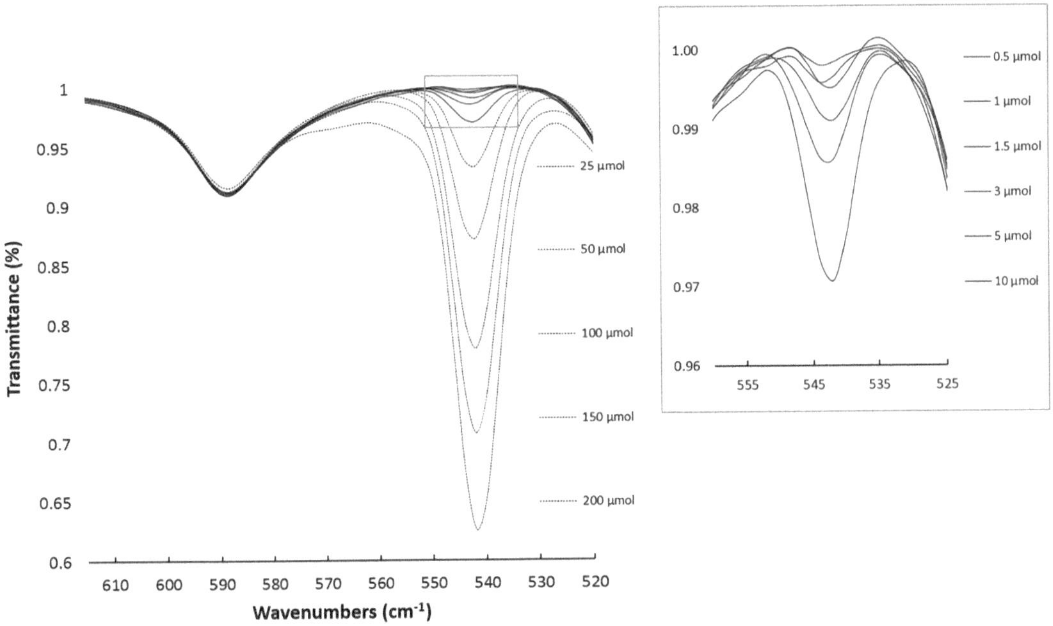

Fig. 3 Example of TPPO peak-to-peak height obtained using a GladiATR™ Single Reflexion Diamond ATR with a DLaTGS (deuterated lanthanum α-alanine-doped triglycine sulfate) detector and MIR source after 16 scans at 24 °C

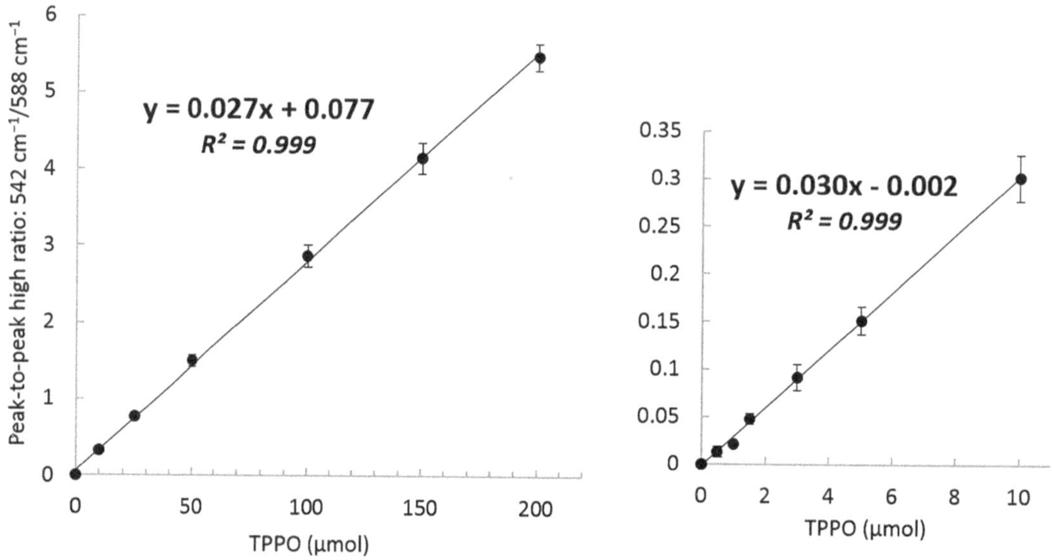

Fig. 4 Example of TPPO calibration curve obtained using a GladiATR™ Single Reflexion Diamond ATR with a DLaTGS (deuterated lanthanum α-alanine-doped triglycine sulfate) detector and MIR source after 16 scans at 24 °C. TPPO amount ranges from 0.5 to 200 μmol

samples of TBHP and CuOOH with concentrations ranging from 1 to 400 mM.

4. Add to separate glass tubes 500 μL from each concentration and 500 μL of nonoxidized oil (e.g., sunflower oil) to target peroxide amounts in the mixture ranging from 0.5 to 200 μmol (0.5 to 200 mM). Prepare a control with 500 μL of TPP solution in butan-2-one (150 mg/mL, ~572 mM) and 500 μL of nonoxidized oil (e.g., sunflower oil).
5. Cap the tube and vortex the samples for 30 s.
6. Allow to stand for 10 min at room temperature in the dark (*see* **Note 6**).
7. Collect and drop 10 μL from 1 mL of the reaction medium on the ATR crystal for FTIR signal acquisition (see Subheading 3.5 for signal acquisition).
8. Measure the peak-to-peak height ratio between the characteristic vibration bands of TPPO and butan-2-one (A542 cm^{-1}/A588 cm^{-1}) and determine the TPPO quantity according to the calibration curve (see Subheading 3.2).
9. Draw a linear plot of the measured peroxides (= μmol TPPO) to the expected peroxide amount ranging from 0.5 to 200 μmol. The relationship has to be close to 1 (R^2 > 0.99) with an excellent reproducibility [13].

3.4 Measurement of PV in Oils by the TPP/TPPO-FTIR Method

1. Prepare a stock solution of TPP in butan-2-one at 150 mg/mL (~572 mM).
2. To a glass tube, add 500 μL of an oil sample (*e.g.* stored sunflower oil) and 500 μL of the TPP stock solution. Note the exact weight (x mg) of 500 μL of the oil sample for further PV calculations (see Subheading 3.6).
3. Cap the tube and vortex for 30 s.
4. Allow to stand for 10 min at room temperature in the dark.
5. Collect and drop 10 μL from the reacting medium on the ATR crystal for FTIR signal acquisition (see Subheading 3.5 for signal acquisition).
6. See Subheading 3.6 for the calculation of the PV of oil after measurement of the TPPO peak-to-peak height intensity.

3.5 ATR-FTIR Signal Acquisition

1. Perform the measurements at room temperature (~25 °C) using an FTIR spectrometer equipped with an ATR accessory, a DLaTGS (deuterated lanthanum α-alanine-doped triglycine sulfate) detector, and an MIR source.
2. Drop 10 μL of the reacting medium on the ATR crystal and immediately cover with a cylindrical object to prevent evaporation (e.g., cap from a Falcon™ 15 mL conical centrifuge tube).
3. Record the FTIR signal acquisition after several scans (e.g., 16) at 4 cm^{-1} resolution with a wavenumber ranging from 400 to 4 000 cm^{-1}.

3.6 TPPO signal treatment and PV calculation

1. Convert the FTIR transmission data of spectra to absorption.
2. Calculate the peak-to-peak height at 542 cm^{-1} (corresponding to its X-substituent-sensitive phenyl vibration) to evaluate the TPPO amount because of its sharp and strong intensity while being easily differentiated from the other peaks. Here is an example of the TPPO signal in sunflower oil (Fig. 5).
3. Calculate the peak-to-peak height of the butan-2-one vibrational band at 588 cm^{-1}.
4. Calculate the peak-to-peak height ratio between the TPPO and butan-2-one (A542 cm^{-1}/A588 cm^{-1}).
5. Convert the peak-to-peak height ratio in μmol of TPPO by means of the TPPO calibration curve (see Subheadings 3.1 and 3.2).
6. 1 mole of TPP is required to convert 1 mole of LOOH in TPPO and LOH derivative (Fig. 1). The TPPO quantity (μmol) corresponds to those that would result from the complete reaction between TPP and LOOH obtained from 0.5 mL (x mg) of oil. The hydroperoxide concentration in the oil from the TPP/TPPO assay can be expressed in mmol/kg = TPPO

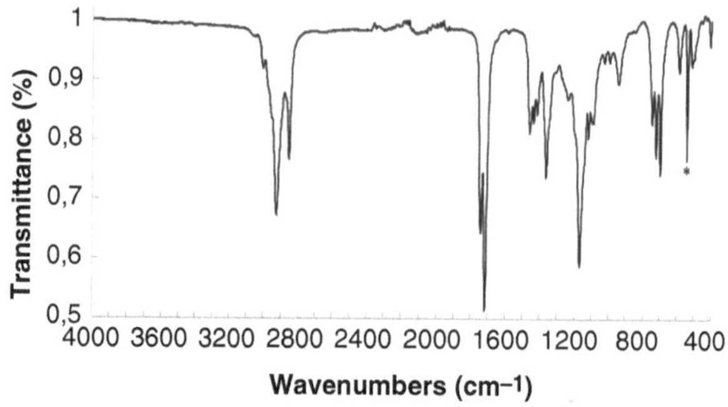

Fig. 5 Example of TPPO spectrum at 100 mM in butan-2-one and sunflower oil (1:1 v:v ratio) obtained using a GladiATR™ Single Reflexion Diamond ATR with a DLaTGS (deuterated lanthanum α-alanine-doped triglycine sulfate) detector and MIR source after 16 scans at 24 °C. *represents the sharp and strong intensity vibrational band at 542 cm^{-1}

quantity (μmol)/oil mass (x mg). If we express the TPP/TPPO assay in terms of the PV in mEq/kg, each millimole of TPPO per kg of the sample corresponds to 1 mmol of LOOH formed and represents a PV of 2 (*see* **Notes** 7 and **8**).

4 Notes

1. Different organic solvents (besides butan-2-one) could be used to assist this process, but they must (i) be nonreactive toward peroxides and TPP, (ii) be free of peroxides (e.g., solvents such as tetrahydrofuran may contain peroxides), (iii) allow a perfect solubilization of chemicals, oils, and LOOH, and (iv) not overlap signals in the FTIR.
2. Precheck the oxidation level of oil to ensure its nonoxidation either by following the current procedure (from 3.4 to 3.6) or using an alternative method [1, 2].
3. The butan-2-one peak at 588 cm^{-1} is perfectly suitable for the calibration procedure because it does not overlap with other signals.
4. Regarding the peroxide detection limits, it turns out that the minimum is about 500 nmol of peroxide formed from 0.5 mL of oil (~1 mmol/kg, or ~2 mEq/kg, or ~2 PV), while the maximum was about 200 μmol of LOOH formed from 0.5 mL of oil (~400 mmol/kg, or ~800 mEq/kg, or ~800 PV), mostly due to TPPO precipitation issues.
5. To increase method sensitivity, TPPO analyses could be done by liquid chromatography using UV detector at 220 nm instead of ATR-FTIR spectrometer to lower the detection limit to around 10 pmol (0.02 PV) [10].

6. The conversion of TPP to TPPO is complete in a short time (~5 min), and TPPO in the reaction medium remains stable over 60 min.
7. The PV, expressed in units of milliequivalents, is defined as the amount of active oxygen per 1 kg of sample (mEq/kg) and corresponds to the amount of iodine (I_2) which is formed by the reaction of LOOH with 2 moles of iodide ion (I^-). In the TPP/TPPO method, 1 mole of TPP is required to convert 1 mole of LOOH in TPPO and LOH derivative. Therefore, PV expressed in mmol/kg with TPP/TPPO will be equal to half of PV expressed in mEq/kg determined by iodine titration.
8. Epoxides, which may be formed in the lipid oxidation reaction [14], could react with TPP to form TPPO. Yet, we have observed that cis-2,3-epoxybutane mixed in sunflower oil/butan-2-one (1/1 v/v) in the presence of TPP did not result in the formation of TPPO, suggesting that TPP was not highly reactive with epoxides, similar to what has been previously shown with H_2O_2 [15].

References

1. Barthel G, Grosch W (1974) Peroxide value determination-comparison of some methods. J Am Oil Chem Soc 51:540–544. https://doi.org/10.1007/BF02636025
2. Lea CH (1952) Methods for determining peroxide in lipids. J Sci Food Agric 3:586–594. https://doi.org/10.1002/jsfa.2740031205
3. Liang P, Chen C, Zhao S et al (2013) Application of fourier transform infrared spectroscopy for the oxidation and peroxide value evaluation in virgin walnut oil. J Spectrosc 1:1–5. https://doi.org/10.1155/2013/138728
4. Sinelli N, Cosio MS, Gigliotti C, Casiraghi E (2007) Preliminary study on application of mid infrared spectroscopy for the evaluation of the virgin olive oil "freshness". Anal Chim Acta 598:128–134. https://doi.org/10.1016/j.aca.2007.07.024
5. Guillén MD, Cabo N (2002) Fourier transform infrared spectra data versus peroxide and anisidine values to determine oxidative stability of edible oils. Food Chem 77:503–510. https://doi.org/10.1016/S0308-8146(01)00371-5
6. Muik B, Lendl B, Molina-Diaz A et al (2007) Two-dimensional correlation spectroscopy and multivariate curve resolution for the study of lipid oxidation in edible oils monitored by FTIR and FT-Raman spectroscopy. Anal Chim Acta 593:54–67. https://doi.org/10.1016/j.aca.2007.04.050
7. Bendini A, Cerretani L, Di Virgilio F et al (2007) Preliminary evaluation of the application of the ftir spectroscopy to control the geographic origin and quality of virgin olive oils. J Food Qual 30:424–437. https://doi.org/10.1111/j.1745-4557.2007.00132.x
8. van de Voort FR, Ismail AA, Sedman J et al (1994) The determination of peroxide value by fourier transform infrared spectroscopy. J Am Oil Chem Soc 71:921–926. https://doi.org/10.1007/BF02542254
9. Walling C, Rabinowitz R (1959) The reaction of trialkyl phosphites with thiyl and alkoxy radicals. J Am Chem Soc 81:1243–1249. https://doi.org/10.1021/ja01514a057
10. Nakamura T, Maeda H (1991) A simple assay for lipid hydroperoxides based on triphenylphosphine oxidation and high-performance liquid chromatography. Lipids 26:765–768. https://doi.org/10.1007/BF02535628
11. Mochida Y, Nakamura S (2006) Determination of total hydroperoxides in oxidized vegetable oils based on triphenylphosphine oxidation using electron ionization mass spectrometry. J Mass Spectrom Soc Jpn 54:235–242. https://doi.org/10.5702/massspec.54.235
12. Ma K, van de Voort FR, Sedman J, Ismail AA (1997) Stoichiometric determination of hydroperoxides in fats and oils by fourier transform infrared spectroscopy. J Am Oil Chem Soc 74:

897–906. https://doi.org/10.1007/s11746-997-0001-8
13. Deyrieux C, Villeneuve P, Baréa B et al (2018) Measurement of peroxide values in oils by Triphenylphosphine/Triphenylphosphine Oxide (TPP/TPPO) assay coupled with FTIR-ATR spectroscopy: comparison with iodometric titration. Eur J Lipid Sci Technol 120: 1800109. https://doi.org/10.1002/ejlt.201800109
14. Schaich KM (2020) Lipid oxidation: new perspectives on an old reaction. edible oil and fat products: chemistry, properties, and safety aspects. In: Shahidi F (ed) Bailey's industrial oil and fat products, 7th edn. Wiley, Hoboken. https://doi.org/10.1002/047167849X.bio067.pub2
15. Nourooz-Zadeh J, Tajaddini-Sarmadi J, Wolff SP (1994) Measurement of plasma hydroperoxide concentrations by the ferrous oxidation-xylenol orange assay in conjunction with triphenylphosphine. Anal Biochem 220:403–409. https://doi.org/10.1006/abio.1994.1357

Chapter 15

Method for Quantifying Oxidized Fatty Acids in Food Samples Using Ultra-High Pressure Liquid Chromatography-Tandem Mass Spectrometry

Marie Hennebelle, Shiva Emami, Zhichao Zhang, Bianca Ferraz Teixeira, Qing Shen, Yurika Otoki, and Ameer Y. Taha

Abstract

While most studies on lipid oxidation in food have focused on the formation of hydroperoxides (primary lipid oxidation products) and volatile compounds (secondary oxidation products), recent studies have pointed toward the formation of additional primary lipid oxidation products, including fatty acid epoxides, ketones, diols, and hydroxides. These products can be formed via autoxidation or enzymatic oxidation. Here, we describe the methods for the extraction and the hydrolysis of 76 oxidized fatty acids from food samples and their quantitation by ultra-high pressure liquid chromatography-tandem mass spectrometry (UPLC-MS/MS) operating in negative ionization mode and dynamic multiple reaction monitoring (dMRM) conditions.

Key words Lipid oxidation, Liquid chromatography-tandem mass spectrometry, Oxidized fatty acids, Polyunsaturated fatty acids, Oxylipins, Oils, Food

1 Introduction

Lipid oxidation in food systems has been described as a radical chain reaction driven by hydrogen abstraction and leading to the formation of hydroperoxides as main primary lipid oxidation products. Their further degradation into volatile compounds, known as secondary oxidation products (e.g., aldehydes), is responsible for the development of off-flavors [1, 2]. While the measurement focus on primary oxidation products has been devoted to hydroperoxides, it has been known for decades that various fatty acid alcohols, epoxides, and ketones can be formed during lipid oxidation [3]. On top of these nonenzymatic pathways, enzymes such as lipoxygenase (LOX) [4–10], cyclooxygenase (COX) [9–11], or cytochrome P450 (CYP450) [12] play a critical role in the formation of

Christelle Lopez et al. (eds.), *Multidimensional Characterization of Dietary Lipids*, Methods and Protocols in Food Science, https://doi.org/10.1007/978-1-0716-3758-6_15,

oxidized fatty acids prior to or during storage and processing of raw food (e.g., plants, fish, and meat).

Due to their multiple biological roles in vivo [13–15], oxidized fatty acids, also known as oxylipins, have been widely studied in biological systems. Their analysis involves solid phase extraction (SPE) followed by liquid chromatography-tandem mass spectrometry [16, 17]. A similar protocol was adapted for the study of food samples, including oils, milk, cream, and French fries [18–22]. Because most oxidized fatty acids in foods are esterified to complex lipid molecules (e.g., triacylglycerol, phospholipids, and sterol esters) [23], a base hydrolysis step is required. Prior SPE separation could be used to separate triacylglycerol from phospholipids and to yield class-specific esterified oxidized fatty acids [24]. A proper optimization of the hydrolysis step is required to avoid low recovery due to incomplete hydrolysis and/or degradation by the hydrolysis base reagent. In biological samples (i.e., plasma), a lower sample volume and an increased base concentration were reported to improve the efficiency of hydrolysis [25]. A similar approach was recently used to improve the recovery of oxidized fatty acids in oils [19]. Following the hydrolysis step, oxidized fatty acids are extracted using SPE columns (typically Oasis HLB from Waters) and analyzed by ultra-high pressure liquid chromatography-tandem mass spectrometry (UPLC-MS/MS) operating in the negative ionization mode and using dynamic multiple reaction monitoring (dMRM).

The protocol described below outlines the published steps for the extraction, detection, and quantification of up to 76 oxidized fatty acids in food samples.

2 Materials

2.1 Preparation of External and Surrogate Standard Stock Solutions

The external standards consist of unlabeled oxidized fatty acid standards used to prepare the standard curves. This is necessary to estimate the instrument response factor for each analyte. The deuterated surrogate standards are used both in the standard curves and in the samples to correct for possible losses during extraction. All standards need to be prepared on ice to minimize their degradation.

The present method includes 76 external standards and 9 surrogate standards (Table 1). *See* **Notes 1** and **2**.

1. Prepare one stock solution for each of the external standards by diluting them in methanol (CAS: 67-56-1; Optima LC-MS grade, Fisher Chemical; Cat#: A456) to reach the final concentration presented in Table 1 (Concentration of stock solution = 100-fold of the highest concentration of the standard curve). Store all the solutions at −80 °C until use.

Table 1
List of standards and concentration of stock solutions

Name	Abbreviation	Surrogate standard[a]	Fatty acid precursor[b]	Cayman Cat#	Concentration stock solution (μM)
12(13)-epoxyoctadecamonoenoic acid	12(13)-EpOME	d_{11}-11(12)-EpETrE	LA	52450	50
9(10)-epoxyoctadecamonoenoic acid	9(10)-EpOME	d_{11}-11(12)-EpETrE	LA	52400	50
13-hydroxyoctadecadienoic acid	13-HODE	d_4-9-HODE	LA	38600	500
9-hydroxyoctadecadienoic acid	9-HODE	d_4-9-HODE	LA	38400	500
13-oxo-octadecadienoic acid	13-oxo-ODE	d_6-20-HETE	LA	38620	25
9-oxo-octadecadienoic acid	9-oxo-ODE	d_6-20-HETE	LA	38420	25
12,13-dihydroxyoctadecamonoenoic acid	12,13-DiHOME	d_{11}-14,15-DiHETrE	LA	10009832	200
9,10-dihydroxyoctadecamonoenoic acid	9,10-DiHOME	d_{11}-14,15-DiHETrE	LA	53400	200
9,12,13-trihydroxyoctadecamonoenoic acid	9,12,13-TriHOME	d_4-PGE2	LA	14-1802-14-39	250
9,10,13-trihydroxyoctadecamonoenoic acid	9,10,13-TriHOME	d_4-PGE2	LA	14-1802-15-39	100
15(S)-hydroxyeicosatrienoic acid	15(S)-HETrE	d_8-5-HETE	GLA	36720	50
Prostaglandin E1	PGE1	d_4-PGE2	DGLA	13010	100
Prostaglandin D1	PGD1	d_4-PGE2	DGLA	12000	50
20-hydroxyeicosatetraenoic acid	20-HETE	d_6-20-HETE	AA	90030	65
Leukotriene C4	LTC4	d_4-LTB4	AA	20210	50
Leukotriene D4	LTD4	d_4-LTB4	AA	20310	100
Leukotriene E4	LTE4	d_4-LTB4	AA	20410	100
6-trans-leukotriene B4	6-trans-LTB4	d_4-LTB4	AA	35250	50
Leukotriene B4	LTB4	d_4-LTB4	AA	20110	50
5-hydroxyeicosatetraenoic acid	5-HETE	d_8-5-HETE	AA	34210	200
20-OH-Leukotriene B4	20-OH-LTB4	d_4-LTB4	AA	20190	50
20-COOH- Leukotriene B4	20-COOH-LTB4	d_4-LTB4	AA	20180	50
6-keto-prostaglandin F1 alpha	6-keto-PGF1a	d_4-6-keto-PFG1a	AA	15210	100
Thromboxane B2	TXB2	d_4-TXB2	AA	19030	100
Prostaglandin F2 alpha	PGF2a	d_4-PGE2	AA	16010	100

(continued)

Table 1
(continued)

Name	Abbreviation	Surrogate standard[a]	Fatty acid precursor[b]	Cayman Cat#	Concentration stock solution (μM)
Prostaglandin E2	PGE2	d_4-PGE2	AA	14010	50
Prostaglandin D2	PGD2	d_4-PGE2	AA	12010	50
Prostaglandin J2	PGJ2	d_4-PGE2	AA	18500	50
Prostaglandin B2	PGB2	d_4-PGE2	AA	11210	200
15-hydroxyeicosatetraenoic acid	15-HETE	d_8-5-HETE	AA	34700	200
11-hydroxyeicosatetraenoic acid	11-HETE	d_8-5-HETE	AA	34500	200
8-hydroxyeicosatetraenoic acid	8-HETE	d_8-5-HETE	AA	34340	200
12-hydroxyeicosatetraenoic acid	12-HETE	d_8-5-HETE	AA	34550	200
9-hydroxyeicosatetraenoic acid	9-HETE	d_8-5-HETE	AA	34400	200
Lipoxin A4	LXA4	d_4-LTB4	AA	90410	20
15-deoxy-prostaglandin J2	15-deoxy-PGJ2	d_4-PGE2	AA	18570	100
14(15)-epoxyeicosatrienoic acid	14(15)-EpETrE	d_{11}-11(12)-EpETrE	AA	10007263	25
8(9)-epoxyeicosatrienoic acid	8(9)-EpETrE	d_{11}-11(12)-EpETrE	AA	50351	25
11(12)-epoxyeicosatrienoic acid	11(12)-EpETrE	d_{11}-11(12)-EpETrE	AA	50511	25
5(6)-epoxyeicosatrienoic acid	5(6)-EpETrE	d_{11}-11(12)-EpETrE	AA	50211	100
15-oxo-eicosatetraenoic acid	15-oxo-ETE	d_8-5-HETE	AA	34730	25
12-oxo-eicosatetraenoic acid	12-oxo-ETE	d_8-5-HETE	AA	34580	25
5-oxo-eicosatetraenoic acid	5-oxo-ETE	d_8-5-HETE	AA	34250	100
14,15-dihydroxyeicosatrienoic acid	14,15-DiHETrE	d_{11}-14,15-DiHETrE	AA	51651	50
11,12-dihydroxyeicosatrienoic acid	11,12-DiHETrE	d_{11}-14,15-DiHETrE	AA	51511	50
8,9-dihydroxyeicosatrienoic acid	8,9-DiHETrE	d_{11}-14,15-DiHETrE	AA	51351	50
5,6-dihydroxyeicosatrienoic acid	5,6-DiHETrE	d_{11}-14,15-DiHETrE	AA	51211	50
8,15-dihydroxyeicosatetraenoic acid	8,15-DiHETE	d_{11}-14,15-DiHETrE	AA	35370	80
5,15-dihydroxyeicosatetraenoic acid	5,15-DiHETE	d_{11}-14,15-DiHETrE	AA	35280	25
13- hydroxyoctadecatrienoic acid	13-HOTrE	d_4-9-HODE	ALA	39620	25
9- hydroxyoctadecatrienoic acid	9-HOTrE	d_4-9-HODE	ALA	39420	50
Leukotriene B3	LTB3	d_4-LTB4	ETA	20109	50

(continued)

Table 1
(continued)

Name	Abbreviation	Surrogate standard[a]	Fatty acid precursor[b]	Cayman Cat#	Concentration stock solution (μM)
17(18)-epoxyeicosateteaenoic acid	17(18)-EpETE	d_{11}-11(12)-EpETrE	EPA	50861	100
14(15)-epoxyeicosatrienoic acid	14(15)-EpETE	d_{11}-11(12)-EpETrE	EPA	10173	25
11(12)-epoxyeicosateteaenoic acid	11(12)-EpETE	d_{11}-11(12)-EpETrE	EPA	10462	25
8(9)-epoxyeicosateteaenoic acid	8(9)-EpETE	d_{11}-11(12)-EpETrE	EPA	10470	100
15-hydroxyeicosapentaenoic acid	15-HEPE	d_8-5-HETE	EPA	32700	100
12-hydroxyeicosapentaenoic acid	12-HEPE	d_8-5-HETE	EPA	32540	100
8-hydroxyeicosapentaenoic acid	8-HEPE	d_8-5-HETE	EPA	32340	100
5-hydroxyeicosapentaenoic acid	5-HEPE	d_8-5-HETE	EPA	32200	100
Prostaglandin E3	PGE3	d_4-PGE2	EPA	14990	30
Prostaglandin D3	PGD3	d_4-PGE2	EPA	12990	50
8,9-dihydroxyeicosatetraenoic acid	8,9-DiHETE	d_{11}-14,15-DiHETrE	EPA	10473	100
11,12-dihydroxyeicosatetraenoic acid	11,12-DiHETE	d_{11}-14,15-DiHETrE	EPA	10466	100
17,18-dihydroxyeicosatetraenoic acid	17,18-DiHETE	d_{11}-14,15-DiHETrE	EPA	10006999	100
14,15-dihydroxyeicosatetraenoic acid	14,15-DiHETE	d_{11}-14,15-DiHETrE	EPA	10006998	100
Resolvin E1	Resolvin E1	d_4-PGE2	EPA	10007848	100
5,6-dihydroxyeicosatetraenoic acid	5,6-DiHETE	d_{11}-14,15-DiHETrE	EPA	10467	100
19(20)-epoxydocosapentaenoic acid	19(20)-EpDPE	d_{11}-11(12)-EpETrE	DHA	10175	25
16(17)-epoxydocosapentaenoic acid	16(17)-EpDPE	d_{11}-11(12)-EpETrE	DHA	10174	25
13(14)-epoxydocosapentaenoic acid	13(14)-EpDPE	d_{11}-11(12)-EpETrE	DHA	10464	25
10(11)-epoxydocosapentaenoic acid	10(11)-EpDPE	d_{11}-11(12)-EpETrE	DHA	10471	25
7(8)-epoxydocosapentaenoic acid	7(8)-EpDPE	d_{11}-11(12)-EpETrE	DHA	10465	25
16,17-dihydroxydocosapentaenoic acid	16,17-DiHDPA	d_{11}-14,15-DiHETrE	DHA	18176	100

(continued)

Table 1
(continued)

Name	Abbreviation	Surrogate standard[a]	Fatty acid precursor[b]	Cayman Cat#	Concentration stock solution (μM)
19,20-dihydroxydocosapentaenoic acid	19,20-DiHDPA	d_{11}-14,15-DiHETrE	DHA	10007001	100
17- hydroxydocosahexaenoic acid	17-HDoHE	d_4-9-HODE	DHA	10005099	30

[a]Oxidized fatty acids using d_4-PGE2 as a surrogate standard are not quantifiable if the samples are hydrolyzed because d_4-PGE2 degrades during base hydrolysis

[b]ALA, α-linolenic acid (C18:3 n-3); ARA, arachidonic acid (C20:4 n-6); DGLA, dihomo-γ-linolenic acid (C20:3 n-6); DHA, docosahexaenoic acid (C22:6 n-3); EPA, eicosapentaenoic acid (C20:5 n-3); ETA, eicosatetraenoic acid (C20:4 n-3); and GLA, γ-linolenic acid (C18:3 n-6)

2. Prepare one stock solution for each of the surrogate standards by diluting them in methanol (CAS: 67-56-1; Optima LC-MS grade, Fisher Chemical; Cat#: A456) to reach a final concentration of 200 μM. The 9 surrogate standards used are d11-11 (12)-EpETrE (Cayman Chemical; Cat#: 10006413), d11-14,15-DiHETrE (Cayman Chemical; Cat#: 1008040), d4-6-keto-PGF1a (Cayman Chemical; Cat#: 315210), d4-9-HODE (Cayman Chemical; Cat#: 338410), d4-LTB4 (Cayman Chemical; Cat#: 320110), d4-PGE2 (Cayman Chemical; Cat#: 314010), d4-TXB2 (Cayman Chemical; Cat#: 319030), d6-20-HETE (Cayman Chemical; Cat#: 390030), and d8-5-HETE (Cayman chemical; Cat#: 334230). Store all the solutions at −80 °C until use.

2.2 Preparation of Surrogate Standard Mix Solution

All standards need to be prepared on ice to minimize their degradation:

1. Prepare a 20 μM solution of surrogate standards (for standard curves): Mix 100 μL of each of the 9 stock solutions of surrogate standards (200 μM) with 100 μL of methanol (CAS: 67-56-1; Optima LC-MS grade, Fisher Chemical; Cat#: A456) to reach a 10-fold dilution. Store the solution at −80 °C until use.
2. Prepare a 2 μM solution of surrogate standards (for samples): Dilute the 20 μM solution of surrogate standards by 10-fold using methanol (CAS: 67-56-1; Optima LC-MS grade, Fisher Chemical; Cat#: A456). Prepare a sufficient volume for your analysis as 10 μL of this 2 μM solution of surrogate standards will be added to each sample. Store the solution at −80 °C until use (*see* **Note 3**).

2.3 Preparation of Standard Mix for Standard Curve

All standards need to be prepared on ice to minimize their degradation:

1. Transfer 15 μL of each of the 76 external standard stock solutions into an amber vial containing 360 μL of methanol (CAS: 67-56-1; Optima LC-MS grade, Fisher Chemical; Cat#: A456). This solution corresponds to the highest concentration of your standard curve.
2. From this solution, use a 4 times dilution series to prepare the 8 other concentrations of your standard curve. At each dilution step, transfer 275 μL of the standard mix to a vial containing 825 μL of methanol (CAS: 67-56-1; Optima LC-MS grade, Fisher Chemical; Cat#: A456). At the end of this step, you will have 9 external standard mixes with concentrations varying by a factor of 4 between each standard mix.
3. Prepare 9 empty amber vials and add 6 μL of the 20 μM deuterated surrogate standard solution to each vial. Dry under nitrogen. Add 600 μL of the 9 external standard mixes (described above) respectively to each vial (i.e., one per vial). At this point, you have a 9-point standard curve, each containing 0.2 μM of surrogate standard mix (*see* **Note 4**). The volume per vial is 600 μL. Store all solutions at −80 °C until use (*see* **Note 3**).

2.4 Hydrolysis Step

The chemicals and solutions used for this step are as follows:

1. Antioxidant mix: Prepare a solution containing 0.2 $mg.mL^{-1}$ triphenylphosphine (TPP; CAS: 603-35-0; Sigma-Aldrich; Cat#: T84409), butylated hydroxytoluene (BHT; CAS: 128-37-0; Sigma-Aldrich; Cat#: W218405), and ethylenediaminetetraacetic acid (EDTA; CAS: 60-00-4; Sigma-Aldrich; Cat#: EDS) in methanol/water (1:1 v/v). HPLC-grade methanol (CAS: 67-56-1; Optima for HPLC, Fisher Chemical; Cat#: A454) and distilled water (*see* **Note 5**). Filter the antioxidant solution with a 0.45 μm Millipore filter (Millipore-EMD; Cat#: SHLN013NL) and a syringe.
2. Extraction solvent: Ice-cold HPLC-grade methanol (CAS: 67-56-1; Optima for HPLC, Fisher Chemical; Cat#: A454) containing 0.1% acetic acid (CAS: 64-19-7; Sigma-Aldrich; Cat#: 695092) and 0.1% BHT (CAS: 128-37-0; Sigma-Aldrich; Cat#: W218405).
3. 0.25 M sodium hydroxide (CAS: 1310-73-2; Sigma-Aldrich; Cat#: S5881) in methanol/water (1:1 v/v). HPLC-grade methanol (CAS: 67-56-1; Optima for HPLC, Fisher Chemical; Cat#: A454) and distilled water are used. *See* **Note 6**.
4. Acetic acid (CAS: 64-19-7; Sigma-Aldrich; Cat#: 695092).
5. Ultrapure water.

The equipment and materials required are as follows:

1. Litmus paper.
2. Heating block.
3. 2 mL centrifuge tubes (USA Scientific; Cat#: 1620-2700).

2.5 Oxidized Fatty Acid Extraction by Solid Phase Extraction (SPE)

The chemicals and solutions used for this step are as follows:

1. SPE buffer: Solution of 0.1% acetic acid (CAS: 64-19-7; Sigma-Aldrich; Cat#: 695092) and 5% HPLC-grade methanol (CAS: 67-56-1; Optima for HPLC, Fisher Chemical; Cat#: A454) in ultrapure water.
2. Ethyl acetate (CAS: 141-78-6; Optima, Fisher Scientific; Cat#: E196).
3. HPLC-grade methanol (CAS: 67-56-1; Optima for HPLC, Fisher Chemical; Cat#: A454).
4. LC-MS-grade methanol (CAS: 67-56-1; Optima LC-MS grade, Fisher Chemical; Cat#: A456).

The equipment and materials required are as follows:

1. 60 mg Oasis HLB SPE columns (Waters; Cat#: WAT094226).
2. SPE manifold connected to a vacuum pump.
3. Nitrogen evaporator or SpeedVac vacuum concentrator.
4. Centrifuge that can hold 2 mL centrifuge tubes equipped with a temperature controller, e.g., Eppendorf centrifuge 5424R.
5. Ultrafree-MC VV centrifugal filter (Millipore; Cat#: UFC30VV00).
6. Low-volume (150 μL) insert with a plastic spring (Waters; Cat#:WAT094171).
7. LC-MS amber vials: Verex™ vial, 9 mm screw top, 2 mL (Phenomenex Inc; Cat#: AR0-3911-13).
8. Preslit caps for LC-MS vials (for analysis): Verex™ cap (preassembled), 9 mm, screw top, with bonded-in PTFE/silicone preslit septa (Phenomenex Inc; Cat#: AR0-8972-13-B).
9. Non preslit caps for LC-MS vials (for storage): Verex™ cap (preassembled), 9 mm, screw top, with bonded-in PTFE/silicone septa (Phenomenex Inc; Cat#: AR0-8952-13-B).

2.6 Liquid Chromatography-Tandem Mass Spectrometry

The analysis of oxidized fatty acids is performed using an ultra-high pressure liquid chromatography-tandem mass spectrometry instrument equipped with an electrospray ionization source (ESI) and a ZORBAX Eclipse Plus C18 reversed-phase column (ID: 2.1 mm; length: 150 mm; particle size: 1.8 μm; Agilent Technologies; Cat#: 959759-902).

1. Mobile phase A: 0.1% acetic acid (CAS: 64-19-7; Sigma-Aldrich; Cat#: 695092) in ultrapure water.
2. Mobile phase B: acetonitrile: methanol (80:15, v:v) containing 0.1 % acetic acid (CAS: 64-19-7; Sigma-Aldrich; Cat#: 695092). LC-MS-grade solvents are used to prepare the mobile phase: LC-MS-grade acetonitrile (CAS: 75-05-8; Optima LC-MS grade; Fisher Chemical; Cat#: A955) and LC-MS-grade methanol (CAS: 67-56-1; Optima LC-MS grade, Fisher Chemical; Cat#: A456).

3 Methods

3.1 Samples

The method can be used for oil solubilized in chloroform/methanol (2:1, v/v) and dried under nitrogen, as the starting material [19]. For other types of food samples (e.g., emulsions and solid food samples), a lipid extraction step (e.g., Folch or Bligh and Dyer method) is required, followed by complete evaporation of the solvents under nitrogen. *See* **Note 7**.

3.2 Sample Preparation and Hydrolysis (Fig. 1)

1. Aliquot 2 μL of oil or lipid extract (amounting to ~2 mg of total lipid) solubilized in chloroform/methanol (2:1, v/v) into a 2 mL centrifuge tube.
2. Dry under nitrogen to remove chloroform:methanol.
3. Add 10 μL of antioxidant mix.
4. Add 10 μL of 2 μM surrogate standard mix.
5. Dry under nitrogen.
6. Add 190 μL of extraction solvent and vortex.
7. Add 200 μL of 0.25 M sodium hydroxide solution and vortex. *See* **Notes 6** and **8**.
8. Heat for 30 min at 60 °C:
   - While the samples are being hydrolyzed, SPE columns can be washed and preconditioned (**steps 1–3** in Subheading 3.3) to limit the waiting time between the hydrolysis step and the extraction of oxidized fatty acids by SPE.
9. Let the samples cool down for approximately 5 min at room temperature.
10. Add 25 μL of acetic acid and vortex. At this step, verify with litmus paper that the pH of the sample is between 4 and 6. The pH should be in the acidic range to ensure that oxidized fatty acids are protonated.
11. Add 1575 μL of ultrapure water and vortex. At this step, a 2 mL sample is obtained and it can be used for SPE.

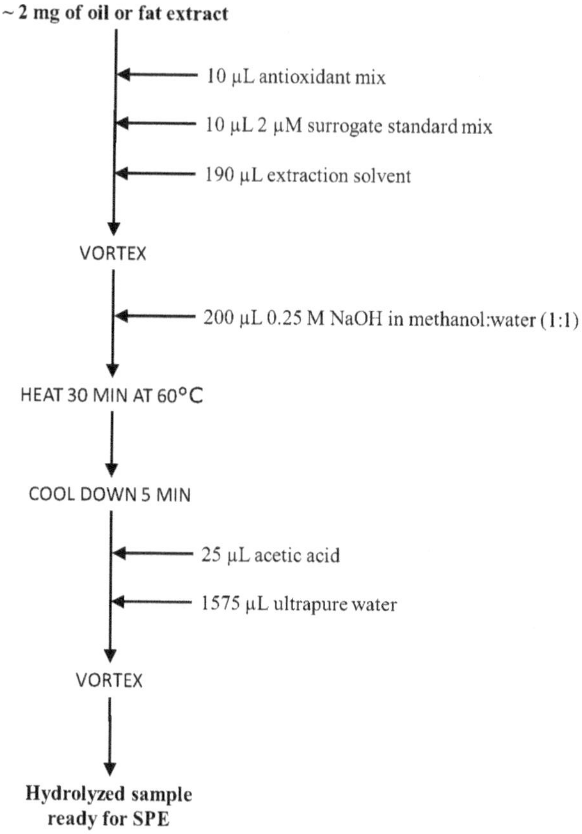

Fig. 1 Overview of the sample preparation and hydrolysis protocols

3.3 Oxidized Fatty Acid Extraction by Solid Phase Extraction (SPE) (Fig. 2)

1. Place the 60 mg Oasis HLB SPE columns on the manifold.
2. Wash the column with one column volume of ethyl acetate, followed by two column volumes of HPLC-grade methanol.
   - For **steps 2–5**, make sure that the columns remain wet at all times.
3. Equilibrate the columns with two column volumes of SPE buffer.
4. Pour the 2 mL samples obtained after the hydrolysis step on the SPE columns (one sample per column).
5. Rinse with two column volumes of SPE buffer.
6. Dry the SPE columns under vacuum (15–20 psi) for 20 min.
7. Elute the oxidized fatty acid extract in 2 mL centrifuge tubes with 0.5 mL HPLC-grade methanol followed by 1.5 mL ethyl acetate.
8. Dry the samples under nitrogen or alternatively with a Speed-Vac vacuum concentrator.
9. Reconstitute in 100 μL LC-MS-grade methanol.

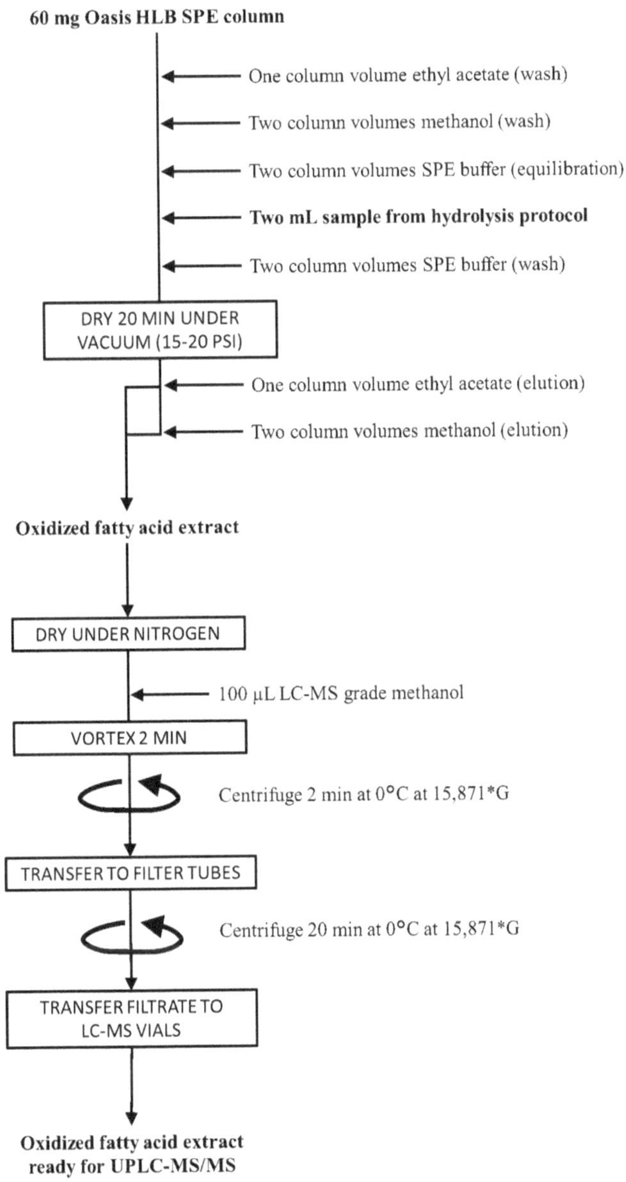

Fig. 2 Overview of the solid phase extraction and reconstitution protocols

10. Vortex for 2 min.
11. Centrifuge at 15,871× *g* and 0 °C for 2 min.
12. Transfer the samples into centrifuge tubes containing an ultrafree-MC VV centrifugal filter.
13. Centrifuge at 15,871× *g* and 0 °C for 20 min.
14. Discard the filter and transfer the sample containing the oxidized fatty acids to LC-MS vials containing 150 µL inserts. Cap with a non preslit cap for storage purpose or with a preslit cap to run on the UPLC-MS/MS. Make sure that there are no

bubbles in the sample. If there are bubbles, remove them by gently tapping the vial on the bench.

3.4 UPLC-MS/MS Analysis

1. Set up the UPLC method following the gradient and flow rate described in Table 2, along with the following specifications:
   - Autosampler temperature: 4 °C.
   - Column compartment temperature: 45 °C.
   - Injection volume: 10 μL.
   - Total run time: 27.3 min per sample.
2. Set up the MS method using the negative ionization mode and the dynamic multiple reaction monitoring (dMRM) conditions described in Table 3. *See* **Note 9**:
   - This protocol was developed using an Agilent 1290 UPLC system coupled to an Agilent 6460 triple quadrupole mass spectrometer (Agilent Corporation, Palo Alto, Ca, USA). ESI is used as an ion source with the following parameters: drying gas temperature, 300 °C; drying gas flow, 10 L.min^{-1}; sheath gas temperature, 350 °C; sheath gas flow, 11 L.min^{-1}; nebulizer, 35 psi; capillary voltage, 4000 V; and nozzle voltage, -1000 V.
3. Install the ZORBAX Eclipse Plus C18 column in the UPLC-MS/MS system.

Table 2
Solvent gradient and flow rate for UPLC

	Time	Solvent gradient	Flow rate
1	0 min	65% A; 35% B	0.25 mL.min^{-1}
2	3 min	60% A; 40% B	0.25 mL.min^{-1}
3	4 min	52% A; 48% B	0.25 mL.min^{-1}
4	10 min	40% A; 60% B	0.25 mL.min^{-1}
5	20 min	30% A; 70% B	0.25 mL.min^{-1}
6	24 min	15% A; 85% B	0.25 mL.min^{-1}
7	24.5 min	15% A; 85% B	0.25 mL.min^{-1}
8	24.6 min	0% A; 100% B	0.35 mL.min^{-1}
9	26 min	0% A; 100% B	0.35 mL.min^{-1}
10	26.1 min	65% A; 35% B	0.35 mL.min^{-1}
11	27 min	65% A; 35% B	0.35 mL.min^{-1}
12	27.3 min	65% A; 35% B	0.30 mL.min^{-1}

Table 3
Retention time, precursor ion, product ion, and collision energy used for the analysis of 76 oxidized fatty acid standards and 9 deuterated surrogate standards by UPLC-MS/MS

Compound abbreviation	Fragmentor voltage (V)	Precursor ion	Product ion	Collision energy (V)	Retention time (min)
d-11-11(12) EpETrE	80	330.2	167.2	7	23.5
d11-14,15-DiHETrE	125	348.2	207.1	10	13.8
d4-6-keto-PGF1a	90	373.3	167.1	19	4.7
d4-9HODE	90	299.2	172.3	13	18
d4-LTB4	80	339.2	197.2	10	11.4
d4-PGE2	90	355.2	275.3	7	7
d4-TXB2	105	373.3	173.2	10	6
d6-20-HETE	110	325.2	281.2	7	16.5
d8-5-HETE	75	327.2	116.1	7	21.2
10(11)-EpDPE	90	343.2	153.2	4	23.6
11(12)-EpETE	90	317.2	167.2	4	20.5
11(12)-EpETrE	105	319.2	167.2	4	23.9
11,12-DiHETE	100	335.5	167	10	12.4
11,12-DiHETrE	120	337.2	167.1	13	14.8
11-HETE	100	319.2	167.2	7	19.5
12(13)-EpOME	95	295.3	195.2	7	22.5
12,13-DiHOME	130	313.2	183.2	16	12.7
12-HEPE	110	317.2	179.2	4	17.2
12-HETE	120	319.2	179.2	7	20.5
12-oxo-ETE	115	317.2	153.1	7	21.1
13(14)-EpDPE	80	343.2	193.2	4	23.4
13-HODE	95	295.2	195.2	13	18
13-HOTrE	125	293.2	195.1	10	15.6
13-oxo-ODE	95	293.2	195.1	13	19.3
14(15)-EpETE	100	317.2	207.2	4	20.2
14(15)-EpETrE	130	319.2	219.3	4	22.8
14,15-DiHETE	95	335.3	207.2	7	12.3
14,15-DiHETrE	130	337.2	207.1	10	13.9
15(S)-HETrE	85	321.2	221.2	7	21.1

(continued)

Table 3
(continued)

Compound abbreviation	Fragmentor voltage (V)	Precursor ion	Product ion	Collision energy (V)	Retention time (min)
15-deoxy-PGJ2	130	315.2	271.2	4	16.1
15-HEPE	90	317.2	219.2	4	16.6
15-HETE	120	319.2	219.2	4	19.1
15-oxo-ETE	125	317.2	113.1	10	20.1
16(17)-DiHDPA	100	361.5	233.1	8	14.5
16(17)-EpDPE	130	343.2	233.2	4	23.3
17(18)-EpETE	130	317.2	215.2	4	19.2
17,18-DiHETE	105	335.3	247.2	7	11.7
17-HDoHE	95	343.2	281.2	4	19.5
19(20)-DiHDPA	100	361.5	273.1	8	13.8
19(20)-EpDPE	130	343.2	241.2	7	22.5
20-COOH-LTB4	120	365.2	347.2	7	4
20-HETE	125	319.2	275.1	10	16.5
20-OH-LTB4	95	351.2	195.2	13	5
5(6)-EpETrE	115	319.2	191.1	4	24.2
5,15-DiHETE	95	335.2	173.2	7	10.83
5,6-DiHETE	90	335.2	115.2	4	10.85
5,6-DiHETrE	85	337.2	145.1	7	16.7
5-HEPE	115	317.2	115.1	4	17.8
5-HETE	90	319.2	115.1	10	21.4
5-oxo-ETE	120	317.2	273.2	7	23.4
6-keto-PGF1α	90	369.3	163.2	22	4.7
6-trans-LTB4	125	335.2	195.1	7	10.8
7(8)-EpDPE	85	343.2	113.1	4	23.8
8(9)-EpETE	115	317.2	127.2	4	20.8
8(9)-EpETrE	90	319.2	167.2	4	23.7
8,15-DiHETE	90	335.2	235.2	7	11
8,9-DiHETE	100	335.5	126.9	12	12.8
8,9-DiHETrE	85	337.2	127.1	13	15.6
8-HEPE	115	317.2	155.2	7	17
8-HETE	120	319.2	155.2	7	20.4

(continued)

Table 3
(continued)

Compound abbreviation	Fragmentor voltage (V)	Precursor ion	Product ion	Collision energy (V)	Retention time (min)
9(10)-EpOME	100	295.3	171.1	7	22.8
9,10,13-TriHOME	110	329.2	171.1	16	6.7
9,10-DiHOME	130	313.2	201.2	16	12.42
9,12,13-TriHOME	125	329.2	211.1	16	6.6
9-HETE	80	319.2	167.2	7	19.8
9-HODE	120	295.2	171.1	10	18
9-HOTrE	110	293.2	171.2	4	15.3
9-oxo-ODE	90	293.2	185.1	13	20
LTB3	120	337.2	195.2	7	14.2
LTB4	125	335.2	195.1	7	11.4
LTC4	70	624.3	272.1	22	9.6
LTD4	55	495.3	177.1	13	7.2
LTE4	90	438.2	333.3	13	9.7
LXA4	95	351.2	115.2	10	8.1
PGB2	125	333.3	175.1	13	10
PGD1	75	353.3	317.2	7	7.2
PGD2	80	351.2	271.3	10	7
PGD3	120	349.3	269.2	7	6
PGE1	75	353.3	317.2	7	7.3
PGE2	80	351.2	271.3	10	7
PGE3	120	349.3	269.2	7	6
PGF2α	120	353.2	309.2	10	6.7
PGJ2	90	333.3	189.2	10	9.9
Resolvin E1	115	349.3	195	10	4.6
TXB2	80	369.2	169.1	10	6

4. Let the system equilibrate with your starting condition (35% B) until the pressure is stable.
5. Run your standard curves and your samples using the UPLC-MS/MS method set-up in **steps 1** and **2**. A representative chromatogram of 76 oxidized fatty acid standards is presented in Fig. 3:

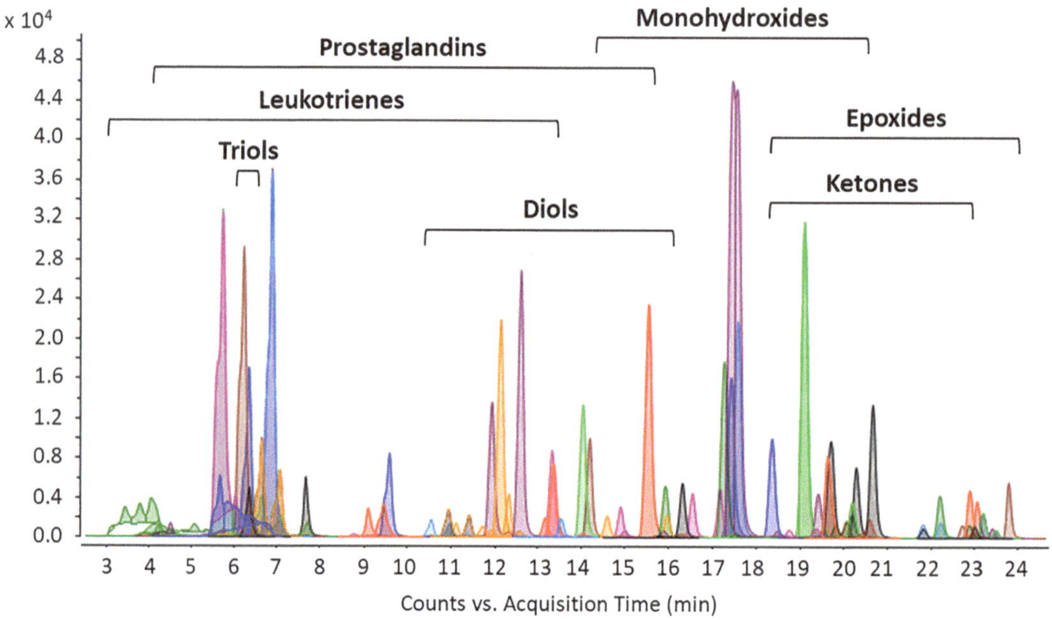

Fig. 3 UPLC-MS/MS chromatogram of 76 oxidized fatty acids

- Start by running a blank (pure LC-MS-grade methanol) to verify the background of your signal and the cleanliness of the system. Depending on the outcome, you might need to run an additional blank and/or to clean some parts of your instrument before running your standard curves and samples.
- Run a standard curve and visually check the chromatograms. You should be able to see a clear peak for all of your compounds and that the peak areas increase with increasing concentrations.
- Run a blank (pure LC-MS-grade methanol) after your first standard curve and every 10 samples on average to verify that no compounds were retained in the system.
- For large batches of samples, it is recommended to run (part of) your standard curves again in the middle and at the end of your run to correct for system fluctuation (e.g., slight decrease in signal over time).
- Once the analysis is completed, wash your column for 15 min with 100% mobile phase B and for 15 min with 100% LC-MS-grade methanol. If the column is dirty, you may need to wash the column with 100% LC-MS isopropanol for at least 1 h (the amount of time required might vary depending on the dirtiness of the column). Store the column in 100% LC-MS-grade methanol. For these washing

steps, make sure that the UPLC is disconnected from the MS and that all solvents go to waste.

3.5 Method Validation

Anytime the method is set up on a new instrument and/or for a new type of sample matrix, it is recommended to perform some validation steps. *See* **Note 10**.

1. Surrogate standard recovery to evaluate losses during hydrolysis and SPE: To do so, run a sample spiked with 0.2 μM surrogate standard and extracted following the protocol described above (i.e., sample preparation, hydrolysis, and SPE), along with a 0.2 μM surrogate standard solution obtained by direct dilution of the surrogate standard mix in LC-MS-grade methanol without the hydrolysis and SPE steps. Each sample is prepared and analyzed in triplicate. The total volume of each of the spiked sample and surrogate standard should be the same; typically 100 μL:
   - Surrogate recovery = ($A_{surrogate}$ in sample/$A_{surrogate}$ in standard mix) × 100, where $A_{surrogate}$ represents the peak area of the surrogate standard.
   - Surrogate recovery should be as close as possible to 100%. *See* **Notes 11** and **12**.
2. Intraday repeatability to verify the stability of your system: Run the same sample (either an experimental sample or a standard mix) multiple times within the same run:
   - Ideally, the full run should be covered, that is, do not inject the same sample multiple times in a row, but divide your multiple injection over the full run (i.e., beginning, middle, and end of the run).
   - Intraday coefficient of variation (CV) = (standard deviation of the replicates/mean of the replicates) × 100.
   - CV should be as low as possible, ideally below 10%.
3. Interday reproducibility to evaluate the interexperimental variability: Using the protocol described above, prepare, hydrolyze, extract, and analyze the same sample on different days:
   - Interday CV = (standard deviation of the means/overall mean) × 100.
   - CV should be as low as possible, ideally below 10%.
4. Matrix effect to verify that matrix-related ion suppression or enhancement is minimal: Prepare, hydrolyze, and extract one sample using the protocol described above. Divide the 100 μL extract obtained into two vials (50 μL each). In one of the vials, bring the volume up to 100 μL by adding 50 μL of LC-MS-grade methanol (unspiked extract); in the other vial, bring the

volume up to 100 μL by adding 50 μL of unlabeled oxidized fatty acid standards and deuterated surrogate standard mix (spiked extract). Analyze the spiked extract, the unspiked extract, and a pure standard mix (50 μL unlabeled oxidized fatty acid standards and deuterated surrogate standard mix + 50 μL LC-MS-grade methanol to obtain the same final concentration as the spiked sample) by UPLC-MS/MS:

- Matrix effect = ($A_{compound}$ in the spiked sample)/[($A_{compound}$ in the unspiked sample) + ($A_{compound}$ in the pure standard mix)] × 100, where $A_{compound}$ represents the peak area of the compound of interest.
- A range between 70% and 130% is considered acceptable.
- When the matrix effect is below 70%, ion suppression is observed; when the matrix effect is above 130%, ion enhancement is observed. *See* **Note 13**.

3.6 Data Processing

1. Integrate your peaks. *See* **Note 14**.
2. For each unlabeled standard, plot a standard curve with the ratio of the amount (in pmol) of the unlabeled standard to the amount (in pmol) of the corresponding deuterated surrogate standard (Table 1) injected in the UPLC-MS/MS on the x-axis and the ratio of the area of unlabeled standard to the area of the corresponding deuterated surrogate standard on the y-axis:
   - Use quadratic regression model and $1/x$ or $1/x^2$ weighting factor to create the standard curve. Typically, $1/x$ works best for oxidized fatty acids. Aim at $R^2 > 0.99$. *See* **Note 15**.
3. Using the standard curve equation and the known amount of surrogate standard in your sample, calculate the concentration of each oxidized fatty acid in your samples:
   - Amount of oxidized fatty acid in the sample (in pmol) = (area of oxidized fatty acid in the sample/area of corresponding deuterated surrogate standard in the sample) × (1/slope of the standard curve) × 2 pmol * 10; 2 pmol corresponds to the amount of the internal standard injected into the UPLC-MS/MS system. The multiplication factor of 10 takes account for the fact that 10 μL out of a 100 μL extract were injected into the UPLC-MS/MS system.
   - Concentration of oxidized fatty acid in the sample = amount (pmol) of oxidized fatty acid in the sample/weight or volume of the starting material (g or mL). *See* **Note 16**.

4 Notes

1. This chapter focuses on the analysis of food samples. However, the method described above can also be used for the analysis of oxidized fatty acids in biological samples (e.g., blood, brain, and liver) with some adjustments. For this reason, the standard list also includes oxidized metabolites derived from enzymatic metabolism (e.g., prostaglandins and thromboxanes).
2. The list of suggested standards is currently used in our laboratory, but it could be extended or reduced to fit for the purpose of the experiment.
3. The surrogate solutions and standard curves can be stored at −80 °C until use. However, we recommend preparing fresh stock solutions after the third freeze-thaw cycle or when the UPLC-MS/MS analysis reveals irregularities in the standard curves. Aliquoting in multiple small volumes could help reducing freeze-thaw cycles of stock solutions.
4. The detection of the lowest concentration in the standard curve depends on the instrument sensitivity. Of the nine standards prepared, likely seven will be above the instrument detection limits in most cases resulting in a 7-point standard curve.
5. It is possible to prepare stock solutions of each of the three antioxidants and store them at −20 °C until use. The three recommended stock solutions are 0.6 mg TPP.mL^{-1} distilled water/HPLC-grade methanol (1:1 v/v), 0.6 mg BHT.mL^{-1} HPLC-grade methanol and 0.6 mg EDTA.mL^{-1} in distilled water. Mix 1 mL of each solution to reach a final concentration of 0.2 mg.mL^{-1} TPP, BHT, and EDTA in methanol/water (1: 1 v/v). The antioxidant mix can also be stored at −20 °C until use.
6. The hydrolysis step was previously tested using a 0.25 M sodium carbonate solution in methanol/water (1:1 v/v) [19]. However, data indicate that hydrolysis with sodium carbonate is not as efficient as sodium hydroxide, leading to possible underestimation of the oxidized fatty acid concentrations.
7. The method has been validated and used for the analysis of oil, milk, and milk cream [18, 19, 23, 24]. Anytime the method is set-up on a new instrument and/or for a new type of sample matrix, you should first test for matrix effects and reproducibility.
8. The protocol described is meant to measure the total oxidized fatty acid and therefore includes a hydrolysis step using a 0.25 M sodium hydroxide solution to release esterified oxidized fatty acids from their backbone (e.g., glycerol). In the

case where there is an interest in measuring free oxidized fatty acids (i.e., not esterified to a bigger molecule, such as triacylglycerol), you can skip the hydrolysis step by adding 1.8 mL of ultrapure water directly after the addition of 190 μL of extraction solvent (**step 4** in Subheading 3.2, Preparation of Samples and Hydrolysis). A 2 mL sample is then obtained and it can be applied directly to the steps described in Subheading 3.3 – Oxidized Fatty Acid Extraction by Solid Phase Extraction (SPE).

9. The retention time and collision energy, as well as the parameters used for the ESI are given for reference, but they need to be adjusted and optimized depending on the UPLC-MS/MS system used.
10. Validation and optimization parameters are instrument and matrix dependent. For example, information about the recovery, the reproducibility and the matrix effect for the analysis of fish and algae oil samples are thoroughly discussed in the study by Emami et al. [19].
11. If the surrogate recovery is low, two options are possible: either to identify which step in the protocol was responsible for the standard loss and adjust the protocol when possible (preferred option) or to proceed with your analysis taking into account the low recovery in the interpretation of your results.
12. All surrogate standards are available in their free form, which can induce a bias in the esterified oxidized fatty acid calculation. Indeed, this calculation assumes that free and esterified compounds will behave in a similar manner during hydrolysis and SPE, which is probably not the case. The lack of esterified standards is an important bottleneck for this method. The recent development of a few methyl ester surrogate standards (e.g., deuterated epoxide methyl esters) is the first step, but more esterified standards will be needed in the future to cover all classes of oxidized fatty acids.
13. Matrix effect (i.e., ion suppression or ion enhancement) is one of the main challenges in quantitative analysis by UPLC-MS/MS due to the presence of coeluting molecules that hinder or increase the ionization of a compound of interest and its detection with a mass spectrometer. Adjusting the mobile phase gradient to allow better separation of the target compound from the interacting molecule may solve the problem. However, in most cases, the interacting molecule has similar properties (e.g., polarity) with the target compound, making it a challenge to properly separate them.
14. Most analytical software give the opportunity to develop and save a processing method which will automatically integrate the peaks of interest. When using such an automatic method,

always verify that the integration is performed properly and in a similar way for all samples and standard curves.

15. A linear fit can be used. However, our experience suggests that using a quadratic curve with a 1/x weighting factor increases the accuracy, especially at low concentrations. The limit of detection (LOD) and the limit of quantification (LOQ) are dependent on the instrument used and can be estimated using a method suggested by the Environmental Protection Agency (EPA) which is based on the region of the calibration curve where a significant change in sensitivity occurs (40 CFR, Appendix B to Part 136 revision 1.11, U.S.). The significant change in the sensitivity ($p < 0.05$) on the calibration curve could be determined by running a *t*-test between the standard curve points. The standard deviation (SD) of the replicates of the first significantly different calibration level is then used to estimate the LOD by multiplying it to the Students *t*-test value appropriate for n-1 degrees of freedom, where n is the number of replicates, and a 95% or 99% one-tail confidence level (1-α) as follows: $LOD = t_{n-1,\ 1-\alpha = 0.95 \text{ or } 0.99} \times SD$. The LOQ will be calculated as follows: $LOQ = 3 \times LOD$. Only compounds with the concentrations above the LOQ and within the standard curve boundaries can be accurately quantified.

16. The formula can be adjusted to consider any additional dilution steps (e.g., hydrolysis and SPE performed on a fraction of a larger lipid extract).

References

1. Frankel EN, Review. (1991) Recent advances in lipid oxidation. J Sci Food Agric 54(4): 495–511
2. Porter NA, Caldwell SE, Mills KA (1995) Mechanisms of free radical oxidation of unsaturated lipids. Lipids 30(4):277–290
3. Schaich KM (2020) Lipid oxidation: new perspectives on an old reaction. In: Bailey's industrial oil and fat products, pp 1–72
4. Bian H, Ma J, Geng Z, Liu T, Sun C, Wang D, Zhang M, Xu W (2019) Changes of hydroxyl-linoleic acids during Chinese-style sausage processing and their relationships with lipids oxidation. Food Chem 296:63–68
5. Lampi AM, Yang Z, Mustonen O, Piironen V (2020) Potential of faba bean lipase and lipoxygenase to promote formation of volatile lipid oxidation products in food models. Food Chem 311:125982
6. Wang T, Hammond EG (2010) 5 – Lipoxygenase and lipid oxidation in foods. In: Decker EA (ed) Oxidation in foods and beverages and antioxidant applications. Woodhead Publishing, pp 105–121
7. Mohammadi F, Marti A, Nayebzadeh K, Hosseini SM, Tajdar-Oranj B, Jazaeri S (2020) Effect of washing, soaking and pH in combination with ultrasound on enzymatic rancidity, phytic acid, heavy metals and coliforms of rice bran. Food Chem 334:127583
8. German JB, Chen SE, Kinsella JE (1985) Lipid oxidation in fish tissue. Enzymic initiation via lipoxygenase. J Agric Food Chem 33(4): 680–683
9. Flaskerud K, Bukowski M, Golovko M, Johnson L, Brose S, Ali A, Cleveland B, Picklo M Sr, Raatz S (2017) Effects of cooking techniques on fatty acid and oxylipin content of farmed rainbow trout (Oncorhynchus mykiss). Food Sci Nutr 5(6):1195–1204
10. Raatz SK, Golovko MY, Brose SA, Rosenberger TA, Burr GS, Wolters WR, Picklo MJ Sr (2011) Baking reduces prostaglandin, resolvin, and hydroxy-fatty acid content of farm-raised

Atlantic salmon (Salmo salar). J Agric Food Chem 59(20):11278–11286

11. Beckman B, Mustafa T (1992) Arachidonic acid metabolism in gill homogenate and isolated gill cells from rainbow trout, Oncorhynchus mykiss: the effect of osmolality, electrolytes and prolactin. Fish Physiol Biochem 10(3):213–222
12. Oleksiak MF, Wu S, Parker C, Qu W, Cox R, Zeldin DC, Stegeman JJ (2003) Identification and regulation of a new vertebrate cytochrome P450 subfamily, the CYP2Ps, and functional characterization of CYP2P3, a conserved arachidonic acid epoxygenase/19-hydroxylase. Arch Biochem Biophys 411(2):223–234
13. Gilroy DW, Edin ML, De Maeyer RP, Bystrom J, Newson J, Lih FB, Stables M, Zeldin DC, Bishop-Bailey D (2016) CYP450-derived oxylipins mediate inflammatory resolution. Proc Natl Acad Sci U S A 113(23): E3240–E3249
14. Morisseau C, Inceoglu B, Schmelzer K, Tsai HJ, Jinks SL, Hegedus CM, Hammock BD (2010) Naturally occurring monoepoxides of eicosapentaenoic acid and docosahexaenoic acid are bioactive antihyperalgesic lipids. J Lipid Res 51(12):3481–3490
15. Gabbs M, Leng S, Devassy JG, Monirujjaman M, Aukema HM (2015) Advances in our understanding of oxylipins derived from dietary PUFAs. Adv Nutr 6(5): 513–540
16. Strassburg K, Huijbrechts AM, Kortekaas KA, Lindeman JH, Pedersen TL, Dane A, Berger R, Brenkman A, Hankemeier T, van Duynhoven J, Kalkhoven E, Newman JW, Vreeken RJ (2012) Quantitative profiling of oxylipins through comprehensive LC-MS/MS analysis: application in cardiac surgery. Anal Bioanal Chem 404(5):1413–1426
17. Yang J, Schmelzer K, Georgi K, Hammock BD (2009) Quantitative profiling method for oxylipin metabolome by liquid chromatography electrospray ionization tandem mass spectrometry. Anal Chem 81(19):8085–8093
18. Richardson CE, Hennebelle M, Otoki Y, Zamora D, Yang J, Hammock BD, Taha AY (2017) Lipidomic analysis of oxidized fatty acids in plant and algae oils. J Agric Food Chem 65(9):1941–1951
19. Emami S, Zhang Z, Taha AY (2020) Quantitation of oxylipins in fish and algae oil supplements using optimized hydrolysis procedures and ultra-high performance liquid chromatography coupled to tandem mass-spectrometry. J Agric Food Chem 68(35):9329–9344
20. Dias FFG, Augusto-Obara TR, Hennebelle M, Chantieng S, Ozturk G, Taha AY, Vieira T, Leite Nobrega de Moura Bell JM (2020) Effects of industrial heat treatments on bovine milk oxylipins and conventional markers of lipid oxidation. Prostaglandins Leukot Essent Fatty Acids 152:102040
21. Teixeira BF, Dias FFG, de Souza Vieira TMF, Taha AY, de Moura Bell JMLN (2022). Early detection of lipid oxidation in infant milk formula by measuring free oxylipins—Comparison with hydroperoxide value and thiobarbituric acid reactive substance methods. J. Food Sci. 87(12):5252–5262. https://doi.org/10.1111/1750-3841.16364
22. Zhang Z, Hennebelle M, Emami S, Taha AY (2021) Effects of Potato Processing and Frying on Oxylipin Concentrations. ACS Food Sci Technol 1(8):1436–1443. https://doi.org/10.1021/acsfoodscitech.1c00139
23. Shen Q, Zhang Z, Emami S, Chen J, Leite Nobrega de Moura Bell JM, Taha AY (2021) Triacylglycerols are preferentially oxidized over free fatty acids in heated soybean oil. NPJ Sci Food 5:7
24. Gan J, Zhang Z, Kurudimov K, German JB, Taha AY (2020) Distribution of free and esterified oxylipins in cream, cell, and skim fractions of human milk. Lipids 55(6):661–670
25. Ostermann AI, Koch E, Rund KM, Kutzner L, Mainka M, Schebb NH (2020) Targeting esterified oxylipins by LC-MS – effect of sample preparation on oxylipin pattern. Prostaglandins Other Lipid Mediat 146:106384

Chapter 16

Quantitative ^{1}H NMR Analysis of Food Lipid Hydroperoxides and Aldehydes

Donny Merkx, Vincent Boerkamp, John van Duynhoven, and Marie Hennebelle

Abstract

^{1}H NMR spectroscopy enables rapid and sensitive one-shot quantification of both lipid hydroperoxides and aldehydes as primary and secondary oxidation products in lipid-based food ingredients and products. The method is straightforward and can be applied to any type of vegetable oils and oil extracts but requires access to a commercial high-field NMR instrument. The sensitivity is improved threefold by using a band-selective pulse excitation, which results in a limit of detection below 0.1 mmol kg^{-1} for lipid hydroperoxides when using a 600 MHz NMR equipped with a cryoprobe. Here, we describe a step-by-step approach for quantification of both hydroperoxide and aldehyde contents in an oil (extract) in one go.

Key words NMR, qNMR, Lipid oxidation, Hydroperoxides, Aldehydes, Oil, Mayonnaise

1 Introduction

By monitoring only a single class of oxidation products, the extent of lipid oxidation could be greatly underestimated [1]. Therefore, to accurately assess the oxidative state of foods, both the primary and secondary oxidation products should be quantified. Various techniques have been put forward to track primary oxidation using the peroxide value (PV), including iodometric titration and colorimetric detection of Fe(II) oxidation [2]. For the analysis of secondary oxidation, thiobarbituric acid reactive substances (TBARS) and the p-anisidine value (p-AV) are frequently used [3]. These methods determine only single species type (peroxides or specific aldehydes) and suffer from high labor intensity, low sensitivity, poor reproducibility, and distortion by contaminants [2–4]. Analysis of oxidation products with chromatographic methods (GC and HPLC) mitigates most issues as it enables identification of a wide range of lipid hydroperoxides (LOOH) or volatiles with high

Christelle Lopez et al. (eds.), *Multidimensional Characterization of Dietary Lipids*, Methods and Protocols in Food Science,
https://doi.org/10.1007/978-1-0716-3758-6_16,

sensitivity. These methods, however, suffer from a low throughput and require pure standards for accurate quantification [2].

In recent years, proton nuclear magnetic resonance (^{1}H NMR) spectroscopy has been used increasingly for the analysis of food lipid oxidation products [5–11]. It is reproducible, is easy to automate, provides high throughput, is able to detect various lipid oxidation species simultaneously, and is independent of pure standards for quantification. The method can be applied directly to vegetable oil samples or after lipid extraction to more complex food samples. The addition of DMSO-d_6 to $CDCl_3$ as solvent allows for accurate quantification of hydroperoxides [6]. Sensitivity issues due to dynamic range limitations can be overcome using selective pulse gradients [7]. This enables ^{1}H NMR measurements of less than 3 min per sample (600 MHz NMR instrument equipped with a cryoprobe) with a limit of detection of 0.1 mmol kg^{-1} for LOOH, which is comparable or even outperforms the classical titration and UV methods. The method also enables simultaneous quantification of nonvolatile aldehydes and hydroperoxides. The quantification of both products is possible without the addition of an internal standard as the triglyceride backbone peak can be used as a reference. Finally, by integration of individual signals, hydroperoxides and aldehydes can be quantified at a molecular substructure level, which can potentially provide valuable information for reaction pathway elucidation [7, 11].

Here, we describe a step-by-step protocol to quantify hydroperoxides and aldehydes in oils (or lipid extracts) using ^{1}H NMR.

2 Materials

2.1 Equipment

This method requires a high-field NMR system, which implies a Larmor frequency of ^{1}H nuclei of at least 400 MHz (9.4 T). It also requires a console which can perform selective gradient pulses, which is common for modern commercial NMR spectrometers. A cryogenic probe head will allow for short measurement times, but it is not a necessity. Optional, but highly recommended, is the use of an automatic sample changer to allow a large sequence of samples to be measured without intervention of the analyst.

2.2 Chemicals

- Deuterated chloroform ($CDCl_3$) + 0.03% TMS, 99.80% D, CAS [865-49-6].
- Deuterated dimethyl sulfoxide (DMSO-d_6), 99.80% D, CAS [2206-27-1].
- Optional: deuterated molecular sieves, 99.80% D, CAS [N1311].
- 5 mm NMR tubes.

- Glass laboratory bottle, 100–500 mL.
- Disposable 1.5 mL PP microcentrifuge tubes.

3 Method

3.1 NMR-Solvent Preparation

(i) Transfer 50 mL of deuterated chloroform ($CDCl_3$) into a glass laboratory bottle.

(ii) Add 10 mL of deuterated dimethylsulfoxide (DMSO-d_6).

(iii) Optional: add 100 mg of deuterated molsieves. This step is advised if the solvent is to be stored for longer than a week. If molsieves are not available, ensure that the solvent attracts as minimal water as possible.

(iv) Close the bottle and mix the solution manually for at least 10 s. The solution will warm up slightly.

(v) Let the solution cool down prior to the first use.

(vi) When using the solution, swirl the bottle briefly (~3 s).

3.2 Sample Preparation

This method requires oil or fat which needs to be devoid of water as this might disturb accurate and precise NMR quantification. For example, oil can be used directly; for high-oil emulsions such as mayonnaise use a freeze-thaw induced phase separation to separate the oil; other types of samples and emulsions may require a lipid extraction first, such as the Folch or Bligh and Dyer method:

(i) Transfer 150 μL of oil or ~135 mg of fat to a small container, e.g., a 1.5 mL micro-centrifuge tube. These are the recommended and optimal quantities; note that the minimal suggested amount is 50 μL oil or 45 mg fat.

(ii) To this, add 450 μL of NMR-solvent (5:1 $CDCl_3$:DMSO-d_6) or fill to a total volume of 600 μL in case less than 150 μL oil was used.

(iii) Mix the solution by closing the container and gentle tumbling of the container (2 s) to ensure that the solution is clear and no phase separation is observed.

(iv) Transfer the full solution (~600 μL) to a 5 mm NMR tube using a glass Pasteur pipette. Cap the NMR tube properly.

3.3 NMR Acquisition

The following steps describe the manual operation of a Bruker NMR spectrometer. Note that most, if not all steps, can be done fully automatically in most modern NMR facilities. The commands as used in the Bruker Topspin software are displayed in brackets, whereas for other data acquisition software, a skilled operator can identify similar commands.

Note that this protocol has been developed using TopSpin 3.6.1, operating a AVANCE III console with a cryoprobe on a 600 MHz system (see **Note 1**). *The suggested pulse values may therefore be different in your system and it is advised to always consult the NMR supervisor for the optimal setup:*

(i) Insert the NMR tube into the magnet.

(ii) Set the internal temperature of the probe to 295 K (*see* **Note 2**) [TE 295]:

 (a) Wait until the sample temperature is stable (minimum 3 min) before continuing to **step vii** (shimming). Meanwhile, you can continue with **steps iii–vi**.

(iii) Create a new experiment [NEW].

(iv) Select a single-pulse proton experiment [zg // zg30].

(v) Tune and match the sample [ATMA].

(vi) Lock the sample to deuterated DMSO [LOCK DMSO]:

 (a) Note that due to the presence of deuterated CDCl3, the system might lock wrongly in this step. If this happens, lock onto deuterated CDCl3 instead [LOCK CDCl3].

Two experiments will be run after each other: a single-pulse experiment to record the full spectrum (including the triglyceride peak for quantification) and a band-selective pulse experiment for the assessment of the oxidation products.

Single-pulse experiment:

(vii) Shim the sample [TOPSHIM or GS].

 (a) Proper shimming is required to ensure sharp quantifiable peaks.

(viii) Determine the length of the 90° pulse [PULSECAL].

(ix) Set the relaxation delay to 5 s [D1 5].

(x) Set the acquisition time to 4 s [AQ 4].

(xi) Set the spectral width to 20 ppm [SW 20].

(xii) Set the offset to 6.175 ppm [O1P 6.175].

(xiii) Determine the ideal receiver gain [RGA].

(xiv) Set the number of scans to a minimum of 4 [NS 4].

(xv) Set the number of dummy scans to 2 [DS 2].

(xvi) Run the experiment [ZG].

 (a) Note if TopSpin reports "receiver overflows" and the receiver gain is already at its lowest value, then the pulse length can be shortened to, e.g., 10° (possible in cryoprobe systems).

The band-selective pulse experiment can be created manually or by an interactive module in TopSpin.

Manual (Skip Steps xxvi–xxxv):

(xvii) Create a new experiment [NEW] or increase the experiment number [IEXPNO].
   (a) Ensure that the first experiment is not overwritten, as both spectra are required for the quantification.

(xviii) Select a selective pulse sequence [PULPROG selgpse].

(xix) Retrieve the powers [GETPROSOL].

(xx) Set the center of the selective pulse on the center of the desired region, e.g., at δ 10.5 ppm [CNST21 10.5].

(xxi) Set the pulse length at the same value as in the single-pulse experiment [e.g., P1 7.6].

(xxii) Set the width of the band-selective pulse by changing the following:
   (a) SPDB2 15
   (b) P12 550

   Note that lower SPDB2 and P12 increase the width of the band-selective pulse area.

(xxiii) Set the digitizer mode to digital [DIGMOD DIGITAL].

(xxiv) Set the zg-options to calculate the properly shaped pulse [ZGOPTNS -DCALC_SPOFFS].

(xxv) Set the correct gradient files:
   (a) GPNAM 1 SMSQ10.100.
   (b) GPZ1 15.

   Optional: Use a REBURP pulse instead of the default Gaussian inverse pulse for a sharper excitation. More details can be found in the work of Merkx et al. [7] The pulse sequences for these selective pulses are available in the libraries of modern NMR spectrometers. TopSpin commands:
   (a) SPNAM2 Reburp.1000.
   (b) P12 2000.

Interactive (Skip Steps xvii–xxv):

(xxvi) Run the single-pulse experiment.

(xxvii) Locate the hydroperoxide and aldehyde regions and integrate [.INT] the area between 13 and 8 ppm, or tighter if possible.

(xxviii) Click on the floppy disk with an "a" next to it on the top of the integration module and select the option "Save Regions to reg."

(xxix) Leave the integration module [.SRET].

(xxx) Locate and click the "Acquire" tab on the top of your TopSpin interface.

(xxxi) Click on the sub-tab "More" and select "Setup Selective 1D Expts."

(xxxii) Click on "Create Datasets."

(xxxiii) Select "Selective Gradient 1H."

(a) Optional: change the shape of the pulse to Reburp.1000.

(xxxiv) Press "Accept."

(xxxv) In the next window, press "cancel" to only create the dataset and not start the acquisition.

Manual and Interactive:

(xxxvi) Set the relaxation delay to 5 s [D1 5].

(xxxvii) Set the acquisition time to 4 s [AQ 4].

(xxxviii) Set the spectral width to 20 ppm [SW 20].

(xxxix) Determine the ideal receiver gain automatically [RGA].

(xl) Set the number of scans to a minimum of 16 [NS 16].

(xli) Set the number of dummy scans to 4 [DS 4].

(xlii) Run the experiment [ZG].

3.4 NMR Preprocessing

(i) Make sure that the size of the spectrum [SI] is equal to the size of the Free Induction Decay (FID) [TD]. Note that the size of the spectrum can be changed after acquisition but the size of the FID cannot.

(ii) Use a Fourier transform on the FID spectrum [FT]:

(a) Optionally, the Fourier transform can be performed with an exponential window function with a line broadening of 0.3 Hz [LB 0.3], followed by [EFP] to obtain smoother peaks.

(iii) Correct the phase of the spectrum manually [.PH] or automatically [APK].

(iv) Correct the baseline manually [.BASL] or automatically [ABSN]. Steps (iii) and (iv) can also be combined with [APBK] in TopSpin v4 and higher.

(v) Calibrate the spectrum to have trimethylsilane (TMS) at 0.00 ppm, either manually [.CAL] or automatically [SREF].

3.5 Integration

Spectral integration is what allows us to quantify the oxidation products. In this protocol, the emphasis will be on the integration of the three peak areas that are required to quantify the LOOHs and aldehydes. In the recorded single-pulse experiment, the triglyceride peaks at 4.30 and 4.13 ppm need to be integrated. In the selective pulse experiment, the LOOH (11.3–10.6 ppm) and aldehyde areas (9.9–9.2 ppm) need to be integrated. The practical steps of integration differ slightly for each software package, but the principle remains largely the same. Here, the integration steps for TopSpin are described:

(i) Open the integration module [.INT].

(ii) On the top left, make sure the small horizontal '['is toggled (If you hover over the correct button, it says "Define new region using cursor. (toggle)"].

(iii) When integrating, include 0.1–0.2 ppm of baseline on both sides of the peak of interest. This is important for the following step.

(iv) To make sure the integral is not biased, make sure that both sides of the integral you take are horizontal (because integrating baseline should be zero). Do this by holding "b" (b stands for *bias* and can be interpreted as baseline offset) and dragging (or scrolling) it up or down to adjust the slope of the beginning of the integral. The same can be done for the end of the integral using "s" (s stands for *slope* and can be interpreted as such). In Fig. 1, integrals with incorrect and correct "b" and "s" values are shown. Figure 2 provides an example of correct integration for all regions of interest.

(v) Save and return when you are satisfied [.SRET].

(vi) Optional:

  (a) If you want to save the integration settings and copy them to another spectrum, type "wmisc intrng *name*", with a name for the integral. The saved integral settings can be found when typing "*rmisc*" and can be used in different datasets.

  (b) Then, select another sample and type "**r**misc intrng *name*" to apply the same integral settings. When you use this option, it is recommended to go back to **step 4** and check the bias and slope of the integral.

(vii) Extract the absolute integral number and go to the tab "integrals" and copy the number under "integral [abs]":

  (a) Note that these absolute integral numbers are often large numbers that should be copied and pasted to a spreadsheet rather than typing manually.

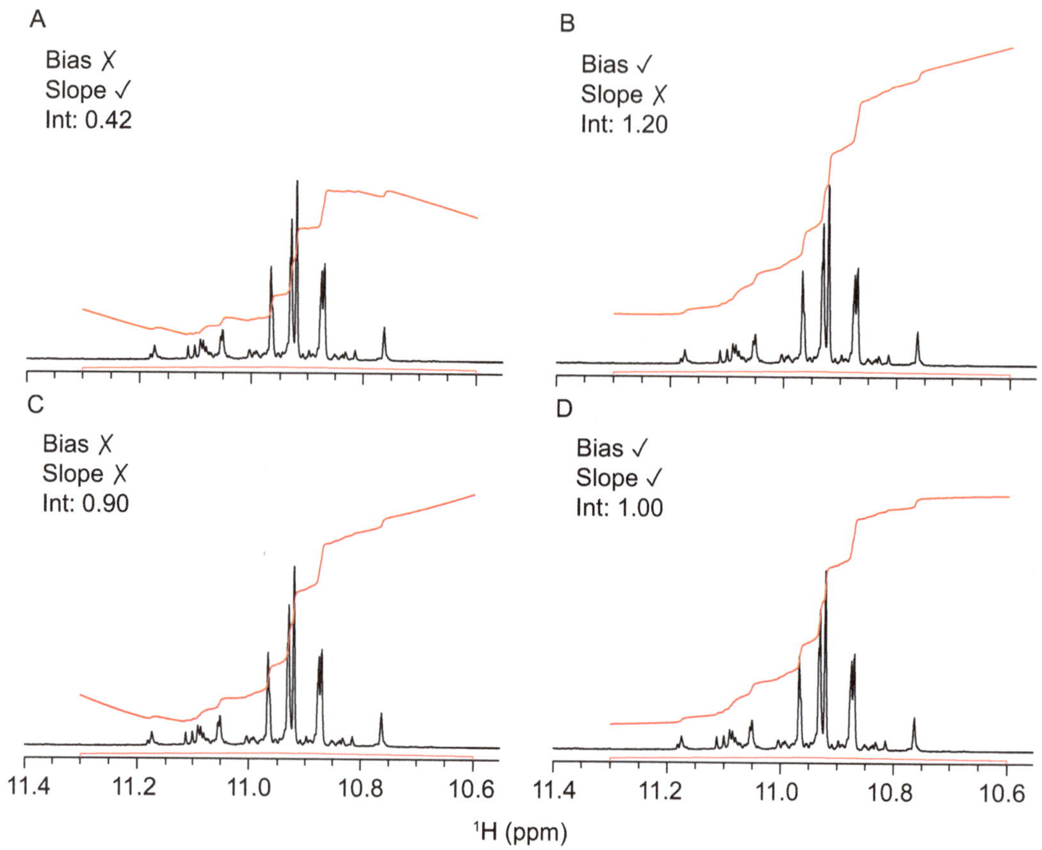

Fig. 1 The impact of Bias (b) and Slope (S) on spectral integration. Both sides of the integral should be horizontal. Modifying the Bias (b) and Slope (S), respectively, impact the steepness of the beginning and end of the integral. Note the impact of the absolute integral numbers when the Bias and Slope are incorrect (**a–c**) versus correct (**d**)

Note that this integration describes the sum of hydroperoxides and the sum of aldehydes. Using the same approach, the individual lipid hydroperoxides and aldehydes can also be quantified. More details on the identification of these oxidation products can be found in the work of Merkx et al. (2018) [7].

3.6 Calculation

Traditionally, oxidation products are expressed as millimoles (mmol) or milliequivalents (meq) of the oxidation product per kilogram oil. Therefore, the standard way this quantification method is set up yields the oxidation products in the "mmol/kg of oil" expression. The method does not require the addition of an internal standard and instead uses the present triglyceride peak and its molar mass. The molar mass of the oil can be estimated easily by looking at the fatty acid composition of the used fat/oil. In the following calculation, a molar mass of 880 g/mol is used.

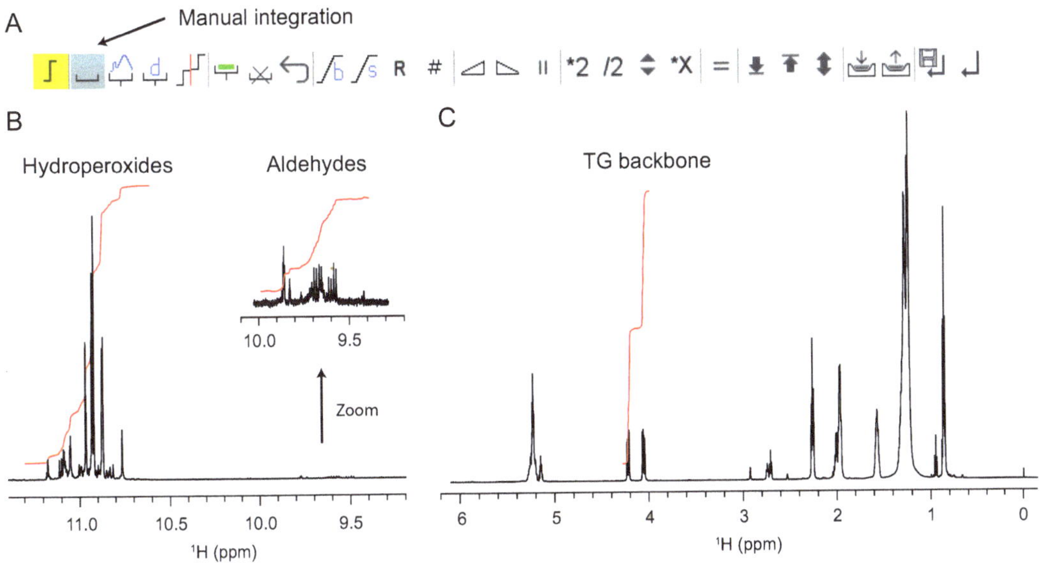

Fig. 2 Example of the correct integration of the desired three regions. (**a**) Selection of manual integration mode in TopSpin v4.1 software. (**b**) The integration of the oxidation products in the selective pulse spectrum. (**c**) The integration of the triglyceride backbone peak in the single-pulse spectrum

The quantification can be done either by using only the single-pulse NMR experiment or by using both the single-pulse and selective pulse experiments. One can use the first approach if the oxidation products are sufficiently present in the single-pulse spectrum, which means that the S/N ratio of the peaks should be above 10. If this is not the case, the selective pulse experiment is advised for accurate quantification.

Using only the single-pulse experiment, the quantification can be performed as follows:

$$c_{ox}\left(\mathrm{mmol}/_{\mathrm{kg}}\right) = \frac{I_{ox}}{I_{TG}} \cdot \frac{N_{TG}}{N_{ox}} \cdot \frac{10^6}{MW_{TG}} \tag{1}$$

Here, I is the NMR signal intensity after integration, N is the number of protons, and MW is the molecular weight of the fat or oil. The index ox stands for the oxidized lipid product and TG for triglyceride. The integrals I need to be extracted from the spectrum as described under *Integration/vii*. The number of protons comes from prior knowledge of the system, so in case of LOOHs and aldehydes, N_{ox} is 1, whereas N_{TG} is 4. MW_{TG} needs to be estimated by the analyst as described before.

When the band selective pulses are used, several terms are added to the equation to correct for the use of different pulse sequences:

	Single Pulse (ZG)	Band-Selective Pulse (SEL)
I_{ox}	407	136119
ns	4	16
RG	2	203
sin(p°)	1	1
Selfactor	$\frac{I_{ox}(zg)}{I_{ox}(sel)} \cdot \frac{ns_{sel}}{ns_{zg}} \cdot \frac{RG_{sel}}{RG_{zg}} \cdot \frac{\sin(p_{sel})}{\sin(p_{zg})}$ =	$\frac{407}{136119} \cdot \frac{16}{4} \cdot \frac{203}{2} \cdot \frac{1}{1}$ = **1.21**

Fig. 3 SelFactor calculation example

$$c_{ox}\left(\text{mmol}/\text{kg}\right) = \frac{I_{ox}}{I_{TG}} \cdot \frac{N_{TG}}{N_{ox}} \cdot \frac{ns_{zg}}{ns_{sel}} \cdot \frac{\sin(p_{zg})}{\sin(p_{sel})} \cdot \frac{RG_{zg}}{RG_{sel}} \cdot \frac{10^6}{MW_{TG}} \cdot \text{SelFactor} \tag{2}$$

Here, ns is the number of scans, RG is the receiver gain, and p is the pulse angle in degrees. The index zg stands for the direct pulse experiment and sel for the band-selective experiment. SelFactor stands for the empirically determined factor which corrects for the signal intensity loss due to relaxation differences between the single-pulse and band-selective experiments. SelFactor will typically have a value between 1.0 and 1.3.

Note that the first time this quantification method is used for a new NMR system, the following steps should be performed to determine SelFactor:

(i) Analyze one or more rancid oils (PV > 10 meq/kg) according to the protocol. Record both single-pulse and band-selective pulse spectra:

  (a) Note that this step is to determine SelFactor for your NMR system.

(ii) Create a new spreadsheet file for calculations.

(iii) Retrieve the number of scans (NS) and receiver gain (RG) for both the single-pulse and selective pulse experiments and put them in the spreadsheet as exemplified in Fig. 3:

  (a) If the single pulse is recorded with a pulse angle other than 90°, also include the sine of this angle in the spreadsheet, e.g., by (SIN(RADIANS(30)) in Microsoft Excel for the sinus of a 30° pulse.

(iv) Determine the ox/TG corrected ratio for both spectra according to Eq. 3 (Eqs. 1 and 2 combined and reduced):

$$\text{SelFactor} = \frac{I_{ox}(zg)}{I_{ox}(sel)} \cdot \frac{ns_{sel}}{ns_{zg}} \cdot \frac{RG_{sel}}{RG_{zg}} \cdot \frac{\sin(p_{sel})}{\sin(p_{zg})} \tag{3}$$

4 Notes

1. The sensitivity of the method depends on the strength of the magnetic field, the sensitivity of the probe, and the number of scans used. Cryogenic probe heads can provide significant sensitivity enhancement but are not commonly available. Increasing the magnetic field strength is sometimes an option. The sensitivity (SNR) increase that is obtained from increasing the magnetic field strength (B_0) is $\left(\frac{B_0-\text{High}}{B_0-\text{Low}}\right)^{1.5}$. The number of scans can always be increased, where increasing the number of scans by x will result in an increase in the SNR by the square root of x. For example, going from 16 to 256 scans will increase the SNR by a factor of $\sqrt{(256/16)} = 4$.
2. At temperatures higher than 290 K, the samples in the NMR tube may partially phase separate after prolonged storage (>16 h). This will yield samples that are hard to shim and reduce the resolution of the NMR spectrum. To prevent this, work with small series, manually tumble the NMR tubes every few hours, or use a cooled sample changer. The latter may however not be available in every facility.

References

1. Schaich KM (2020) Lipid oxidation: new perspectives on an old reaction. In: Shahidi F (ed) Bailey's industrial oil and fat products, 7th edn. Wiley
2. Dobarganes MC, Velasco J (2002) Analysis of lipid hydroperoxides. Eur J Lipid Sci Technol 104(7):420–428
3. Barriuso B, Astiasarán I, Ansorena D (2013) A review of analytical methods measuring lipid oxidation status in foods: a challenging task. Eur Food Res Technol 236(1):1–15
4. Antolovich M, Prenzler PD, Patsalides E, McDonald S, Robards K (2002) Methods for testing antioxidant activity. Analyst 127(1): 183–198
5. Skiera C, Steliopoulos P, Kuballa T, Holzgrabe U, Diehl B (2012) 1H NMR approach as an alternative to the classical p-anisidine value method. Eur Food Res Technol 235:1101–1105
6. Skiera C, Steliopoulos P, Kuballa T, Holzgrabe U, Diehl B (2012) 1H-NMR spectroscopy as a new tool in the assessment of the oxidative state in edible oils. J Am Oil Chem Soc 89:1383–1391
7. Merkx DWH, Hong S, Ermacora A, van Duynhoven J (2018) Rapid quantitative profiling of lipid oxidation products in a food emulsion by 1H NMR. Anal Chem 90(7):4863–4870
8. Goicoechea E, Guillen MD (2010) Analysis of hydroperoxides, aldehydes and epoxides by 1H nuclear magnetic resonance in sunflower oil oxidized at 70 and 100 °C. J Agric Food Chem 58(10):6234–6245
9. Guillen MD, Goicoechea E (2009) Oxidation of corn oil at room temperature: primary and secondary oxidation products and determination of their concentration in the oil liquid matrix from 1H nuclear magnetic resonance data. Food Chem 116(1):183–192
10. Guillén MD, Ruiz A (2008) Monitoring of heat-induced degradation of edible oils by proton NMR. Eur J Lipid Sci Technol 110(1): 52–60
11. Boerkamp VJP, Merkx DWH, Wang J, Vincken J-P, Hennebelle M, van Duynhoven JPM (2022) Quantitative assessment of epoxide formation in oil and mayonnaise by 1H-13C HSQC NMR spectroscopy. Food Chem 390: 133145

Chapter 17

Quantification of Secondary Lipid Oxidation Products in Oils, Fats, and Food Products: Malondialdehyde by HPLC-DAD and TBA-RS by UV-Visible Spectrophotometry

Lucie Ribourg-Birault and Anne Meynier

Abstract

This chapter presents an accurate method to prepare, analyze, and quantify secondary lipid oxidation products namely malonaldehyde MDA (as MDA-$(TBA)_2$) and TBA reactive substances (TBA-RS). Each step of the extraction of the targeted compounds, preparation of MDA standard, external calibration, and finally analysis of the reaction media are presented. For the latest, we present two alternative procedures: the first is HPLC paired with a diode array detector (DAD) to specifically quantify MDA and the second is UV-visible spectrophotometry for the quantification of TBA-RS. The limit of quantification is 0.25 μM (250 nM) or 1 nmoles eq MDA g^{-1} lipids.

Key words Malondialdehyde, MDA, TBA-RS, Thiobarbituric acid, Lipid oxidation, HPLC, Spectrophotometry

1 Introduction

Despite decades of research on lipid oxidation, this subject is still present in the literature. The objectives are to get a better understanding of the lipid oxidation mechanisms, to identify and quantify a larger array of oxidation products, and to optimize food processes, storage conditions, and the usage of antioxidants [1]. Classically, among the products formed, a distinction is made between primary products such as peroxides, hydroperoxides, and conjugated dienes and volatile or nonvolatile secondary products, such as aldehydes. These molecules affect the organoleptic qualities of the products and may be deleterious for health [2]. The test of thiobarbituric acid reactive substances (TBA-RS) is one of the oldest and most frequently used method to evaluate oxidation and quantify secondary lipid oxidation products in foods. The reaction of unsaturated aldehydes with 2-thiobarbituric acid (TBA), giving a pink chromophore red at about 532 nm, is known since the late

Christelle Lopez et al. (eds.), *Multidimensional Characterization of Dietary Lipids*, Methods and Protocols in Food Science,
https://doi.org/10.1007/978-1-0716-3758-6_17,

40s. It was initially employed to assess oxidative stress [3] and applied since the early 1960s to quantify secondary products of lipid oxidation in food products [4, 5]. Four major types of procedures exists: test on the whole sample, test on aqueous or acid extract of the sample, steam distillate or on lipid extract [6].

As mentioned above, the principle of measurement of TBA-RS is based on the reactivity of low molecular weight aldehydes, especially malonaldehyde (MDA), with thiobarbituric acid followed by the colorimetric measurement of TBA-RS. The quantification is achieved by applying the specific extinction coefficient of MDA-TBA adduct or by calibration with authentic MDA standard curves. The results are expressed as equivalent MDA, as other aldehydes such as 2,4-alkadienals, or other interfering substances may have participated to the reaction and contributed to the final value. Unlike most TBA-reactive aldehydes, MDA results specifically from the decomposition of the primary oxidation products of polyunsaturated fatty acids containing at least two double bonds or by further decomposition of some aldehydes such as 2,4 alkadienals. To specifically measure MDA, a liquid chromatography separation step can be applied [5].

The method described in this chapter was adapted from previous works [7–10]. It allows quantification of both TBA-RS and MDA from the same reaction medium. The one or the other protocol, or both, can be used depending on the available equipment, resources, and research question.

It is based on the extraction under acidic conditions of MDA and other TBA-RS. Then, the extract is reacted with TBA to form the colored complex (Fig. 1). The reaction mixture can be further analyzed either by high-performance liquid chromatography (HPLC) to quantify specifically the MDA(TBA)$_2$ complex or by spectrophotometry to quantify TBA-RS. The quantification is achieved due to an external quantification with MDA prepared from the acidic hydrolysis of tetra-hydroxy-propane (TEP).

Depending on the samples, noticeable differences (up to threefold) could be observed between MDA assessed by chromatographic methods and TBA-RS assessed by spectrophotometry, particularly for biological samples, as a result of various interferences [11]. When starting a new set of experiments on unknown matrices, it could be of interest to perform both

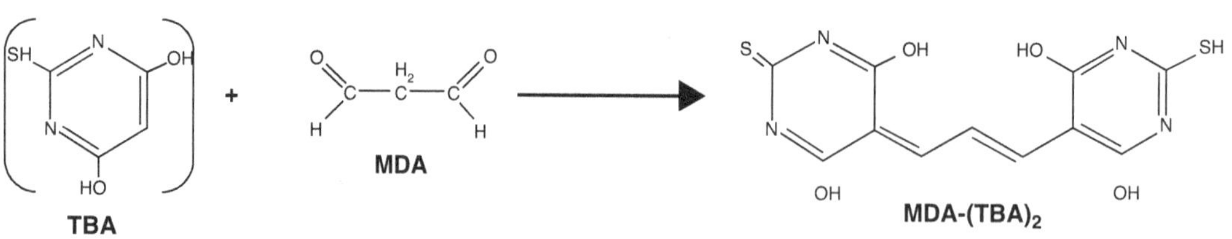

Fig. 1 The chemical structure of MDA and TBA and the reaction between these molecules

determination of MDA after HPLC separation and TBA-RS by UV-visible spectrophotometry to compare the results and then to choose the methods as regard to the research topics and to the lipid composition of the targeted samples. Indeed high oleic content in the samples could lead an underestimation of secondary lipid oxidation products when assessed by MDA only [12].

The described protocol was successfully applied to a wide range of samples and food matrices from oils varying in the PUFA content and composition [13] to in vitro digesta [14, 15], mouse diet and plasma [16], and ham treated at high pressure [17]. The quantification of MDA and/or TBA-RS could be completed, in particular for a better knowledge of the oxidation mechanism, by that of volatile compounds (see Chap. 18). Other molecules can be considered and analyzed such as isoprostanes or prostaglandins (see Chap. 19) or hydroxyalkenals [13] even if their quantification requires specific equipment and highly trained staff. The latter methods could provide relevant information on the relative oxidation of n-3 and n-6 fatty acids.

2 Materials

2.1 Chemical Products

Prepare solutions using ultrapure water (conductivity = 18 MΩ. cm^{-1} at 25 °C) and/or HPLC-grade reagents. Diligently follow safety instructions and waste disposal regulations when disposing waste materials (Table 1).

2.2 Mandatory Equipment

- Analytical precision balance, sensitivity: 0.01 mg.
- Weighing shoes.
- Volumetric flask (10 mL, 20 mL, 100 mL, 200 mL, 250 mL, 500 mL, and 1 L).
- 20 mL glass bottle.
- Pipettes (200 μL, 1 mL, and 5 mL) and cones.
- 1 milliliter graduated glass pipette.
- Transfer pipettes.
- 10 mL (or more) glass screw tube fitted with a cap with Teflon seal.
- Magnetic bar.
- pH meter.
- Water bath capable of heating up to 100 °C.
- Water bath with temperature controlled at 70 ± 1 °C.
- UV-visible spectrophotometer and quartz semimicro cells (optical path length: 10 mm).
- Vortex.

Table 1
Chemicals required for the MDA and TBA-RS methods

Chemical	Chemical formula	CAS number	Molecular weight (g. mol^{-1})	Quality/ reference	Safety instructions
2-thiobarbituric acid (TBA)	$C_4H_4N_2O_2S$	504-17-6	144.15		
Butyled hydroxytoluene (BHT)	$C_{15}H_{24}O$	128-37-0	220.35		
Ethylenediaminetetraacetic acid disodium salt (EDTA)	$C_{10}H_{14}N_2Na_2O_8.2H_2O$	6381-92-6	372.23		
Sodium hydroxide	NaOH	1310-73-2	40.00	Reagent grade	
Trichloroacetic acid (TCA)	Cl_3CCOOH	76-03-9	163.39	ACS reagent	
Chlorydric acid	HCl	7647-01-0	36.46	Reagent grade, 37%	
Sodium phosphate monobasic	NaH_2PO_4	7558-80-7	119.98		
Sodium phosphate dibasic	Na_2HPO_4	7558-79-4	141.96		
1,1,3,3-tetraethoxypropane (TEP)	$(C_2H_5O)_2CHCH_2CH(OC_2H_5)_2$	122-31-6	220.31		

Methanol	CH_3OH	67-56-1	32.04	HPLC grade	
Acetonitrile	C_2H_3N	75-05-8	41.05	HPLC grade	
Ethanol	C_2H_6O	64-17-5	46.07	For analysis	

- Ice.
- Ultrasonic bath.
- Positive displacement pipettes and disposable capillary piston tips.
- Fridge or temperature-controlled room at +4 °C.
- 1 L glass bottle.
- Alternative equipment:
  - Fluorimeter detector.
  - 96 well plates and UV-star® microplates recommended.

2.3 Preparation of Solutions and Reagents

2.3.1 Solutions for Extraction

- 0.3 mol. L^{-1} TCA: Weigh 10 g of TCA in a weighing shoe, introduce it into a 200 mL volumetric flask, and make up to 200 mL with water. Then, store it at 4 °C.
- 1 μg.μL^{-1} BHT: Weigh 10 mg of BHT in a weighing shoe, introduce it into a 10 mL volumetric flask, and make up to 10 mL with ethanol. Then, store it at 4 °C.
- 0.1 mol. L^{-1} EDTA: Weigh 745 mg of EDTA in a weighing shoe, introduce it into a 20 mL volumetric flask, and make up to 20 mL with water. Then, store it at 4 °C.

2.3.2 Reagent

- 0.05 mol. L^{-1} NaOH: Weigh 500 mg of NaOH (pellets, anhydrous) in a weighing shoe, introduce it into a 250 mL volumetric flask, and make up to 250 mL with water. Then, store it at 4 °C.
- 55.5 mmol. L^{-1} TBA reagent: Weigh 800 mg of TBA in a weighing shoe, introduce it into a 100 mL volumetric flask, and make up to 100 mL with NaOH solution. Prepare this reagent solution extemporaneously (*see* **Note 1**).

2.3.3 Preparation and Checking of the Stock Solution of MDA

TEP Hydrolysis

- 0.1 N HCl (*see* **Note 2**): Introduce about 50 mL of water into a 100 mL volumetric flask and add 0.97 mL of HCl 37% with a glass graduated pipette. Make up to 100 mL with water. Then, store it at 4 °C.
- 5 mmol. L^{-1} TEP: Introduce 25 μL of TEP 97% into a 20 mL volumetric flask and make up to 20 mL with HCl solution. Transfer the solution to a bottle and heat to 100 °C for 5 min. The acid hydrolysis of the TEP solution makes it possible to obtain a standard MDA solution with a concentration of 5 mmol. L^{-1} (if all TEP is hydrolyzed). Then, cool and store it at 4 °C (*see* **Note 3**).

Measurement of the MDA Concentration of Stock Solution

After TEP hydrolysis, it is crucial to determine the actual concentration of the MDA standard solution:

- Carry out dilutions of the MDA standard solution in water by 100, 200, and 500.
- Measure the absorbance of the three diluted solutions with an UV-visible spectrophotometer at 245 nm. As an indication, the absorbance is around 0.75 for the solution diluted by 100 and around 0.15 for the solution diluted by 500.
- Calculate the actual MDA concentration (C, mmol.L^{-1}) of the standard solution using the Beer–Lambert law by taking into account the dilutions. Take the average of the three values obtained by the following formula:

$$C = \frac{\text{Abs245nm} \times \text{dilution} \times 1000}{\varepsilon \times l}$$

where ε = Molar extinction coefficient of MDA at 245 nm = 13700 L.mol^{-1}.cm^{-1} [18] and l = optical path of the cell (cm).

- *See* **Note 3** for storage duration and use recommendations.

2.3.4 Preparation of Dilution Standard Solution (0.3 mol. L^{-1} TCA, 0.004 g.L^{-1} BHT, and 0.8 mol. L^{-1} EDTA)

- Weigh 10 g of TCA in a weighing shoe, introduce it into a 200 mL volumetric flask, and add 800 μL of BHT solution and then 1.6 mL of EDTA solution. Make up to 200 mL with water. Then, store it at 4 °C.

2.3.5 Preparation of MDA Calibration Solutions (10 mol.L^{-1} and 0.5 mol. L^{-1})

- From the actual MDA concentration of the stock standard solution, calculate the volume to be withdrawn so as to prepare 10 mL of 10 μmol.L^{-1} MDA solution in a flask. Introduce this volume of solution (in the order of 20 μL) with a suitable pipette and make up to 10 mL with the solution of standard dilution.
- Finally, dilute this solution by 20 with the solution of standard dilution to obtain a solution at 0.5 μmol. L^{-1} (Fig. 2).

3 Method

3.1 Calibration

See Subheading 2.

*3.1.1 Determination of the Concentration of MDA in the Standard Solution (See **Note 3**)*

3.1.2 Preparation of Calibration Tubes

- Prepare solutions of calibration in glass screw tubes fitted with a cap with Teflon seal with a capacity of 10 mL. Each tube is prepared by adding the diluted MDA standard solution at

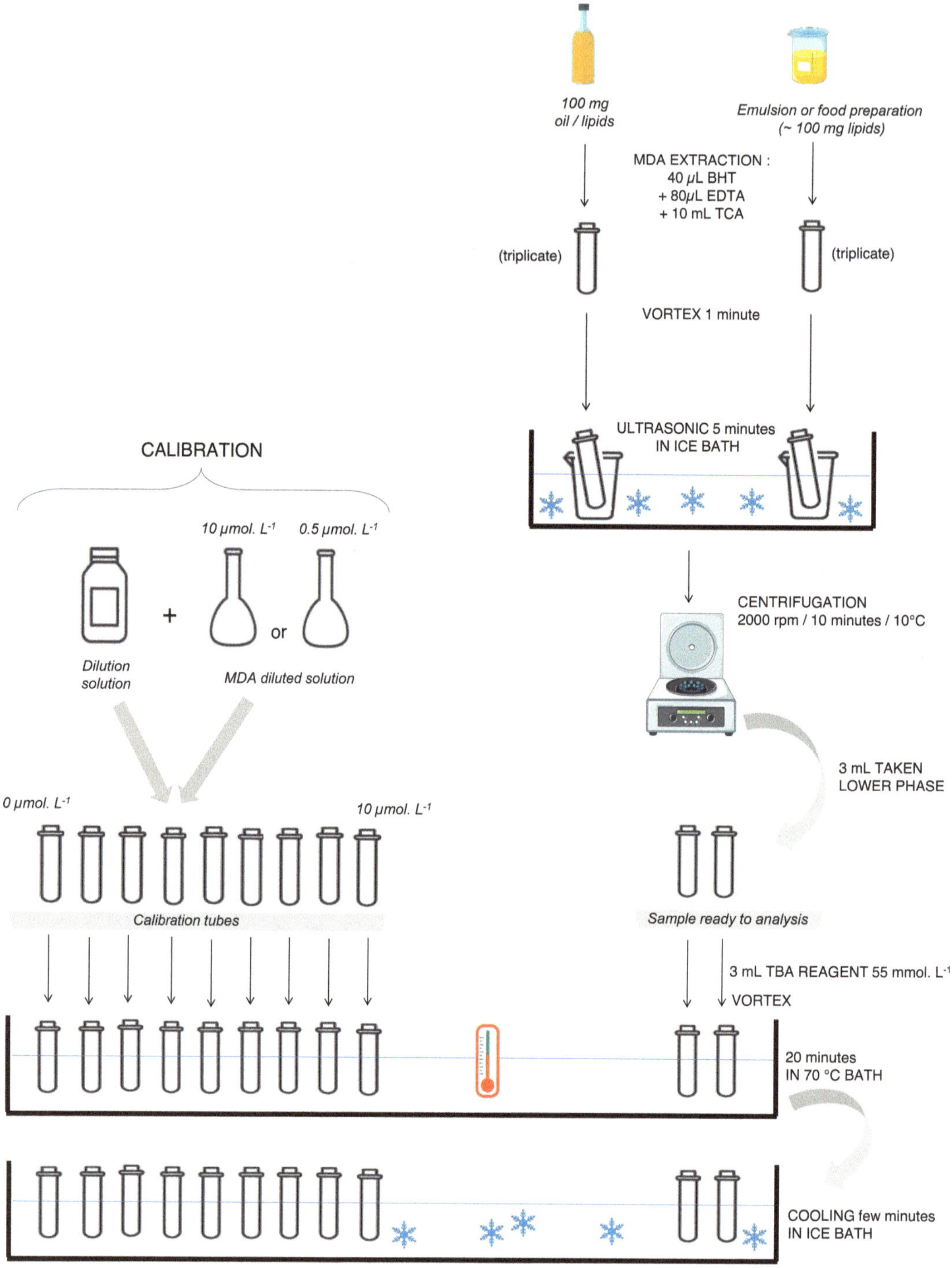

Fig. 2 Flowchart of the main steps in the preparation of MDA standard solutions

Table 2
Preparation of calibration tubes

Volume of reactant (μL)	Calibration tube – Concentration of MDA (μmol. L^{-1}) 0	0.025	0.05	0.1	0.5	1	2	5	7	10
Volume of diluted standard solution 0.5 μmol.L^{-1} (μL)	0	150	300	600						
Volume of diluted standard solution 10 μmol.L^{-1} (μL)	0				150	300	600	1500	2100	3000
Volume of solution of dilution standard (μL)	3000	2850	2700	2400	2850	2700	2400	1500	900	0

10 μmol.L^{-1} or at 0.5 μmol.L^{-1} for the least concentrated points to the solution of standard dilution by following the volumes indicated in Table 2. The total volume is 3 mL per tube and the concentrations range from 0 to 10 μmol.L^{-1} (Fig. 3).

3.2 Samples

3.2.1 Extraction of MDA in Samples

Into a glass screw tube fitted with a cap with Teflon seal, introduce the test sample, then 40 μL of 1 μg.μL^{-1} BHT, 80 μL of 0.1 mol. L^{-1} EDTA, and finally, 10 mL of 0.3 mol. L^{-1} TCA (*see* **Note 4**).

The order of magnitude of the test sample is ideally 100 mg of lipids and is to be adjusted according to the type of sample, its lipid content, and its state of oxidation. For an oxidized sample, this test sample can be reduced. For a sample of low fat content such as flour, up to 5 g of flour to which 10 mL of TCA is added (*see* **Note 5**) can be used. In this case, the concentration of TCA can be increased to 0.45 mol. L^{-1} to promote protein precipitation and allow optimal extraction of substances reactive with thiobarbituric acid (*see* **Note 6**).

Vortex each tube for at least 15 s and place the tubes, previously plunged into an ice bath, in an ultrasonic bath for 5 min.

3.2.2 Preparation of Sample Tubes

Introduce 3 mL of the extract into a new screw tube fitted with a cap with Teflon seal, taking care not to bring the lipid film on the surface or the pellet.

3.3 Reaction Between MDA and Related Substances and TBA Reagent

- To each tube (calibration and sample) add 3 mL of TBA reagent (*see* **Notes 4** and 7).
- Vortex and then immerse the tubes in a water bath at 70 °C for 20 min.

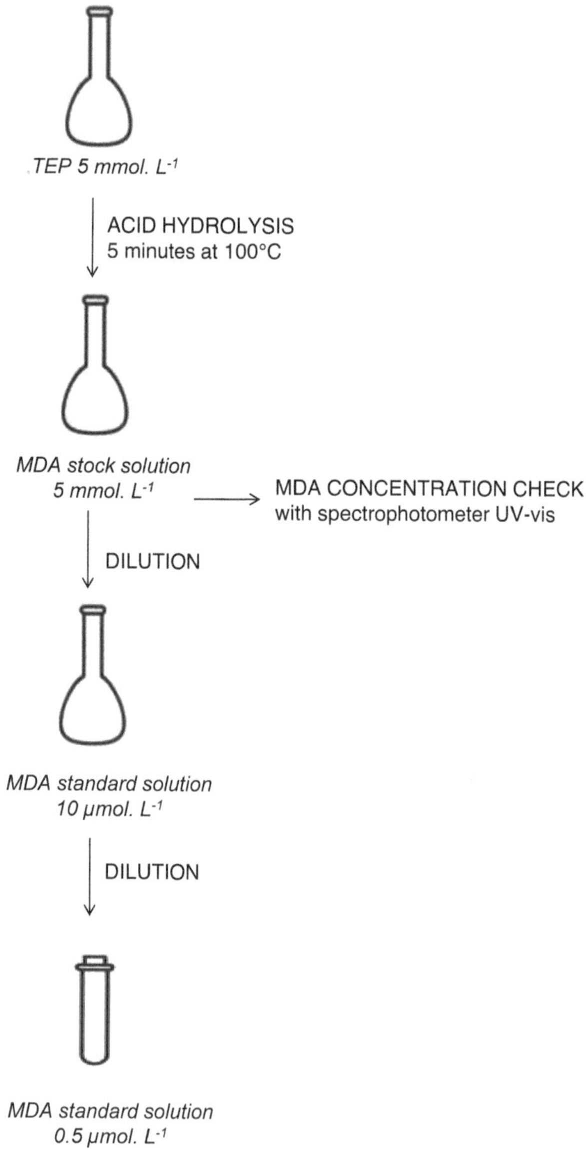

Fig. 3 Flowchart of the main steps in the preparation of calibration tubes and samples for the quantification of MDA and/or TBA-RS in oils, fats, emulsions, and complex food products by HPLC and UV-visible spectrophotometry

- After the ending time, put the tubes in an ice water bath for about 5 min. The calibration tubes are colored from pale pink to more intense pink depending on their MDA and TBA-RS concentrations.

It is important that all the tubes (calibration and sample) are immersed and taken out at the same time so that the reaction time is exactly the same for all.

3.4 Analysis

After the reaction, it is possible to specifically quantify malondialdehyde (MDA) by HPLC paired with a diode array detector or fluorimeter or to measure all TBA-RS by spectrophotometry in cuvettes or on a plate.

It is important to carry out the assays within a few hours of the end of the reaction and to keep the tubes at room temperature; otherwise, crystals may appear in the tubes.

3.4.1 Separation of MDA by HPLC-DAD or HPLC-FLD (See **Note 8**)

1. Mandatory equipment
   - Graduated cylinders.
   - Vacuum flask and hydrophilic filter 0.22 μm.
   - C18 HPLC column, 150 × 4.6 mm and particle size, 5 μm.
   - HPLC with a diode array detector or a fluorescence detector.
   - Syringes and filter membrane PTFE 0.45 μm.
   - HPLC glass vials.
2. Preparation of the mobile phase

   NaH_2PO_4 buffer 50 mmol. L^{-1}, pH 6.8.

   Monobasic solution: Weigh 3 g of NaH_2PO_4 in a weighing shoe, introduce it into a 500 mL volumetric flask, and make up to 500 mL with water.

   Dibasic solution: Weigh 7.1 g of Na_2HPO_4 in a weighing shoe, introduce it into a 1 L volumetric flask, and make up to 1 L with water.

   Measure the pH of the dibasic solution. Then, slowly add the monobasic solution to it under magnetic stirring until it reaches pH 6.8.
3. HPLC analysis:

   Inject 20 μL of each calibration point in the HPLC apparatus equipped with an isocratic solvent delivery system (*see* **Note 9**) and a C18 (150 cm × 4.6 mm ID, 5 μm) column maintained at 20 °C. The mobile phase is pH 6.8 sodium phosphate buffer 50 mmol. L^{-1}/methanol/acetonitrile (72/17/11, v/v/v) at a flow rate of 1 mL.min^{-1}. The temperature of sampler is also set at 20 °C (*see* **Note 10**). The detector can be a diode array detector (DAD) with the wavelength set at 532 nm or a fluorescence detector (FLD) with the excitation wavelength set at 535 nm and the emission wavelength set at 555 nm.

   The chromatogram shows one peak corresponding to the MDA-$(TBA)_2$ complex (Figs. 4 and 5).

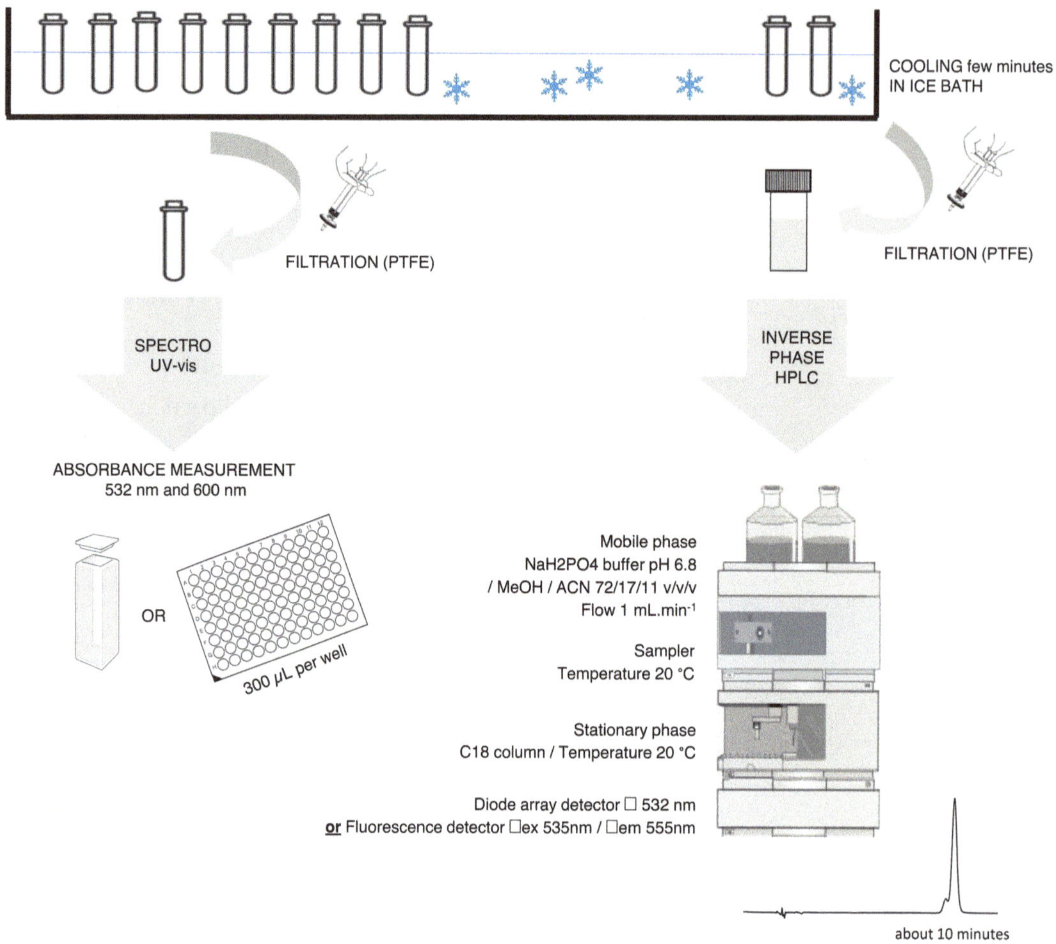

Fig. 4 Flowchart of the main analytical steps of standards and samples for the quantification of MDA and TBA-RS in oils, fats, emulsions, and complex food products by HPLC and UV-visible spectrophotometry

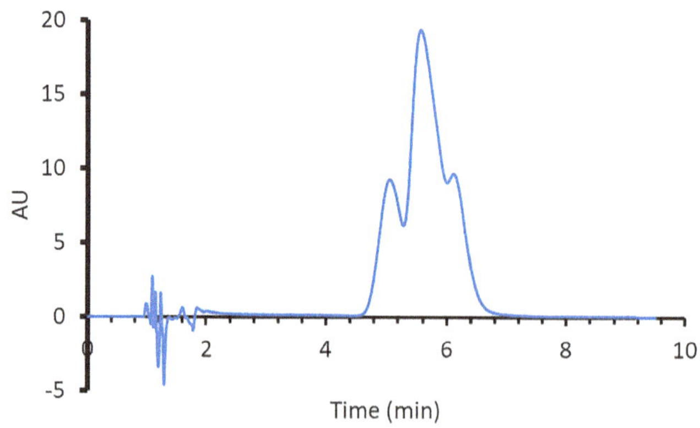

Fig. 5 Example of the chromatogram of the most concentrated calibration tube (concentration, 10 μmol.L^{-1}) with DAD set at 532 nm

3.4.2 Analysis of TBA-RS by UV-Visible Spectrophotometry

This spectrophotometric assay can be carried out in a sample-by-sample tank (glass cuvettes) or on a plate. We recommend to use UV-star® microplates which are adapted to the acidity of the solutions analyzed (*see* **Note 11**).

The absorbance of the calibration points is measured at 532 and 600 nm, which corresponds to the baseline in order to subtract the possible turbidity of the sample.

3.5 Quantification of MDA and TBA-RS in Samples

3.5.1 Determination of the Sample MDA Content by HPLC-DAD or HPLC-FLD

Integrate the peak of the MDA complex in each chromatogram (calibration tube or sample) and record the values of the areas.

There is a linear relationship between the MDA concentration of the injected solutions and the area of the MDA peak. Establish the calibration curve $y = ax + b$, where y is the peak area and x is the concentration of MDA of injected solution in $\mu mol.L^{-1}$.

With these parameters, calculate the concentration of MDA (x in the formula) in the reaction tubes in $\mu mol.\ L^{-1}$ and then calculate the number of moles of MDA in the samples, *n*, using the formula

$$n = \text{concentration of MDA} \times \text{extraction volume}$$

where the concentration of MDA in the sample is expressed in $\mu mol.L^{-1}$ and the extraction volume in L, which corresponds to the volume of the sample, BHT solution, EDTA solution, and TCA solution for taking into account the dilution of MDA during the extraction step.

In the case of varied samples with different lipid contents, for interpreting the data, it is generally relevant to express the results both in nmol MDA per gram of sample and per gram of lipids.

$$\text{MDA content} = \frac{n \times 10^3}{\text{mass of sample or lipids in the sample}}$$

where the MDA content is expressed in $nmol.\ g^{-1}$, n in µmol, and the mass of the sample or lipids in mg.

The calibration curve exhibited a linear fitting from 0.1 and 10 $\mu mol.L^{-1}$, which corresponds to a limit of quantification of 10 $nmol.g^{-1}$ lipids. If a lower value is expected (in the case of very slightly oxidized samples), it is possible to carry out a specific calibration between 0 and 0.1 $\mu mol.L^{-1}$ and the linearity is validated in this specific zone. The limit of quantification drops accordingly to 1 $nmol.g^{-1}$ lipids.

3.5.2 Determination of Sample TBA-RS Content by UV-Visible Spectrophotometry

Establish the calibration curve $y = ax + b$, where y is the absorbance at 532 nm–absorbance at 600 nm and x is the concentration of MDA of the injected solution in $\mu mol.L^{-1}$.

With these parameters, calculate the content of TBA-RS in the samples in equivalent MDA, $\mu mol.L^{-1}$. The value can then be expressed in nmol eq. MDA per gram of suspension or nmol eq. MDA per gram of lipid.

The linearity with 1 cm pathway cuvettes is validated between 0.25 and 10 $\mu mol.L^{-1}$ which corresponds to a limit of quantification of 20 $nmol.g^{-1}$ lipids. If a lower value is expected (in the case of fresh or slightly oxidized samples), it is possible to carry out a calibration range between 0 and 0.25 $\mu mol.L^{-1}$. In that case, the linearity is validated in this specific zone and the limit of quantification drops to 1 nmol. g^{-1} lipids (as for the HPLC-DAD method).

When using plates, the linearity is validated between 0 and 10 $\mu mol.L^{-1}$ which corresponds to a limit of quantification of 1 $nmol.g^{-1}$ lipids.

4 Notes

1. 100 mL of TBA reagent allows the preparation of the calibration tubes and triplicates for seven samples.
2. For safety reasons, it is recommended to introduce the acid into water and not the other way around.
3. The stock solution can be stored at least for 6 months at 4 °C after preparation but its MDA concentration must be checked every 2 months and before any large series of analyses. This regular verification is essential to guarantee the accuracy of the standard range and the reliability of the quantitative results.
4. For these steps, we recommend to use a dispenser pipette.
5. Take care to choose a tube with a volume adapted to the final volume (sample test portion + 40 μL of BHT solution + 80 μL of EDTA solution + 10 mL of TCA).
6. In the case of these low-fat samples, a significant turbidity can be observed in the extraction tube if the TCA concentration is too low.
7. After the reaction, the volume of the medium is 6 mL. This volume is necessary to be able to filter it and get enough filtrate for the HPLC MDA assay and spectrophotometry TBA-RS assay by UV-visible spectrophotometry in cuvettes. If a single analysis is planned, for example, the determination of TBA-RS on a plate, it is possible to reduce the quantities by adding only 1 mL of TBA reagent to 1 mL of extract or calibration solution.
8. For the detection and quantification of MDA by HPLC, two types of detectors can be used: a diode array detector or a fluorescence detector. The limit of quantification can change depending on the type of detector used.
9. The injection volume can be changed (e.g., it can be reduced if the area of the MDA peak is outside the calibration range) but be careful that this parameter is then taken into account in the calculations.

10. After the reaction, the samples are maintained at room temperature and then at 20 °C in the HPLC sample compartment to prevent crystallization of the complex.
11. Only the UV-star® plates have been tested. If you wish to use another type of plate, we advise you to test them within the calibration range.

Acknowledgement

We would like to give special thanks to Hernan-Brice Kenmogne-Domguia for his work on the development of the method for the determination of malondialdehyde by HPLC carried out during his Master 2 training in 2009 and then widely used during his thesis. The initial setup of the presented methods have been realized ANR (French National Research Agency) program "Alimentation et industries alimentaires—Food and food industry": ANR-08-ALIA-002 AGEcaninox project.

References

1. Schaich KM (2013) Challenges in elucidating lipid oxidation mechanismswhen, where, and how do products arise? In: Logan A, Nienaber U, Pan XS (eds) Lipid oxidation. Challenges in food systems. AOCS Press, Urbana, pp 1–52
2. Schaich KM (2013) Challenges in analyzing lipid oxidation: are one product and one sample concentration enough? When, where, and how do products arise? In: Logan A, Nienaber U, Pan XS (eds) Lipid oxidation. Challenges in food systems. AOCS Press, Urbana, pp 53–128
3. Tsikas D (2017) Assessment of lipid peroxidation by measuring malondialdehyde (MDA) and relatives in biological samples: analytical and biological challenges. Anal Biochem 524: 13–30. https://doi.org/10.1016/j.ab.2016.10.021
4. Frankel EN (2005) Lipid oxidation, 2nd edn. The Oily Press, Bridgwater
5. Shahidi F, Zhong Y (2005) Lipid oxidation: measurement methods. In: Bailey's industrial oil and fat products. Wiley, Hoboken
6. Hoyland DV, Taylor AJ (1991) A review of the methodology of the 2-thiobarbituric acid test. Food Chem 40:271–291. https://doi.org/10.1016/0308-8146(91)90112-2
7. Botsoglou NA, Fletouris DJ, Papageorgiou GE et al (1994) Rapid, sensitive and specific thiobarbituric acid metrhod for measuring lipid peroxidation in animal tissue, food and feedstuff samples. J Agric Food Chem 42:1931–1937
8. Grau A, Guardiola F, Boatella J et al (2000) Measurement of 2-thiobarbituric acid values in dark chicken meat through derivative spectrophotometry: influence of various parameters. J Agric Food Chem 48:1155–1159
9. Seljeskog E, Hervig T, Mansoor MA (2006) A novel HPLC method for the measurement of thiobarbituric acid reactive substances (TBARS). A comparison with a commercially available kit. Clin Biochem 39:947–954
10. Mendes R, Cardoso C, Pestana C (2009) Measurement of malondialdehyde in fish: a comparison study between HPLC methods and the traditional spectrophotometric test. Food Chem 112:1038–1045
11. Dasgupta A, Klein K (2014) Methods for measuring oxidative stress in the laboratory. Antioxidants Food Vitam Suppl:19–40. https://doi.org/10.1016/b978-0-12-405872-9.00002-1
12. Frankel EN (1998) Lipid oxidation. The Oily Press, Dundee
13. Viau M, Genot C, Ribourg L, Meynier A (2016) Amounts of the reactive aldehydes, malonaldehyde, 4-hydroxy-2-hexenal, and 4-hydroxy-2-nonenal in fresh and oxidized edible oils do not necessary reflect their peroxide and anisidine values. Eur J Lipid Sci Technol

118:435–444. https://doi.org/10.1002/ejlt.201500103

14. Kenmogne-Domguia HB, Meynier A, Boulanger C, Genot C (2012) Lipid oxidation in food emulsions under gastrointestinal-simulated conditions: the key role of endogenous tocopherols and initiator. Food Dig 3: 46–52. https://doi.org/10.1007/s13228-012-0026-9
15. Kenmogne-Domguia HB, Moisan S, Viau M et al (2014) The initial characteristics of marine oil emulsions and the composition of the media inflect lipid oxidation during in vitro gastrointestinal digestion. Food Chem 152:146–154. https://doi.org/10.1016/j.foodchem.2013.11.096
16. Awada M, Soulage CO, Meynier A et al (2012) Dietary oxidized n-3 PUFA induce oxidative stress and inflammation: role of intestinal absorption of 4-HHE and reactivity in intestinal cells. J Lipid Res 53:2069–2080. https://doi.org/10.1194/jlr.M026179
17. Rakotondramavo A, Ribourg L, Meynier A et al (2019) Monitoring oxidation during the storage of pressure-treated cooked ham and impact on technological attributes. Heliyon 5. https://doi.org/10.1016/j.heliyon.2019.e02285
18. Esterbauer H, Schaur RJ, Zollner H (1991) Chemistry and biochemistry of 4-hydroxynonenal, malonaldehyde and related aldehydes. Free Radic Biol Med 11:81–128

Chapter 18

Analysis of Secondary Volatile Oxidation Products by Headspace GC-MS Techniques

Charlotte Jacobsen

Abstract

Lipid oxidation leads to the formation of secondary volatile oxidation products, which can be determined by a range of analytical techniques. This chapter will provide an introduction to different headspace gas chromatography-mass spectrometry (GC-MS) techniques and will discuss their advantages and disadvantages. Moreover, a detailed description of the headspace GC-MS technique used in the author's laboratory is provided. This includes details on headspace sampling conditions for different food matrices (oil, oil-in-water emulsions, mayonnaise, fish and muscle food, and fish feed plus fish meal samples). The use of different multivariate data analytical techniques for statistical analysis of volatile oxidation products is also discussed.

Key words Lipid oxidation, Volatile oxidation products

1 Introduction

Lipid oxidation leads to the formation of secondary volatile oxidation compounds such as aldehydes, ketones, alcohols, acids, and hydrocarbons. These compounds are responsible for the formation of undesirable off-odors and off-flavors as a result of lipid oxidation [1]. Some aldehydes such as malondialdehyde, hydroxy-trans-2-nonenal (HNE), and 4-hydroxy-trans-2-hexenal (HHE) have potential carcinogenic properties and should be avoided in high concentrations [2–4]. Oxidative flavor deterioration can be determined by sensory analysis, chemical analysis, or a combination of both. In case both methods are used, it is possible to correlate the results from the two methods, e.g., by multivariate data analytical techniques as will be discussed in this chapter. The most comprehensive method for analysis of secondary volatile oxidation compounds is currently the collection of volatile compounds by a headspace technique followed by the analysis of the collected volatile compounds by gas chromatography coupled with a mass

Christelle Lopez et al. (eds.), *Multidimensional Characterization of Dietary Lipids*, Methods and Protocols in Food Science,
https://doi.org/10.1007/978-1-0716-3758-6_18,

spectrometer (GC-MS). A large number of volatile oxidation compounds can be identified and quantified by this method. For example, Hartvigsen et al. identified and quantified 148 compounds in fish oil-enriched mayonnaise [5]. When using headspace GC-MS techniques, it is not necessary to extract the lipids from the food matrix before analysis. This is an advantage compared to other simpler methods such as determination of aldehydes by spectrophotometric analysis of anisidine values or other more advanced methods such as ^{1}HNMR [6] (see also Merkx et al. (Chap. 16)). Another advantage of the headspace GC-MS technique is its ability to identify the chemical structure of individual volatile compounds, which cannot be obtained to the same degree by other methods such as determination of anisidine values or TBARS. This is particularly important in studies, which aim at understanding the oxidation reaction pathways. The biggest drawbacks of headspace GC-MS techniques are as follows: (1) the time-consuming data analysis if more than a few compounds are quantified; (2) only few costly deuterated standards are available, meaning that quantification usually has to be done without deuterated standards, which adds some uncertainty to the absolute concentrations obtained by this method; and (3) the complexity of the GC-MS equipment, which requires skilled personnel to operate. The lack of deuterated standards is particularly a problem in complex systems, where it can be difficult to obtain exactly the same recovery of external standards added for quantification as for endogenously formed lipid oxidation products even if the compounds are exactly the same.

Different techniques can be used to collect the volatile oxidation products (static headspace, solid phase micro-extraction (SPME), and manual and automated dynamic headspace). This chapter will provide a detailed description of the manual dynamic headspace (purge and trap) GC-MS method including data treatment. This will be followed by a short comparison of the different headspace techniques, including a discussion of their advantages and drawbacks.

2 Materials for Manual Dynamic Headspace (Purge and Trap) Analysis

This section provides an overview of the materials and equipment used in the author's laboratory. Alternative materials and equipment can be used, particularly for those materials/equipment in the list below for which it is emphasized that they are examples.

2.1 Chemicals and Detergents

- Distilled or ion-exchanged water.
- Dish washing detergent without citrus aromatics.
- Ethanol 96% p.a.

Table 1
Examples of sampling conditions

	Fish	Fishmeal and feed	Oil	10% oil-in-water emulsion
Purge temperature	37 °C	45 °C	60 °C	45 °C
Sample amount	10.0 g	0.5 g	4.0 g	4.0 g
Water	25.0 mL	15.0 mL	–	5.0 mL
Antifoam	1.0 mL		–	1.0 mL
Flow	340 mL/min	150 mL/min	150 mL/min	150 mL/min
Collection time	30 min	15 min	30 min	30 min
Flow of final flush	50 mL/min	50 mL/min	–	50 mL/min
Time of final flush	20 min	20 min	–	20 min

- Internal standard (IS), e.g., 4-methyl-1-pentanol, 30 μg/g in water or rape seed oil, depending on the type of sample to be analyzed.
- Control standard, e.g., 4-methyl-1-pentanol, 1.5 mg/g ethanol.
- Liquid nitrogen—only for muscle food and products hereof.
- Antifoam (e.g., Synperonic) 400 μL/500 mL water—only if necessary (see Table 1).
- Potassium hydroxide p.a.—only for products containing vinegar (e.g., mayonnaise).
- Glass wool—only for products containing vinegar (e.g., mayonnaise).
- Standards for the calibration curves (different aldehydes, ketones, alcohols, furans, etc.).
- Nitrogen purge gas quality: 5.0.

2.2 Headspace Collection Equipment for Purge and Trap Methodology

- Purge bottle (100 mL pear-shaped bottle) (other dimensions can also be used).
- Purge head with a 1/4″ fitting for the Tenax tubes (see Fig. 1).
- Tenax tubes: Perkin Elmer 1/4″ stainless steel tubes, packed with 225 ± 3 mg Tenax GR$^{(R)}$ 60–80 mesh, closed with Teflon caps with O-ring. Other Tenax materials can also be used.
- Water bath.
- Sphere flowmeter.
- 1–2 stop watches.
- Analytical balance (for weighing standards).
- Technical balance (for weighing samples).
- Whirlmixer.

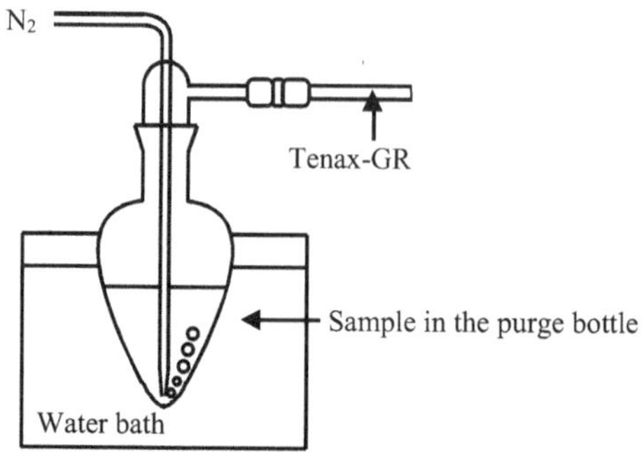

Fig. 1 System for the collection of volatiles by manual dynamic headspace (purge and trap). When the s-tube is used, it is inserted before the Tenax tube using the same types of fittings as shown above

- Drying oven for glass items (temp: 140 °C).
- 5μl Hamilton syringe.
- A regeneration oven for the Tenax-tubes is also recommended if many samples are to be analyzed, but Tenax tubes can also be regenerated in most thermal desorption units (see Subheading 4.1).

2.2.1 Additionally for Muscle Foods

- Dewar container.
- Cut-off Eppendorf syringe as a weighing tool.
- Waring blender.

2.2.2 Additionally for Samples Containing Vinegar (e.g., Mayonnaise)

- Glass-S-tubes.

3 Dynamic Headspace (Purge and Trap) GC-MS Analysis Method

3.1 Preparation Step Before the Collection of the Volatiles

Tenax tubes should be regenerated after each run and before the first use. Most thermal desorption units will have a feature where this can be done. Otherwise it can be done in a separate regeneration oven designed for this purpose. A temperature of 340 °C for 1 h is recommended. After regeneration, Tenax tubes should be stored in a desiccator until use. Right before use, a short regeneration of the Tenax tube (3 min at 340 °C) is recommended.

3.1.1 Packing of S-Tubes (Only for Products Containing Vinegar, e.g., Mayonnaise) [5]

Heat a mortar in the drying oven. Crush approximately 10 KOH pearls in the mortar—use gloves. Weigh approximately 100 mg KOH and use a small funnel to pour it down into the S-tube. Knock the glass tube carefully to make the KOH fall down to the bottom. Now put a small amount (2 mg) of Chrompack glass wool down into the tube. Mount the S-tube carefully between the purge head and the Tenax tube and make sure that the glass wool faces the Tenax tube. The S-tube can be stored in the desiccator until use.

Cleaning of S-Tubes After the Collection of Volatiles

First, remove the glass wool and then flush the S-tubes on the inside with hot water and then soak them separately, which means not together with the purge head and the bottle.

3.2 Collection of Volatiles from the Sample Products

Due to the volatility of the substances, it is important to quickly handle the samples and if necessary store them on ice. Run the analysis as a triple determination. Remember to write down the weighing numbers.

Switch on the water bath (see the temperature in Table 1). Take the glass equipment out of the drying oven and let it cool to room temperature.

The collection procedure needs to be optimized for different types of products. Some examples are given below.

3.2.1 Fish and Other Muscle Food Products

Cut the sample into squares (approximately 1 cm × 1 cm × 1 cm) and freeze them in liquid nitrogen for at least 1 min. The entire sample must be covered with liquid nitrogen. Now blend the frozen pieces in a small steel container with the Waring blender until all the sample has been pulverized.

Weigh approximately 10 g of the powdered sample (4 decimal places) in a 100 mL pear-shaped bottle (purge bottle) using a cut-off Eppendorf syringe.

Weigh approximately 30 mg internal standard (4 decimal places) in the purge bottle. Insert a plug into the bottle. Add 25 mL of water and antifoam as needed to the bottle and shake it using the whirl mixer to mix the powder with water.

Continue as described in Subheading 3.2.5.

3.2.2 Fishmeal and Fish Feed

Pulverize the fish feed in the blender.

Weigh approximately 0.5 g fishmeal/feed (4 decimal places) in a pear-shaped bottle (purge bottle).

Weigh approximately 30 mg internal standard (4 decimal places) in the purge bottle. Insert a plug into the bottle. Add 15 mL of water to the bottle and shake it using the whirl mixer to mix the fishmeal with water.

Continue as described in Subheading 3.2.5.

3.2.3 Oil

Weigh approximately 4 g of the sample (4 decimal places) in a 100 mL pear-shaped bottle (purge bottle) using an automatic pipette. Add approximately 30 mg of internal standard (4 decimal places) to the bottle and insert a plug into it.

Continue as described in Subheading 3.2.5.

3.2.4 10% Oil-in-Water Emulsions

Weigh approximately 4 g sample (4 decimal places) in a 100 mL pear-shaped bottle (purge bottle) and add 5 mL of water, and in most cases 1 mL of antifoam is also needed. Use an automatic pipette.

Adding water to the emulsion will increase the polarity of the matrix and thereby the release of the volatiles will be more efficient. This will increase the sensitivity of the method for most volatiles.

Add approximately 30 mg internal standard (4 decimal places) to the bottle and insert a plug into it.

Continue as described in Subheading 3.2.5.

3.2.5 Procedure for All Products

Open the nitrogen flow on the main tap (approximately 3 bars) and roughly adjust the flow with the sphere flowmeter (see Table 1).

Mount the purge head in the outlet of the sphere flowmeter on the gas supply. Make sure that the purge head faces the right direction (see Fig. 1) and take care not to break the glass tube.

Mount the Tenax tube on the purge head. Face the tube with the two grooves toward the glass tube and press it in entirely. In case that mayonnaise is analyzed, place the S-tube between the purge head and the Tenax tube (face the S-tube with glass wool toward the Tenax tube).

Neither write on the Tenax tubes nor put a label on them. You may mark them with clips that can be fastened to the tubes.

Mount the purge bottle on the purge head with a clip, place the bottle in the water bath, and start the stop watch. Use the flowmeter to check that the system is tight.

For each product, a specific flow (±10 mL/min) in a specific time interval must be used during the collection (see the collection conditions in Table 1).

Check all the joints in case the flow is too low—even if the sphere flowmeter is correctly set, especially critical places such as the mountings of the ground stopper and the collection tube.

Beware of changes in flow if more than one sphere flowmeter is opened during the collection.

3.3 Disassembling the Equipment

When the collection time is over, it is important to disassemble the equipment in the right order.

First, disassemble the Tenax tube and close it with the white plugs. Thereafter, disassemble the bottle. If the test material on the tip of the purge head and water drops under the "collar" can be wiped off with a piece of paper towel, the purge head can be used for the subsequent sample. Dismount the purge head while there is still nitrogen flow in the system.

3.4 Removing Water from Tenax Tubes

For some samples (see Table 1), it is necessary to finally flush nitrogen through the tube after the collection to remove water. For that purpose, place the Tenax tube directly on the outlet of the ball flowmeter in the same direction as that during the collection, which means the grooves face toward the ball flowmeter. Then, the Tenax tube is ready for further analysis by GC-MS.

3.5 Cleaning the Bottles

Rinse the bottles thoroughly. Use liquid soap if necessary before cleaning in a laboratory dish washer. Alternatively, the bottles can be cleaned manually. After washing, rinse them with ethanol.

Wash the purge heads and fittings thoroughly, rinse with ion-exchanged water, and finally with ethanol from a spray bottle.

After rinsing with ethanol, place all parts in the drying oven for at least 60 min at 140 °C. To avoid an ethanol peak in the chromatogram, dry overnight.

3.6 Control Standard

To follow the performance of the GC-MS system, it is recommended to analyze a control standard for each test series. This is done by pipetting 1 μL of the standard directly to a Tenax tube followed by flushing for 5 min using a flow of 50 mL/min. The peak area obtained for the control standard should be plotted on a control chart, which provides an overview of the peak areas obtained for the control standard for each batch of analyses per month and per year.

3.7 Calibration Curve

Prepare a standard mixture (and dilutions thereof) of selected reference substances in appropriate solvent depending on the expected/identified oxidation products and the product.

For GC analysis, there are basically two options for making a calibration curve, namely by using the principle of standard addition to a sample or by adding the standards directly on the Tenax tube. The latter method is mostly used when it is difficult to obtain homogeneous mixing of the standards with the sample (e.g., fish samples).

3.7.1 Calibration Curve as a Standard Addition

The calibration curve is analyzed as a "standard addition," i.e., a specific weight of all standard mixtures is added to a matrix representing the sample.

12.5 mg/g stock solution in 96% ethanol or rapeseed oil:[1]

Weigh approximately 40 mg (4 decimal places) of each selected standard in one 4 mL screw cap glass. *Include 4-methyl-1-pentanol (internal standard) in the mixture.* Weigh ethanol or rapeseed oil in the same glass to a total weight of 3.2 g. Note the weight (4 decimal places) of standards and ethanol/rapeseed oil as the weight will be used for the construction of the calibration curve.

As an example, the following procedure can be used. Make the following dilutions in 4 mL screw cap glasses (in the same solvent as stock solution). Weigh approximately an amount shown in Table 2

[1] Select rapeseed oil when the matrix (samples) has high oil content. Select ethanol for other samples.

Table 2
Examples of concentrations used for calibration curves for standard addition

Conc. (μg/g)	250	100	50	25	10	5	1	0,5
Weigh from:	Stock sol.	250	250	250	100	100	10	10
mg standard mixture	60	800	400	200	200	100	200	100
mg ethanol/rapeseed oil	2940	1200	1600	1800	1800	1900	1800	1900

Table 3
Examples of concentrations used for calibration curves for pipetting directly into the Tenax tubes

Conc. (mg/mL)	2.0	1.0	0.5	0.2	0.1	0.05	0.01	0.005
Pipette from:	Stock	Stock	Stock	2.0	2.0	2.0	0.1	0.1
mL standard mix	1.60	0.80	0.40	0.40	0.20	0.10	0.40	0.20
mL ethanol	2.40	3.20	3.60	3.60	3.80	3.90	3.60	3.80

and note down the weight (4 decimal places) of both the standard mixture and ethanol/rapeseed oil.

Select 5–7 appropriate dilutions for the calibration curve, depending on the concentration in the samples.

Weigh approximately 30 mg (4 decimal places) standard mixture in a purge bottle in which a matrix representing the sample has already been weighed. Note that no internal standard must be added because it is already in the standard mixture. Perform triple determinations in the same way as for the samples. Analyze the Tenax tubes (standards) using the same GC-MS method as the samples.

Store the stock solution and dilutions in a freezer (−20 °C) in 4 mL screw cap glasses.

3.7.2 Standard Mixtures Are Pipetted Directly into Tenax Tubes

5 mg/mL stock solution in 96% ethanol:

Weigh approximately 50 mg (4 decimal places) of each selected standard in one 10 mL measuring flask. *Include 4-methyl-1-pentanol (internal standard) in the mixture.* Fill up to the mark with ethanol.

Make the dilutions in 96% ethanol in 4 mL screw cap glasses, as shown in Table 3.

Select 5–7 appropriate dilutions for the calibration curve, depending on the concentration in the samples.

Pipette 1 μL standard mixture directly into Tenax tubes.[2] Perform triple determinations. Flush the Tenax tubes with N_2 for

[2] Start with the lowest concentration and rinse the syringe with the next concentration.

5 min at a flow of 50 mL/min. Analyze the Tenax tubes (standards) with the same GC-MS method as the samples.

Store the stock solution and dilutions in a freezer (−20 °C) in 4 mL screw cap glasses.

4 Analysis of the Collected Volatile Compounds Using GC-MS

4.1 Equipment for the Analysis of Volatile Compounds

- Gas chromatograph linked with MS detector. Alternatively, GC with a flame ionization detector (FID) can be used, but this will make identification of peaks less trustworthy. There is also a problem with coeluting compounds as they cannot be distinguished and correctly identified by GC-FID.
- Automatic thermal desorption system (ATDS) with autosampler (different brands are available).
- GC column: e.g., J&W Scientific DB-1701 or DB Wax (both 30 m, I.D. 0.25 mm, film 0.5 μm).

4.2 Procedure

Volatile compounds are transferred from the Tenax tube to the GC-MS system with an ATDS. Briefly, the steps are as follows: The Tenax tubes are placed in the ATDS, which heats up the tubes to 200 °C. With a flow of helium (99.9995%) at 60 mL/min the volatiles are transferred to a cold trap (temperature: −30 °C), where the volatiles are concentrated. Then, the cold trap is rapidly heated to 220 °C and with a flow of helium the volatiles are transferred into the GC where they are separated on the column according to their polarity and mass. Thereafter, they are transferred into the MS. In the MS, the volatile compounds are fragmented by, for example, electron ionization, whereafter the mass of each fragment is determined by the mass detector. Thereby, a spectrum of each compound can be obtained, which can be used for the identification of the compound, as described later. Figure 2

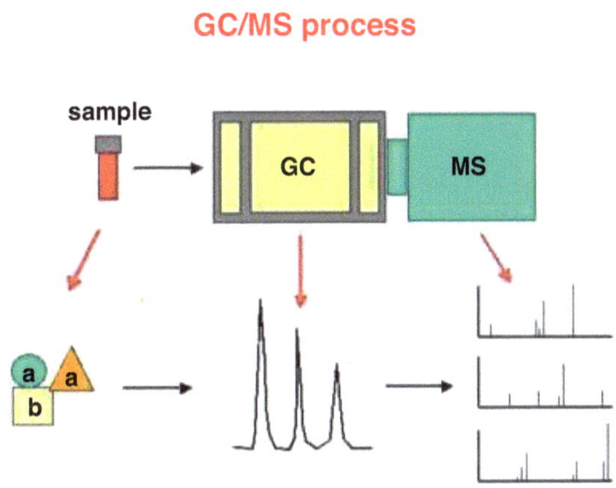

Fig. 2 The overall principle of GC-MS analysis

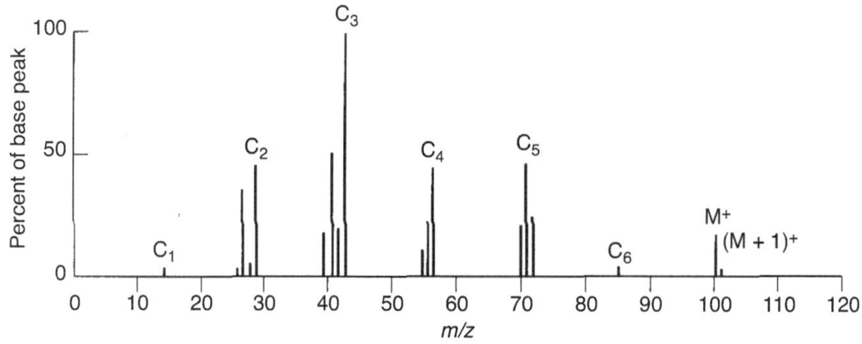

Fig. 3 The spectrum of n-hexanal: $CH_3CH_2CH_2CH_2CH_2CHO$, Mw = 100.16 g/mol

shows the overall principle of GC-MS analysis. As an example, the MS spectrum of hexanal is shown in Fig. 3.

4.2.1 GC-MS Conditions

GC-MS conditions must be optimized to obtain the best possible separation and quantification of volatile compounds. Some examples of different GC oven programs are given below using a DB1701 column (30 m × ID 0.25 mm × 0.5 μm film thickness, J&W Scientific, Folsom, CA, USA) and helium gas flow (1.3 mL/min). In all cases, the MS settings used in our laboratory are as follows: electron ionization mode, 70 eV, mass-to-charge ratio (m/z) scan between 30 and 250 amu. In the author's lab, we usually do not scan below 30 amu whereby we avoid undesirable signals from water molecules which may have entered the system. In cases where very short-chain volatile compounds are searched for, the scan can be changed to search for smaller masses.

GC Oven Program

Examples of different oven programs:

Milk samples with 0.5% fish oil [7]: The initial temperature is set to 45 °C for 5 min, then the temperature is increased by 1.5 °C/min to 55 °C, and again increased by 2.5 °C/min to 90 °C. Finally, it is increased by 12 °C/min to 220 °C and maintained for 4 min.

Fish oil samples [7]: The initial temperature is set to 35 °C for 3 min, then the temperature is increased by 3 °C/min to 120 °C, and again increased by 7 °C/min to 160 °C. Finally, it is increased by 15 °C/min to 200 °C and maintained for 4 min.

Mayonnaise samples with 80% oil [5]: The temperature is set to 35 °C for 5 min, increased at 1.5 °C/min to 55 °C, at 2.5 °C/min to 90 °C, and at 15 °C/min to 220 °C, and kept at 220 °C for 4 min.

Fish samples [8]: The temperature is set to 40 °C for 1 min, 40–175 °C at 4 °C/min, 175–240 °C at 20 °C/min, and finally maintained at 240 °C for 7 min.

4.2.2 Identification and Quantification

Strategy for Selection of Compounds for Quantification

Different strategies can be followed in lipid oxidation research with respect to analysis of secondary volatile oxidation products. Some laboratories choose only to quantify and report one or two compounds, e.g., hexanal in the case of samples rich in omega-6 fatty acids and propanal for samples rich in omega-3 fatty acids [9]. However, the use of dynamic headspace GC-MS offers the possibility to quantify and report many more compounds, which can provide a more complete picture of the oxidation reactions taking place. This may be of particular importance in the case of complex food systems, where lipid oxidation products can further react with proteins, amino acids, or phospholipids and give rise to the formation of Strecker aldehydes and nonenzymatic browning products [10]. In addition, some studies have shown that antioxidants can reduce the formation of some volatiles and at the same time increase the formation of others [11]. This effect is overlooked if only one compound is analyzed. Furthermore, to correlate results from sensory profiling with results obtained from GC-MS analysis of volatile oxidation products it is necessary to determine more than one volatile compound. Examples of such an approach are discussed in Subheading 5.

Identification of Volatile Compounds

Volatile oxidation compounds can be tentatively identified by their mass spectra using the MS library, which is integrated with the MS system. Confirmation of the identity of the volatile compounds can be done by comparison of the obtained retention times with those of external standards including those used for the calibration curves. For FID, volatiles are identified by the retention times of external standards only.

For the analysis of sample types that the laboratory have not analyzed before or for the analysis of the effect of antioxidants, it is recommended to make a visual comparison of the chromatograms of the different samples and select all peaks, which differ markedly between the samples for identification. If standards are available for these compounds, they should be quantified (see below).

For the analysis of sample types, which the lab has experience with, it is recommended to select the 10–15 most relevant compounds for quantification.

Quantification of Volatile Compounds

To quantify the selected compounds, calibration curves should be constructed. The first step is to perform the dynamic headspace sampling of selected standards, as described in Subheading 3.7. Thereafter, the Tenax tubes are analyzed using the same GC-MS conditions as for the samples. Then, calibration curves are constructed using the MS software. Subsequently, concentrations of the samples are calculated with the help of the calibration curves and the MS software. Specific procedures for these two steps depend on the MS software used. Concentrations are usually calculated in ng per g of sample. Therefore, it is important to note down the precise amount of sample used for the headspace collection as this number is used for this purpose.

It is necessary to carefully check that the right peaks have been identified by the MS software and that the peaks have been properly integrated by the software. This is a time-consuming procedure, which includes steps such as evaluation of the relative standard deviation of three measurements of the same sample. In most cases, the relative standard deviation is around or below 10%, but in some cases it can be higher. In our lab, we allow a relative standard deviation of up to 20% for a few sampling points for a few compounds if none of the replicates are obviously wrong and if no reason for the high standard deviation can be found. In case of storage experiments, another check point is whether the development during storage looks reasonable. Assume that a storage experiment of 10 days with sampling points at day 0, 4, 7, and 10 is carried out. If, for example, one sample has a much lower concentration of one compound after day 7 than after days 4 and 10 and if this is not the case for other compounds, the results obtained after 7 days should be checked again to ensure that the peak has been correctly identified and integrated.

5 Data Analysis

Data obtained from the headspace GC-MS analysis can either be analyzed by traditional statistical analysis or by multivariate data analysis techniques. Depending on the experimental design, differences between samples for individual volatile compounds can be determined by one-way or two-way analysis of variance (ANOVA) followed by a post-test to determine the rank order (e.g., the Tukey or Duncan tests). For storage experiments, ANOVA followed by the Bonferroni test is appropriate for ranking samples at a specific time point and ranking the same sample code at different time points.

If more than a few volatile compounds have been analyzed in a storage experiment, multivariate data analysis is recommended. The most simple multivariate data analysis is a principal component analysis (PCA) of all data, which can be used to provide an overview of all data. A PCA can also be used to show different development patterns of volatiles depending on the sample. When performing a PCA on data from a storage experiment, data can be organized in two different ways: in a short thick matrix or in a long thin matrix. For example, assume that a storage experiment with six different sample codes and five sampling times have been performed and 10 volatile compounds have been analyzed. In a short thick matrix, the six samples are the objects and the 10 volatiles × 5 sampling points will give 50 variables (Fig. 4a). This will result in a PCA

A

	Var no. 1			*Var no. 5*		*Var no. 45*		*Var no.50*
	Vol1_T1	Vol1_T2	....	Vol1_T5	...	Vol10_T1	...	Vol10_T5
Sample 1								
Sample 2								
...								
Sample 6								

B

		Vol 1	Vol 2	....	Vol 10
Object no. 1	Sample 1_T1				
Object no. 2	Sample 1_T2				
	...				
Object no. 5	Sample 1_T5				
	...				
Object no. 25	Sample 6_T1				
	...				
Object no. 30	Sample 6_T5				

Fig. 4 Illustration of the data structure for multivariate data analysis in (**a**) Short thick matrix with 6 samples (objects) and 50 variables (300 data points), and (**b**) Long thin matrix with 30 objects and 10 variables (300 data points). The values of the data points are not shown. Vol Volatile compound, T Time point 1, and Var Variable. In **b**, there are only 10 variables. In **a**, there are only 6 objects

where the effect of the storage time for each volatile compound is visualized in the loadings plot. In a long thin matrix, the 6 samples × 5 sampling points will give 30 objects and the 10 volatiles will give 10 variables (Fig. 4b). This will result in a PCA where the effect of storage time for each sample across the 10 volatiles is visualized in the scores plot. Both methods have their advantages and whether to choose one or the other *depends* on the outcome of the experiment. For example, if all volatiles have the same development pattern during storage it may be an advantage to choose the long thin matrix as this will make it easier to visualize the differences between samples. On the other hand, the short thick matrix makes it easier to visualize different development patterns of the volatile compounds during storage.

More advanced multivariate methods namely, different types of partial least squares regression (PLSR) are available. In the conventional PLSR method, one set of experimental data, e.g., instrumental data are used as the X-variables and another set of experimental data, e.g., sensory data are used as Y-variables. The data can be structured in a similar way as shown in Fig. 4a, b.

In the so-called ANOVA PLSR, experimental design variables are used as X-variables and the experimental data, e.g., data for the volatile compounds as Y-variables. The experimental design variables are entered as so-called category variables. For example, if three different antioxidants in two different concentrations have been tested, an antioxidant category variable with three levels and a concentration category variable with two levels can be used. Likewise, a category variable for storage time can be introduced. Similar to a traditional ANOVA, an ANOVA PLSR makes it possible to determine whether the effect of an experimental variable (e.g., the type of antioxidant or concentration of the antioxidant) is significant or not. This can be done for all the measured variables, e.g., volatile compounds, included in the model. An example of this approach can be found in [12].

The ANOVA PLSR method is a "hard" regression that provides information about the experimental design only. Thus, it is not able to differentiate between noise and unexplainable variation in the data and the latter will be regarded as noise. In the so-called discriminant PLSR (DPLSR), the design variables are used as Y-variables and the experimental data as X-variables. This method is a "soft" regression, which provides information about the structure in the empirical (experimental) data in addition to information about the design. Thus, this method is able to differentiate between noise and unexplainable variation in the data. If the variation in the empirical data is caused by the design employed, this will be reflected in the model [13]. An example of the use of DPLSR can be found in [14]. The author's experience is that if the same data set is analyzed by both methods and if the two models obtained show the same pattern in the data, which is often the case, then the model is strong and can be trusted.

6 Comparison of Different Methods for Collecting Volatile Oxidation Products

As mentioned previously, different methods can be used to collect volatile compounds from a sample matrix. The simplest method is static headspace. In this method, the sample is poured into a vial leaving some headspace above the sample. Then, the vial is closed and the sample is equilibrated for some time at a specific temperature before a needle penetrates the septum and collects the volatiles in the headspace. Thereafter, the volatiles are injected into the GC. The advantage of this method is that it is simple to perform, can be automated, and has a good sensitivity for short-chain aldehydes. The disadvantage is lack of sensitivity for longer-chain aldehydes. Thus, usually it cannot be used for the analysis of aldehydes with a longer chain length than C7.

For SPME, the sample is also left in a vial until the equilibrium of volatile compounds between sample and headspace has been obtained. The SPME fiber is then introduced into the headspace of the vial and stays there until the equilibrium has been reached between headspace and fiber or till a well-defined time period where a certain percentage of uptake of volatile compounds has been reached [15]. After extraction, the fiber is removed from the vial and inserted into the GC-injector and the volatiles are thermally released into the GC. The SPME method can also be automated, which is one of the advantages of this method. A disadvantage is that some volatile compounds may interfere with the absorption and cause a reduction in other volatile compounds extracted from the sample [16]. This is because some compounds have a higher affinity for the fiber coating, and thereby the interfering volatile compounds are present in higher concentration in the fiber than in the sample. The same problem was also observed in a study by [17], who studied lipid oxidation reactions during storage in emulsions prepared with two types of marine phospholipids. They found that the formation of high concentrations of a Strecker aldehyde (3-methylbutanal) in one of the emulsions at the end of the storage period led to a huge underestimation of concentrations of other volatile oxidation products in that emulsion. The same problem was not encountered when the emulsions were analyzed by manual dynamic headspace GC-MS. The problem with SPME is that adsorption is a competitive process where the composition of the matrix and the extraction conditions can affect the amount of individual volatile compounds extracted. Another challenge with SPME is that different batches of fibers provide different results when used for the analysis of the same samples. Therefore, a new calibration curve must be made if one changes from one batch of fibers to another batch in the middle of an experiment.

It is also possible to automate the dynamic headspace method. In this case, the sample is poured into a vial and the volatile compounds are extracted and concentrated on the dynamic headspace station located on the auto sampler unit. In this case, only the headspace of the sample is purged with nitrogen, and not the sample matrix as is the case in the manual dynamic headspace method illustrated in Subheading 2. However, the volatile compounds are collected on an adsorbing material, which can be the same as the Tenax material described above. The sample and the adsorbent are thermostated and the vial is agitated in the DHS station. After collection, residual water is removed by purging with nitrogen through a drying vial. Hereafter, the adsorbent is transferred and injected into the GC–MS system through a thermal desorption unit. Distances (the number of transfers and storage time) that volatile compounds have to pass from the sample to injection are very short compared to the manual dynamic headspace method described in Subheading 2. Another advantage is that the risk of losing volatile compounds during transfer in the automated dynamic headspace unit is lower than that in the manual method.

Thomsen et al. [7] compared the automated and manual dynamic headspace collection methods with that of SPME for three different sample types (oil, milk, and emulsions). The same

GC-MS conditions were used for each of the products after the different collection methods.

For oil samples, the linearity range and the detection limit for volatile standards collected with the three methods varied depending on the chain length of the volatile standard. When collected with the manual dynamic headspace, it was possible to detect volatile standards with a short chain length (up to C6) when diluted to approximately 15 ng/g. With a long chain length (C7–C9), detection was not possible with the low concentrations. Thus, the detection limit was higher for these volatile standards. In contrast to the manual dynamic headspace, SPME and the automated dynamic headspace could detect volatile standards with a chain length up to C8–9 in concentrations from 50 ng/g. However, for volatile standards with a chain length above C8 and more than one double bond for all three methods, the detection limit increased dramatically. The linearity range of the volatile standards diluted in oil was the smallest for a short chain length and the largest for a long chain length using all three collection systems. The automated dynamic headspace method was found to have the largest linearity area compared to the other two methods irrespective of the chain length of the volatile standard.

Similar to the study by Lu et al [17], some challenges were also observed for solid-phase microextraction, namely competition problems for space on the SPME fiber. Therefore, it is recommended mainly to apply solid-phase microextraction on simple matrices and to be cautious with more complex matrices such as enriched milk and highly oxidized oils.

7 Conclusions

Lipid oxidation of foods leads to the formation of a large number of volatile oxidation products. Headspace GC-MS provides several advantages for the analysis of volatile oxidation products compared to simpler methods based on spectrophotometric analysis. Different headspace techniques are available and each of them has their advantages and drawbacks. Therefore, the choice of headspace technique and the conditions used for collecting the volatile oxidation products should be carefully considered. Combining GC with an MS detector instead of a flame ionization detector (FID) results in a more certain identification of each volatile compound and enables quantification of compounds that are overlapping in the GC-chromatogram. GC-FID cannot distinguish between compounds that are overlapping and therefore erroneous results may be obtained by this method. Dynamic headspace or SPME coupled with GC-MS analysis offers the possibility of identifying and quantifying many volatile compounds. Multivariate data analytical techniques such as PCA, PLSR, ANOVA-PLSR, and

Discriminant-PLSR are useful statistical tools, which also provide a visual overview of the patterns in the data. This can make it easier to interpret the results compared to univariate statistical analysis of the data obtained for each individual compound.

References

1. Frankel EN (2005) Lipid oxidation, 2nd edn. PJ Barnes and Associates, Bridgwater
2. Guillén MD, Goicoechea E (2008) Toxic oxygenated α,β-unsaturated aldehydes and their study in foods: a review. Crit Rev Food Sci Nutr 48:119–136
3. Nieva-Echevarría B, Goicoechea E, Guillén MD (2020) Food lipid oxidation under gastrointestinal digestion conditions: a review. Crit Rev Food Sci Nutr 60:461–478
4. Esterbauer H, Schaur RJ, Zollner H (1991) Chemistry and biochemistry of 4-hydroxynonenal, malonaldehyde and related aldehydes. Free Rad Biol med 11:81–128
5. Hartvigsen K, Lund P, Hansen LF, Hølmer G (2000) Dynamic headspace gas chromatography/mass spectrometry characterization of volatiles produced in fish oil enriched mayonnaise during storage. J AgricFood Chem 48: 4858–4867
6. Merkx DWH, Hong GTS, Ermacora A, van Duynhoven JPM (2018) Rapid quantitative profiling of lipid oxidation products in a food emulsion by 1H NMR. Anal Chem 90:4863–4870
7. Thomsen BR, Yesiltas B, Sørensen A-DM, Larsen DB, Glastrup J, Jacobsen C (2016) Comparison of three methods for extraction of volatile lipid oxidation products from food matrices for GC-MS Analysis. J Am Oil Chem Soc 93:929–942
8. Baron CP, Svendsen GH, Lund I, Jokumsen A, Nielsen HH, Jacobsen C (2013) Organic vegetable ingredients in the diet of Rainbow trout (Oncorhynchus mykiss). Impact on fish muscle composition and oxidative stability. Eur J Lipid Sci Technol 115:1367–1377
9. Nuchi CD, McClements DJ, Decker EA (2001) Impact of Tween 20 hydroperoxides and iron on the oxidation of methyl linoleate and salmon oil dispersions. J Agric Food Chem 49
10. Lu FSH, Bruheim I, Haugsjerd BO, Jacobsen C (2014) Effect of temperature toward lipid oxidation and non-enzymatic browning reactions in krill oil upon storage. Food Chem 157: 398–407
11. Timm-Heinrich M, Xu X, Nielsen NS, Jacobsen C (2003) Oxidative stability of structured lipids produced from sunflower oil and caprylic acid. Eur J Lipid Sci Technol 105:436–448
12. Sørensen A-DM, Baron CP, Let MB, Brüggemann D, Pedersen LRL, Jacobsen C (2007) Homogenization conditions affects the oxidative stability of fish oil enriched milk emulsions: oxidation linked to changes in the protein composition at the oil-water interface. J Agric Food Chem 55:1781–1789
13. Martens M, Martens H (1986) Partial least squares regression. In: Piggott JR (ed) Statistical procedures. Elsevier Applied Science Publ., UK, pp 293–359
14. Jacobsen C, Hartvigsen K, Lund P, Thomsen MK, Skibsted L, Adler-Nissen J, Hølmer G, Meyer AS (2000) Oxidation in fish oil enriched mayonnaise: 3. Assessment of the influence of the emulsion structure on oxidation by discriminant partial least squares regression analysis. Eur Food Res Technol 211:86–98
15. Zhang Z, Pawliszyn J (1993) Headspace solid-phase microextraction. Anal Chem 65:1843–1852
16. Górecki T, Yu X, Pawliszyn J (1999) Theory of analyte extraction by selected porous polymer SPME fibres. Analyst 124:643–649
17. Lu FSH, Nielsen NS, Jacobsen C (2013) Comparison of two methods for extraction of volatiles from marine PL emulsions. Eur J Lipid Sci Technol 115:246–251

Chapter 19

LC-MS/MS Analysis of Isoprostanoids in Edible Oils, Vegetables, and Animal Tissues

Claire Vigor, Bingqing Zhou, Guillaume Reversat, Valérie Gros, Amandyne Linares-Maurizi, Anis Elloumi, Valérie Bultel-Poncé, Alexandre Guy, Camille Oger, Jean-Marie Galano, and Thierry Durand

Abstract

Nonenzymatic oxidation of polyunsaturated fatty acids (NEO-PUFA) generates isoprostanoids. These lipid peroxidation products are formed in vivo, *post mortem*, and during food processes, through lipid autoxidation. They exhibit biological activities and could be indicators or markers of the oxidation status of foodstuff. This chapter describes a simple method for measuring isoprostanoids in edible oils and vegetable or animal tissues using the LC-MS/MS approach. Protocols have been applied in our laboratory to measure the levels of isoprostanoids in various products; examples of data obtained on microalgae and different seeds are presented in annexes.

Key words Lipid peroxidation, NEO-PUFA, Isoprostanoids, Liquid chromatography-tandem mass spectrometry

1 Introduction

Free radical-induced lipid oxidation is implicated in a number of human diseases [1]. The products resulting from this oxidative process, in particular, isoprostanoids, are considered excellent biomarkers. Isoprostanoids can be classified into different categories, depending on the polyunsaturated fatty acid from which they are derived, namely, phytoprostanes/phytofurans coming from α-linolenic acid (ALA) oxidation, isoprostanes/isofurans of the series F2 coming from arachidonic acid (AA) oxidation, isoprostanes/isofurans of the series F3 coming from eicosapentaenoic oxidation (EPA), neuroprostanes/neurofurans derived from docosahexaenoic acid (DHA) oxidation but also the lesser known metabolites dihomo-IsoPs and dihomo-IsoFs derived from adrenic acid (AdA) or NeuroPs from the series F3 derived from docosapentaenoic acid (DPA) (*see* Tables 1 and 2). Many studies have

Christelle Lopez et al. (eds.), *Multidimensional Characterization of Dietary Lipids*, Methods and Protocols in Food Science,
https://doi.org/10.1007/978-1-0716-3758-6_19,

Table 1
Some nonenzymatic oxidized metabolites classified on the basis of parent ω-3 PUFAs

Parent 3 PUFAs	Isoprostanoids compounds	Isofuranoid compounds
Linolenic acid (ALA)	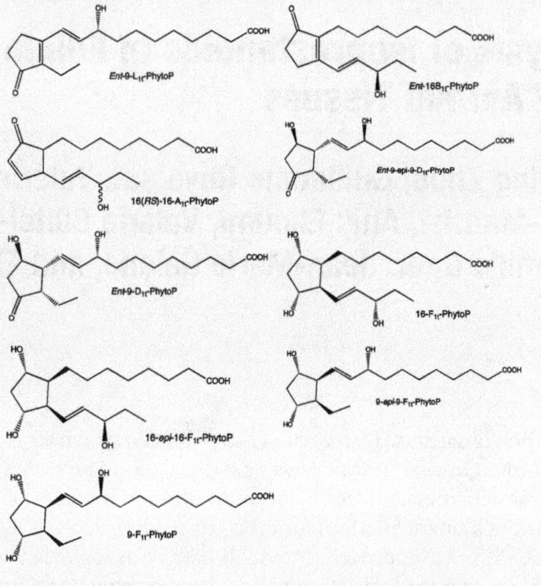 *Ent*-9-L_{1t}-PhytoP; *Ent*-16B_{1t}-PhytoP; 16(*RS*)-16-A_{1t}-PhytoP; *Ent*-9-*epi*-9-D_{1t}-PhytoP; *Ent*-9-D_{1t}-PhytoP; 16-F_{1t}-PhytoP; 16-*epi*-16-F_{1t}-PhytoP; 9-*epi*-9-F_{1t}-PhytoP; 9-F_{1t}-PhytoP	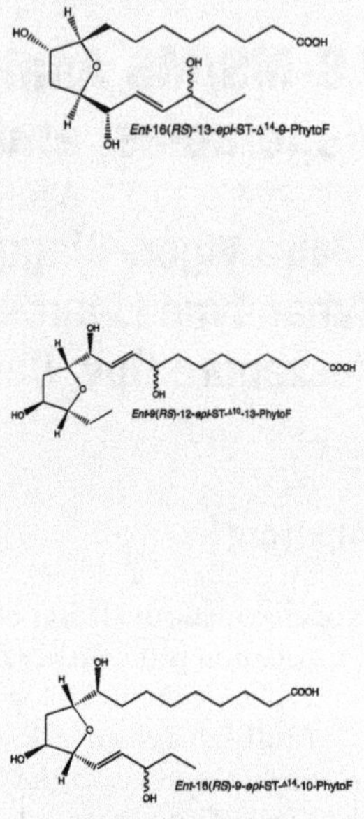 *Ent*-16(*RS*)-13-*epi*-ST-Δ^{14}-9-PhytoF; *Ent*-9(*RS*)-12-*epi*-ST-Δ^{10}-13-PhytoF; *Ent*-16(*RS*)-9-*epi*-ST-Δ^{14}-10-PhytoF
Eicosapentaenoic acid (AdA)	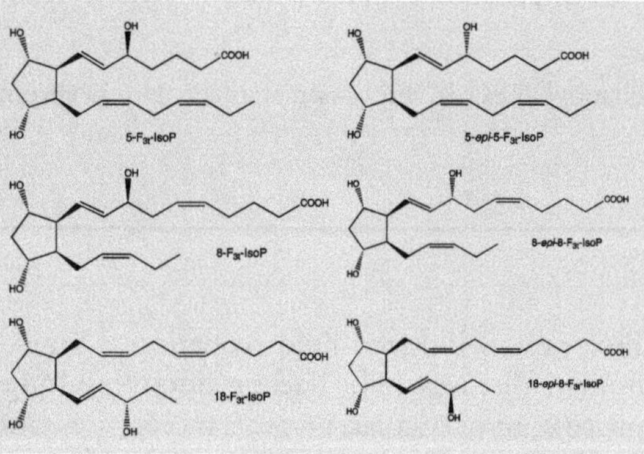 5-F_{3t}-IsoP; 5-*epi*-5-F_{3t}-IsoP; 8-F_{3t}-IsoP; 8-*epi*-8-F_{3t}-IsoP; 18-F_{3t}-IsoP; 18-*epi*-8-F_{3t}-IsoP	
Docosahexaenoic acid (DPA)	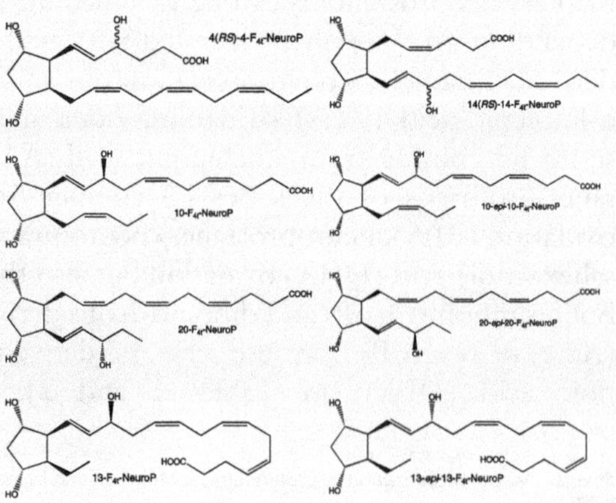 4(*RS*)-4-F_{4t}-NeuroP; 14(*RS*)-14-F_{4t}-NeuroP; 10-F_{4t}-NeuroP; 10-*epi*-10-F_{4t}-NeuroP; 20-F_{4t}-NeuroP; 20-*epi*-20-F_{4t}-NeuroP; 13-F_{4t}-NeuroP; 13-*epi*-13-F_{4t}-NeuroP	

Table 2
Some nonenzymatic oxidized metabolites classified on the basis of parent ω-6 PUFAs

Parent 6 PUFAs	Isoprostanoids compounds	Isofuranoid compounds
Arachidonic acid (AA)	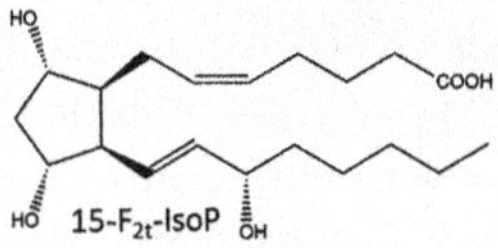 15-F_{2t}-IsoP 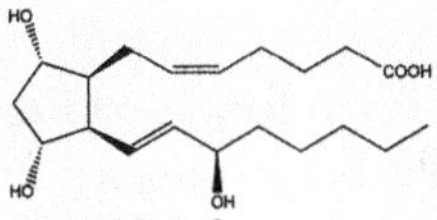 15-*epi*-15-F_{2t}-IsoP 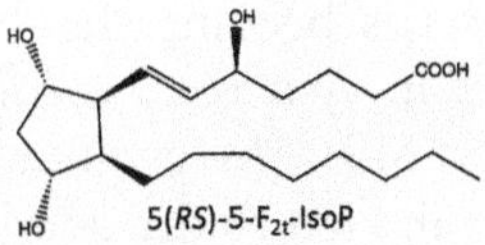 5(*RS*)-5-F_{2t}-IsoP 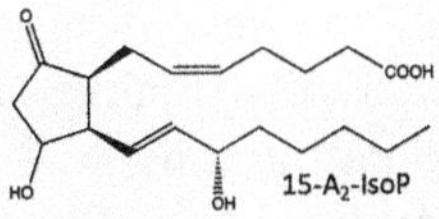 15-A_2-IsoP 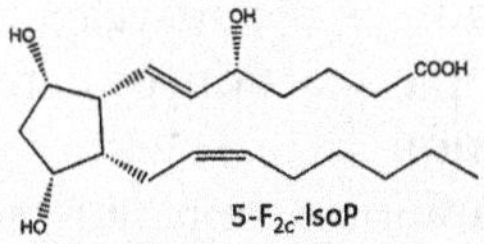 5-F_{2c}-IsoP	
Adrenic acid (AdA)	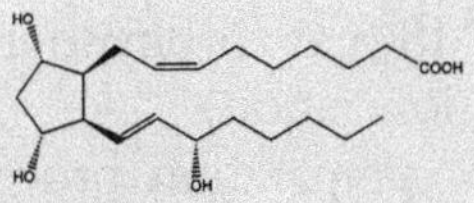 17-*epi*-17-F_{2t}-dihomo-IsoP 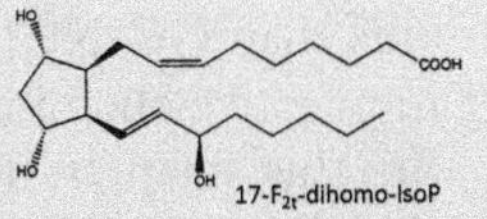 17-F_{2t}-dihomo-IsoP 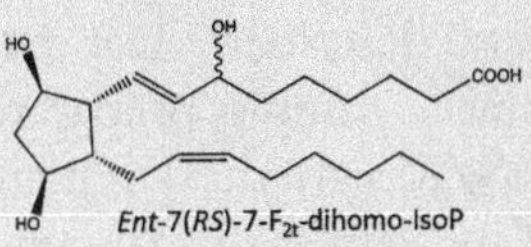 *Ent*-7(*RS*)-7-F_{2t}-dihomo-IsoP	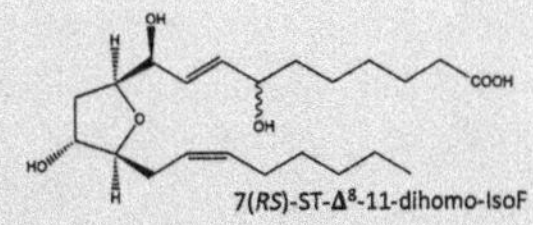 7(*RS*)-ST-Δ^8-11-dihomo-IsoF 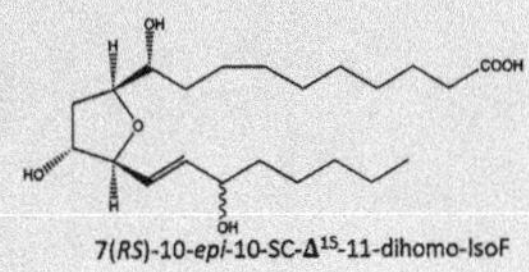 7(*RS*)-10-*epi*-10-SC-Δ^{15}-11-dihomo-IsoF
Docosapentaenoic acid (DPA)	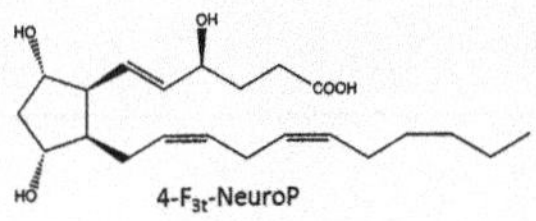 4-F_{3t}-NeuroP 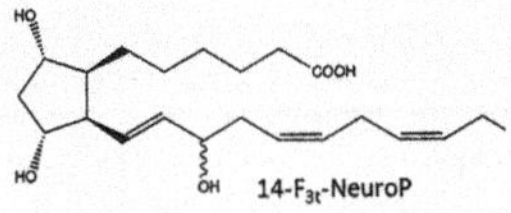 14-F_{3t}-NeuroP	

demonstrated that such compounds are the most reliable available biomarkers of lipid peroxidation in vivo, and recent work examining isoprostanoids formation has provided valuable information about the pathogenesis of numerous human diseases [2, 3]. Oxidative stress does not only occur in animals during the development of pathologies. Numerous studies have shown that plants can also be subject to a change in oxidative status and more particularly to radical attack under the effect of environmental stress for instance (water, heat, or pollution). By assimilation, it is therefore conceivable to consider isoprostanoids as good markers of the oxidative status of tissues of every living organism. These compounds can also be produced after vegetable harvesting or animal death and during the processes that convert living organisms into foods as the radical reaction including lipid autoxidation continues to take place. We are thinking in particular of dietary ingredients and foods containing high amounts of the precursors of isoprostanoids, i.e., PUFAs. Even if no study has yet used these products to evaluate the oxidation status of foodstuff, their measurement would be probably very relevant, considering furthermore their demonstrated biological activities in the inflammatory, neurological, or cardiovascular fields. Knowing the contents of these metabolites in food products to constitute an index of biological interest could be promising for food quality control [4, 5].

Measurement by mass spectrometry-based methods of the levels of these derivatives has become the "gold standard" biomarker of oxidative stress in vivo [6, 7]. Initially studied in gas chromatography (GC-MS/MS), the evolution of liquid chromatography (LC) techniques has led to the development of LC-MS/MS quantification methods in recent years. Notably, the use of LC-MS/MS in the multiple reaction monitoring (MRM) mode has received great attention [8]. Together with the less time-consuming sample preparation compared to GC-MS, LC-MS/MS analysis gives more information related to the identity of compounds than GC-NCI-MS (gas chromatography-mass spectrometry in negative chemical ionization).

In this chapter, the preparation procedure and analysis by LC-MS/MS of isoprostanoids [9] in edible oils and vegetable or animal tissue samples are described. Depending on the type of starting material, a suitable sample preparation protocol is presented, with the LC-MS/MS method remaining unchanged (*see* Fig. 1). Some results of measurements obtained on edible oils and seeds are presented in the annexes (*see* Figs. 3 and 4).

2 Materials

Our analytical process only differs regarding the preparation of the extract to be analyzed by LC-MS/MS depending on the material studied (edible oils, vegetables, and animal tissues). Thus, in the following paragraphs, we present the chemicals and equipment necessary for the treatment of oil, vegetal tissues, or animal tissues and then those needed for the LC-MS/MS analysis.

2.1 Chemicals

Prepare all solutions using ultrapure water. All solvents should be at least of high-performance liquid chromatography (HPLC) grade. All solvents prepared should be stored at room temperature. Phosphate buffer and chloroform are placed at 4 °C.

Table 3 presents the chemicals required to prepare the lipid extracts from oils, animal tissues, and vegetable tissues.

Table 4 presents the solvents required to analyze isoP by LC/MS-MS.

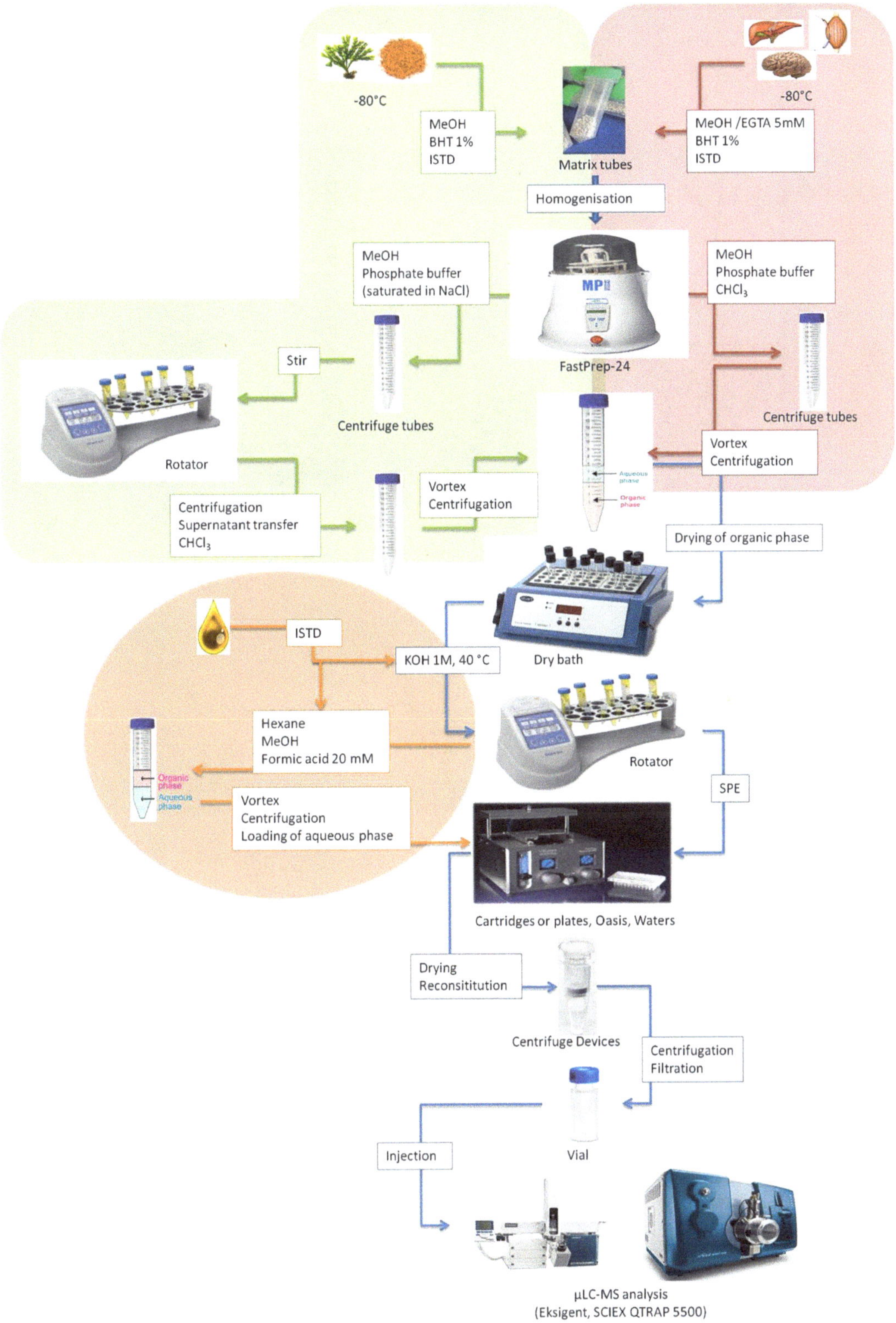

Fig. 1 Overview of the experimental procedure highlighting the sample preparation of oil, vegetable, or animal tissues: green arrows indicate sample preparation part before SPE for oil, red arrows show steps for animal tissue, orange arrows the oil pretreatment, and blue arrows the common extraction steps

Table 3
Chemicals required to prepare the lipid extracts from oils, animal and vegetal tissues

Chemical	Chemical formula	CAS number	Molecular weight (g.mol^{-1})	Safety Instructions
Acetic acid	$C_2H_4O_2$	64-19-7	60.05	 H226, H314
Ammonium hydroxide (NH_4OH) 28-30%	NH_4OH	1336-21-6	17.03	 H302, H314, H335 H400
Butylated hydroxytoluene (BHT)	$C_{15}H_{24}O_2$	128-37-0	220.35	 H400, H410, H411
Formic acid	CH_2O_2	64-18-6	46.03	 H226, H302, H331, H314, H318, H402
Hydrogen chloride 37%	HCl	7647-01-0	36.46	 H280, H314, H331

Ortho-phosphoric acid	H_3O_4P	7664-38-2	98.00	 H290, H314
Potassium hydroxide (KOH)	KOH	1310-58-3	56.11	 H302, H314
Sodium chloride (NaCl)	NaCl	7647-14-5	58.44	
Sodium dihydrogen phosphate	NaH_2PO_4	7558-80-7	119.98	
Sodium Hydroxyde (NaOH)	NaOH	1310-73-2	40.00	 H290, H314

Table 4
Solvents required to analyze isoP by LC/MS-MS

Chemical	Chemical formula	CAS number	Molecular weight (g.mol^{-1})	Quality	Safety Instructions
Acetonitrile (ACN)	C_2H_3N	75-05-8	41.05	LC-MS	 H225, H302, H312, H319, H332
Chloroform	$CHCl_3$	67-66-3	119.38	HPLC	 H302, H315, H319, H331, H336, H351, H361d, H372
Ethanol	C_2H_6O	64-17-5	46.07	LC-MS	 H225, H319
Ethyl acetate	$C_4H_8O_2$	141-78-6	88.11	HPLC	 H225, H319, H336

Hexane	C_6H_{14}	110-54-3	86.18	HPLC	 H225, H302, H305, H315, H336, H361fd, H373, H411
Methanol (MeOH)	CH_4O	67-56-1	32.04	LC-MS	 H225, H301, H311, H331, H370
Water (H_2O)	H_2O	7732-18-5	18.02	LC-MS	
2-propanol	C_3H_8O	67-63-0	60.10	LC-MS	 H225, H319, H336

2.2 Equipment Needed to Prepare Lipid Extracts (LE)

2.2.1 LE of Oil

- 15 mL polypropylene centrifuge tube (Fisher Scientific).
- Classic Vortex Mixer.
- Refrigerated benchtop centrifuge.

2.2.2 LE of Vegetable or Animal Tissue

In addition to the material described in Subheading 2.2.1, it will be necessary to have the following at one's disposal:

- Mixer bench top unit: Mill MM400 (Retsch®).
- Grinding matrix tubes: lysing matrix D (MP Biochemicals, Illkirch, France).
- High-speed benchtop homogenizer: FastPrep-24 (MP Biochemicals).
- Variable-speed, variable-angle vertical rotators: PTR-35 Grant®.

2.3 Solid-Phase Extraction (SPE)

- Mixed-mode reversed-phase/strong anion-exchange (MAX) cartridges or plates (Oasis, Waters).
- Positive or negative pressure filtration systems such as vacuum manifolds, Visiprep (VWR); Waters positive pressure-96 processor; or Biotage® Extrahera™ (automation system for supported SPE).
- Stuart sample concentrator constituted with block heaters in combination with the blown gas (N_2).
- Reacti-vial™ small glass reaction vials (Thermo Scientific).
- Nanosep MF Centrifugal Devices with wwPTFE Membrane 0.45 μm.
- Microcentrifuges Micro Star 17 (VWR).

2.4 LC-ESI-MS/MS Analysis

- HALO C18 analytical column (100 * 0.5 mm, 2.7 μm) (Eksigent Technologies, CA, USA).
- Micro-Liquid-chromatography equipment: MicroLC 200 Plus (Eksigent Technologies, CA, USA).
- Triple quadrupole/linear ion trap mass system: SCIEX Triple Quad™ 5500+ System.

3 Methods

3.1 Sample Preparation

A mixture of internal standards (ISTD) needs to be added to the sample at the beginning of the experiment to compensate for the loss of analytes occurring during the sample preparation and analysis. To prepare this mixture, dilute accurately concentrated stock ISTD (e.g., d_4-15-F_{2t}-IsoP, d_4-4-F_{4t}-NeuroP, d_4-10-F_{4t}-NeuroP, C19-16-F_{1t}-PytoP, C21-15-F_{2t}-IsoP) to 1 ng/μL in absolute methanol. Store at −20 °C. The solution is measured and added with a precision syringe (for instance, eVol® XR, syringe 50 (2–50) μL).

3.1.1 Preparation of Lipid Extracts

Oil Pretreatment

It may be worthwhile to carry out hydrolysis. Preliminary tests may indicate this by comparing the results obtained with or without hydrolysis. Thus, depending on the type of oil, basic hydrolysis before the LE provides different isoprostanoid profiles.

1. Prepare the following solutions before extraction:
   - Potassium hydroxide (KOH) 1 M by dissolving 5.6 g of KOH pellets per 100 mL of ultrapure water.
   - 20 mM formic acid (pH 4.5): add 75 μL of formic acid to ultrapure water per 100 mL final volume. Adjust to pH 4.5 by adding some drops of 1 M KOH solution.
2. Weigh 200 mg of oil in a 15-mL centrifuge tube.
3. In hydrolysis conditions, add 950 μL of KOH 1 M and 4 μL of the ISTD mix with a 100-μL eVol® syringe in the tube. Stir the tube using a PTR-35 Grant® rotator: 360° vertical rotation for 45 s and 90° rotation for 15 s for 30 min, at 40 °C. In nonhydrolysis conditions, this step can be skipped.
4. Add 5 mL of hexane, 2 mL of MeOH using a 5-mL micropipette, and 4 μL of internal standards with a 100 μL-eVol® syringe in the tube and vortex.
5. Add 2 mL of 20 mM formic acid (pH 4.5) in the tube and vortex.
6. Centrifuge the mixture at 2000 rpm for 5 min at room temperature.
7. Eliminate the upper organic phase and charge the lower phase on the conditioned SPE cartridges or plates with a glass pipette. (*see* Subheading 3.1.2 for next steps.)

Modified Folch Extraction of Vegetable and Animal Tissues

Vegetable or animal tissues (*see* **Note 1**) need pretreatment to extract the lipids before obtaining the extract ready for LC/MS-MS analysis. The pretreatment is slightly different between the two matrices.

Vegetable Matrix

Vegetable matrix such as seeds, leaves, flowers, stems, or monocellular and pluricellular algae should be precrushed coarsely using a grinder to obtain a homogenized raw material and ensure a more representative sample. Material will be stored at −80 °C.

1. Prepare the following solutions before extraction:
   - Butylated hydroxytoluene (BHT) 1% (w/v): dissolve 1 g BHT per 100 mL of methanol.

- Phosphate buffer pH 2.1: dissolve 6 g sodium dihydrogen phosphate (NaH_2PO_4), and 3.4 mL phosphoric acid 85%, in 800 mL of ultrapure water. Then adjust to pH 2.1 with HCl 37% in water and top up 1 L. For the salt-saturated buffer, add sodium chloride (NaCl) until crystals persist at the bottom of the bottle (average 300 g/L).

2. Weigh 200 mg of fresh vegetable matrix, 100 mg of dry vegetable matrix, or 50 mg of lyophilized powder in a grinding matrix tube. Add 1 mL of MeOH, 25 µL of BHT 1% in methanol (*see* **Note 2**), and 4 µL of the ISTD mix using a 100-µL eVol® syringe in the tube and vortex.
3. Grind the sample with a FastPrep-24 machine at a speed of 6.5 m/s for 30 s.
4. Transfer the mixture into a 15-mL centrifuge tube.
5. Add 1 mL MeOH in the grinding matrix tube to rinse the tube. Vortex and transfer the liquid into the 15-mL centrifuge tube.
6. Repeat step 4 twice with a total volume of 1.5 mL of phosphate buffer (pH 2.1) saturated with NaCl.
7. Shake the centrifuge tube with the rotator (360° vertical rotation for 45 s and 90° rotation for 15 s) for 1 h.
8. Centrifuge at 5000 rpm for 5 min at room temperature.
9. Collect the supernatant into a 15-mL centrifuge tube.
10. Add 4 mL of 4° $CHCl_3$.
11. Vortex the mixture for 30 s (*see* **Note 3**).
12. Centrifuge the mixture at 2000 rpm for 5 min at 4 °C.
13. Eliminate the upper aqueous phase and isolate the lower organic phase containing lipid fraction in a Pyrex tube.
14. Dry the organic phase under a stream of nitrogen at 40 °C (*see* **Note 4**).

Animal Tissue

1. Prepare the following solutions before the extraction:
   - Butylated hydroxytoluene (BHT) 1% (w/v): dissolve 1 g BHT per 100 mL of methanol.
   - Ethylene glycol-bis(β-aminoethyl ether)-*N*,*N*,*N*′,*N*′-tetraacetic acid (EGTA) 5 mM: dissolve 9.51 g of d'EGTA in 10 mL of 10 M NaOH and 10 mL of ultrapure water. Adjust the pH to 7.4 with HCl 37% in water and top up 50 mL with ultrapure water to obtain a solution of 0.5 M EGTA. Dilute 100 times in MeOH/ H_2O (2:1; v/v).
   - Phosphate buffer pH 2.1: dissolve 6 g sodium dihydrogen phosphate (NaH_2PO_4), and 3.4 mL phosphoric acid 85%, in 800 mL of ultrapure water. Then adjust to pH 2.1 with HCl 37% in water and top up 1 L.

2. Weigh 200 mg of tissue in a grinding matrix tube and store it at −80 °C until extraction. If the mass is not weighed before freezing, excise a portion from frozen tissue if possible. If not, thaw the tissue (*see* **Note 1**).
3. On the day of analysis, add 50 μL of BHT 1%, 1 mL of a mixture of MeOH and 5 mM EGTA in water (2:1; v/v), and 4 μL of ISTD into the grinding tube, in succession, and vortex.
4. Homogenize the sample with a FastPrep-24 instrument for 30 s at 6.5 m/s.
5. Transfer the homogenized tissue into a centrifuge tube.
6. Rinse the grinding tube with (1) 830 μL of MeOH and (2) 800 μL of phosphate buffer (pH 2). Vortex and transfer each rinsing solvent into the 15-mL centrifuge tube.
7. Add 3 mL of 4° $CHCl_3$ to the mixture.
8. Vortex the mixture for 30 s (*see* **Note 3**).
9. Centrifuge the mixture at 2000 rpm for 5 min at 4 °C.
10. Eliminate the upper aqueous phase and isolate the lower organic phase containing lipid fraction in a Pyrex tube.
11. Dry the organic phase under a stream of nitrogen at 40 °C (*see* **Note 4**).

3.1.2 Isoprostanoids Extraction

Before SPE, the extracted lipid fraction needs to be hydrolyzed to release esterified fatty acid chains and isoprostanoids and, thus, to quantify total isoprostanoids: free and esterified analytes. Then, targeted compounds are purified and concentrated by SPE.

Hydrolysis

1. Prepare the solution of potassium hydroxide (KOH) 1 M by dissolving 5.6 g of KOH pellets per 100 mL of ultrapure water.
2. Add 0.95 mL of 1 M KOH and stir the Pyrex tube using the rotators (360° vertical rotation for 45 s and 90° rotation for 15 s) at 40 °C for 30 min.
3. Cool the samples at room temperature. Add 1 mL of 40 mM formic acid and vortex (*see* **Note 5**).

SPE

1. Prepare SPE mounting and the following mixture:
   - 40 mM formic acid (pH 4.5): add 150 μL of formic acid to ultrapure water per 100 mL final volume. Adjust to pH 4.5 by adding some drops of 1 M KOH solution.
   - 20 mM formic acid (pH 4.5): add 75 μL of formic acid to ultrapure water per 100 mL final volume. Adjust to pH 4.5 by adding some drops of 1 M KOH solution.
   - Ammonium hydroxide (NH_4OH) 2% (v/v): dilute 7.7 mL of 28% (v/v) stock NH_4OH in ultrapure water made up to 100 mL.

- Formic acid 20 mM:MeOH (70:30, v/v): for 250 mL, mix 175 mL of 20 mM formic acid and 75 mL of MeOH.
- Hexane:ethyl acetate (70:30, v/v): for 250 mL, add 175 mL of hexane to 75 mL of ethyl acetate.
- Hexane:ethanol:acetic acid (70:29.4:0.6, v/v/v): for 250 mL, mix 175 mL of hexane with 73.5 mL of ethanol and 1.5 mL of acetic acid.

2. Precondition SPE cartridges or plates with 2 mL of MeOH and 2 mL of 20 mM formic acid.
3. Load the hydrolyzed sample onto the preconditioned cartridges or plates.
4. Let the loaded sample pass through the cartridge or plate sorbent at about one drop every two seconds in order to maximize analyte–sorbent interactions (*see* Fig. 2).
5. Wash the cartridges or plates with four solvents in succession at a flow of one drop per second: (1) 2 mL of NH_3 (2% (v/v)); (2) 2 mL of a formic acid 20 mM:MeOH (70:30, v/v) mixture; (3) 2 mL of hexane; and (4) 2 mL of a hexane:ethyl acetate (70:30, v/v) mixture. The aim of this step is to remove weakly bound compounds from sorbent.
6. Increase nitrogen flow to dry the cartridges or plate for about 2 min (*see* **Note 6**).

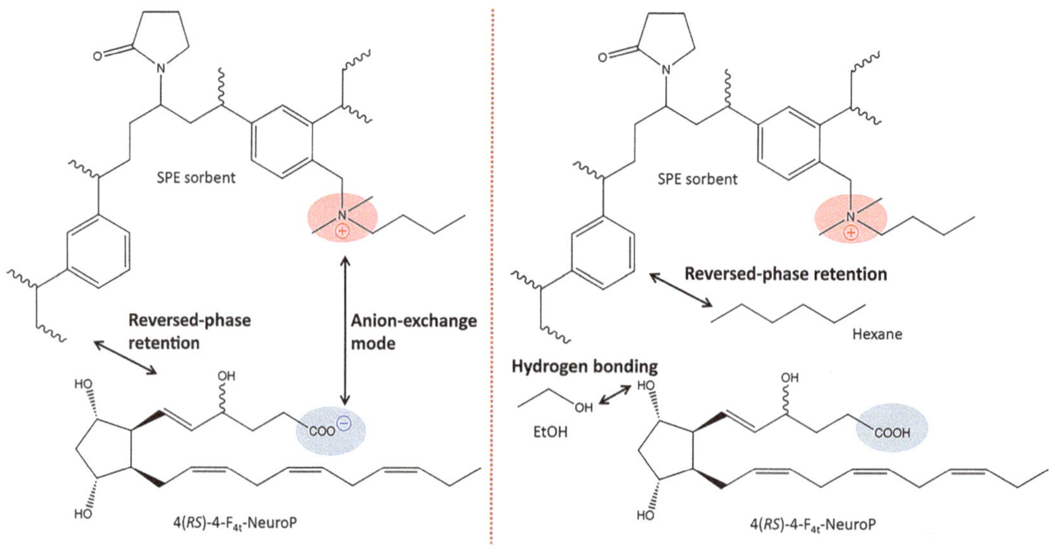

Fig. 2 SPE mechanism: example of 4(*RS*)-4-F_{4t}-NeuroP. (**a**) analyte retention by sorbent; (**b**) analyte elution in acidified solvent

7. Elute isoprostanoids trapped by the sorbent by adding 2 mL of a mixture of hexane:ethanol:acetic acid (70:29.4:0.6, v/v/v) into a glass vial or a collection plate, at a flow of one drop every 2 s.
8. Dry the eluted samples under a nitrogen flow at 40 °C.

Sample Reconstitution

1. Prepare the rework solvent: a mixture of water and acetonitrile (83:17, v/v). For 100 mL, add 17 mL of acetonitrile to 83 mL of ultrapure water.
2. Add 100 μL of the prepared mixture in dried samples and vortex.
3. Let molecules solubilize for 15 min and vortex (*see* **Note 7**).
4. Transfer samples into a centrifuge filter (Nanosep centrifugal devices). Centrifuge at 10,000 rpm for 1 min at room temperature (*see* **Note 8**).
5. Transfer the filtrate into a plastic injection vial.
6. Pool a few volumes of each sample to constitute a "quality control vial" which is spiked by the standards (4 μL of a 0.5 ng/μL standard mixture). This quality control sample will help in molecular identification.

3.2 LC-MS/MS Analysis

Over 45 isoprostanoids can be quantified simultaneously using the following method within 16 min:

1. Prepare the mobile phase constituted by binary solvents A (H_2O/0.1% formic acid; v/v) and B (ACN/MeOH (8:2; v/v); 0.1% formic acid). Prepare also two wash solvents to rinse the autosampler syringe: ultrapure water and a ternary solvent mixture of isopropanol:acetonitrile:methanol (1:1:1; v/v/v).
2. Purge and flush the C18 micro-column, respectively, for 10 cycles at 30 μL/mL.
3. Equilibrate the column with the acquisition method for 3 min before starting the analysis.
4. The elution gradient is as follows: 17% B at 0 min, 17% B at 2.6 min, 21% B at 2.85 min, 25% B at 7.3 min, 28.5% B at 8.8 min; 33.3% B at 11 min; 40% B at 15 min, 95% B at 16.5 min for 1.5 min of equilibration, and then returns to the initial conditions to equilibrate for the next run for 4.5 min.
5. Set the flow rate to 0.03 mL/min and the injection volume to 5 μL. Maintain the column temperature at 40 °C.
6. SCIEX QTRAP 5500 (Sciex Applied Biosystems, ON, Canada) mass spectrometry is operated in the negative mode of electrospray (ESI) (*see* **Note 9**). The source is kept at −4.5 kV and 0 °

C, nitrogen flow is used as curtain gas at 30 psi and also assists the nebulization at 20 psi.

7. Scheduled MRM mode detection was chosen to analyze the targeted compounds with a detection window of 90 s. Two transitions for quantification (T1) and specification (T2) were predetermined by MS/MS analysis of standards. Four potential parameters were optimized for each transition: declustering potential (DP), entrance potential (EP), cell entrance potential (CEP), and cell exit potential (CXP) (*see* Table 5).

3.3 Integration and Quantification

Peak integration and peak quantification of analytes are processed by MultiQuant 3.0 software.

Molecules are identified by comparison of retention time with standards (with ±0.2 min), the presence of the molecular ion, and their two product ions. Automatic integration of peaks is encouraged for integration repeatability.

All peaks with a signal-to-noise ratio above 10 are quantified. An external calibration was used to quantify analytes. Twelve concentrations of standard mixture ranging from 4 to 400 pg/μL (0, 4, 8, 12, 16, 20, 40, 80, 120, 160, 200, 300, and 400 pg/μL) are injected. The linear regression with 1/x weighting factors is elaborated with the height of them (selected concentrations close to the concentration of metabolites in the matrix), based on the area ratio of standards and IS. Standard mixtures are prepared in the reconstitution solution (H_2O/ACN, 83:17, v/v) by diluting a stock solution of 1 g/mL. The calibration curve of each analyte is evaluated by correlation coefficient ($R^2 > 0.99$), and the standard concentration accuracy ranges from 80% to 120%. The appropriate quantification ISTD was chosen for each analyte family: C21-15-F_{2t}-IsoP for neuroprostanes/neurofurans and isoprostanes/isofurans, and C19-16-F_{1t}-PhytoP for phytoprostanes and phytofurans. This type of calibration using ISTD normalizes the extraction efficiency and instrument deviation. The concentration calculated by the software corresponds to the amount of analytes in the injected volume and has to be normalized (*see* **Note 10**).

3.4 Data Processing

Different statistical analyses can be carried out according to the project aim and the number of samples, n. In case of raw data, some statistical parameters, such as the means, standard deviations, and coefficients of variation, are calculated and provide important information. Raw data can be visualized just by graphs or tables, for example, bar charts. In case of comparison to a control group or between independent groups, parametric tests can be performed when the number of samples is large ($n > 30$), if not, nonparametric tests are done (*see* **Note 11**). Parametric tests are always privileged because more information is provided, for example, the mean and the confidence interval.

Table 5
Optimized parameters of a scheduled MRM method

Q1 Mass (Da)	Q3 Mass (Da)	ID	DP (Volts)	EP (Volts)	CEP (Volts)	CXP (Volts)
341,3	297,2	C19-PhytoP_T1	−149	−7	−29,7	−16
341,3	279,1	C19-PhytoP_T2	−149	−7	−32,4	−14
367,2	193,2	C21_15-F_{2t}-IsoP_T1	−174	−3,1	−34	−27
367,2	261,3	C21_15-F_{2t}-IsoP_T4	−174	−3,1	−34	−27
381,2	157	d4-10-epi-10-F_{4t}-NeuroP_T1	−155	−4,8	−27	−10,8
381,2	110	d4-10-epi-10-F_{4t}-NeuroP_T2	−155	−4,8	−32,1	−5,9
357,2	197	d4-15-F_{2t}-IsoP_T1	−65	−6	−35,1	−12
357,2	251	d4-15-F_{2t}-IsoP_T2	−65	−6	−35,1	−12
357,2	213	d4-15-F_{2t}-IsoP_T3	−100	−10	−37	−27
343,2	201	ent-16(RS)-13-epi-ST-Δ^{14}-9-PhytoF_T1	−157,5	−6,37	−34,9	−12
343,2	127	ent-16(RS)-13-epi-ST-Δ^{14}-9-PhytoF_T2	−157,5	−6,4	−42,5	−7,1
343,2	209,1	ent-16(RS)-9-epi-ST-Δ^{14}-10-PhytoF_T1	−151	−7,6	−30,5	−6,9
343,2	82,9	ent-16(RS)-9-epi-ST-Δ^{14}-10-PhytoF_T2	−151	−7,6	−37	−7
343,2	237,1	ent-9(RS)-12-epi-ST-Δ^{10}-13-PhytoF_T1	−161	−6,7	−32,6	−27
343,2	86,9	ent-9(RS)-12-epi-ST-Δ^{10}-13-PhytoF_T2	−161	−6,7	−36	−27
327,22	283,2	ent-16-epi-16-F_{1t}-PhytoP_T1	−100	−4,5	−26,7	−14,9
327,2	251,2	ent-16-epi-16-F_{1t}-PhytoP_T2	−100	−4,5	−26,7	−14,9
327,21	283,2	9-F_{1t}-PhytoP_T1	−100	−4,5	−26,7	−14,9
327,21	171,2	9-F_{1t}-PhytoP_T2	−100	−4,5	−26,7	−14,9
327,22	283,2	ent-16-F_{1t}-PhytoP_T1	−100	−4,5	−26,7	−14,9
327,2	251,2	ent-16-F_{1t}-PhytoP_T2	−100	−4,5	−26,7	−14,9
327,2	283,2	9-epi-9-F_{1t}-PhytoP_T1	−100	−4,5	−26,7	−14,9
327,2	171,2	9-epi-9-F_{1t}-PhytoP_T2	−100	−4,5	−26,7	−14,9
325	289	ent-9-epi-9-D_{1t}-PhytoP_T1	−121,1	−9,52	−22,2	−18,4
325	123	ent-9-epi-9-D_{1t}-PhytoP_T2	−121,1	−9,52	−33,78	−7,47
325	134,9	ent-9-epi-9-D_{1t}-PhytoP_T3	−121,1	−9,52	−28,54	−7,25
325,01	289	ent-9-D_{1t}-PhytoP_T1	−121,1	−9,52	−22,2	−18,4

(continued)

Table 5 (continued)

Q1 Mass (Da)	Q3 Mass (Da)	ID	DP (Volts)	EP (Volts)	CEP (Volts)	CXP (Volts)
325,01	123	ent-9-D_{1t}-PhytoP_T2	−121,1	−9,52	−33,78	−7,47
325,01	134,9	ent-9-D_{1t}-PhytoP_T3	−121,1	−9,52	−28,54	−7,25
307,2	235	16-B_1-PhytoP_T1	−130	−5	−27,5	−18
307,2	223	16-B_1-PhytoP_T2	−130	−5	−27,5	−18
307,2	185,1	9-L_1-PhytoP_T1	−152	−9	−27,9	−10,2
307,2	109	9-L_1-PhytoP_T2	−152	−9	−29,2	−6
307,2	289,2	16(RS)-16-A_1-PhytoP_T1	−115	−8,9	−15,4	−16,5
307,2	249	16(RS)-16-A_1-PhytoP_T2	−115	−8,9	−21,9	−17,9
351,2	289,1	18-epi-18-F_{3t}-IsoP_T1	−167	−6,89	−27,36	−15,9
351,2	249,2	18-epi-18-F_{3t}-IsoP_T2	−167	−6,89	−29,3	−12,7
351,3	127,1	8-F_{3t}-IsoP_T1	−151	−7,6	−30,5	−6,9
351,3	155	8-F_{3t}-IsoP_T2	−151	−7,6	−26,6	−10
351,2	307	5-F_{3t}-IsoP_T1	−140	−2,5	−23,7	−21,9
351,2	115	5-F_{3t}-IsoP_T2	−140	−2,5	−27	−7
351,21	289,1	18-F_{3t}-IsoP_T1	−167	−6,89	−27,36	−15,9
351,21	249,2	18-F_{3t}-IsoP_T2	−167	−6,89	−29,3	−12,7
351,21	307	5-epi-5-F_{3t}-IsoP_T1	−140	−2,5	−23,7	−21,9
351,21	115	5-epi-5-F_{3t}-IsoP_T2	−140	−2,5	−27	−7
351,31	127,1	8-epi-8-F_{3t}-IsoP_T1	−151	−7,6	−30,5	−6,9
351,31	155	8-epi-8-F_{3t}-IsoP_T2	−151	−7,6	−26,6	−10
353,2	193	15-epi-15-F_{2t}-IsoP_T1	−147	−5	−32	−10
353,2	247	15-epi-15-F_{2t}-IsoP_T2	−100	−10	−37	−27
353,21	193	15-F_{2t}-IsoP_T1	−147	−5	−32	−10
353,21	247	15-F_{2t}-IsoP_T2	−100	−10	−37	−27
353,2	115	5(RS)-5-F_{2t}-IsoP_T1	−203	−5	−25,3	−36,9
353,2	309	5(RS)-5-F_{2t}-IsoP_T2	−203	−5	−25,1	−36,9
353,201	115	5-F_{2c}-IsoP_T1	−149	−4,8	−28,6	−12
353,201	309	5-F_{2c}-IsoP_T2	−149	−4,8	−26,8	−20
397,4	201,1	7(RS)-ST-Δ^8-11-dihomo-IsoF_T1	−185	−5	−34,2	−10,5
397,4	183	7(RS)-ST-Δ^8-11-dihomo-IsoF_T2	−185	−5	−36,1	−10,2
381,3	319,2	17-F_{2t}-dihomo-IsoP_T1	−177	−4,3	−31,6	−19,1

(continued)

Table 5
(continued)

Q1 Mass (Da)	Q3 Mass (Da)	ID	DP (Volts)	EP (Volts)	CEP (Volts)	CXP (Volts)
381,3	337,1	17-F_{2t}-dihomo-IsoP_T2	−177	−4,3	−33,8	−14,7
381,3	221	17-F_{2t}-dihomo-IsoP_T3	−168	−4	−37,9	−6
381,3	237	17-F_{2t}-dihomo-IsoP_T4	−168	−4	−31,8	−16,6
381,3	143	ent-7(RS)-F_{2t}-dihomo-IsoP_T1	−161	−5,93	−31,6	−7,8
381,31	327,2	ent-7(RS)-F_{2t}-dihomo-IsoP_T2	−161	−5,93	−28,2	−17,5
381,31	319,2	17-epi-17-F_{2t}-dihomo-IsoP_T1	−177	−4,3	−31,6	−19,1
381,31	337,1	17-epi-17-F_{2t}-dihomo-IsoP_T2	−177	−4,3	−33,8	−14,7
381,31	221	17-epi-17-F_{2t}-dihomo-IsoP_T3	−168	−4	−37,9	−6
381,31	237	17-epi-17-F_{2t}-dihomo-IsoP_T4	−168	−4	−31,8	−16,6
333,2	271	15-A_2-IsoP_T1	−141	−8	−17	−17
333,2	189	15-A_2-IsoP_T2	−141	−8	−23	−17
397,4	155	17(RS)-10-epi-SC-Δ^{15}-11-dihomo-IsoF_T1	−185	−5	−34,2	−10,5
397,4	221	17(RS)-10-epi-SC-Δ^{15}-11-dihomo-IsoF_T2	−185	−5	−36,1	−10,2
379,3	207	14-F_{3t}-NeuroP_T1	−169,8	−3,8	−29,8	−11,8
379,3	179	14-F_{3t}-NeuroP_T2	−169,8	−3,8	−38,2	−9,2
379,3	233	14-F_{3t}-NeuroP_T3	−169,8	−3,8	−26,1	−11,9
379,2	101	4(RS)-4-F_{3t}-NeuroP_T1	−174	−4	−28,9	−7,9
379,2	273,3	4(RS)-4-F_{3t}-NeuroP_T2	−174	−4	−32,1	−12
377,2	193	13(R/S)-13-F_{4t}-NeuroP_T1	−149	−3,8	−27,2	−7,8
377,2	315,2	20-epi-20-F_{4t}-NeuroP_T1	−164	−3	−26,22	−18,4
377,2	323	20-epi-20-F_{4t}-NeuroP_T2	−164	−3	−26,22	−18,4
377,2	259	20-epi-20-F_{4t}-NeuroP_T3	−164	−3	−27	−14
377,2	153	10-F_{4t}-NeuroP_T1	−149	−3,8	−27,2	−14,25
377,2	110	10-F_{4t}-NeuroP_T2	−149	−3,8	−27,2	−7,8
377,21	315,2	20-F_{4t}-NeuroP_T1	−164	−3	−26,22	−18,4
377,21	323	20-F_{4t}-NeuroP_T2	−164	−3	−26,22	−18,4
377,21	259	20-F_{4t}-NeuroP_T3	−164	−3	−27	−14
377,21	153	10-epi-10-F_{4t}-NeuroP_T1	−149	−3,8	−27,2	−14,25
377,21	110	10-epi-10-F_{4t}-NeuroP_T2	−149	−3,8	−27,2	−7,8

(continued)

Table 5 (continued)

Q1 Mass (Da)	Q3 Mass (Da)	ID	DP (Volts)	EP (Volts)	CEP (Volts)	CXP (Volts)
377,2	179	14(RS)-14-F_{4t}-NeuroP_T1	−166	−4,5	−28,4	−9,7
377,2	205	14(RS)-14-F_{4t}-NeuroP_T2	−166	−4,5	−28,5	−12
377,2	101	4(RS)-4-F_{4t}-NeuroP_T1	−161	−4,69	−28,3	−6,9
377,2	271,3	4(RS)-4-F_{4t}-NeuroP_T2	−161	−4,69	−27	−16
393,3	187,2	4(RS)-ST-Δ^5-8-NeuroF_T1	−161	−4,69	−29	−6,9

Abbreviations: *CXP* cell exit potential, *CEP* cell entrance potential, *DP* Declustering potential, *EP* entrance potential, *IsoF* isofuran, *NeuroP* Neuroprostane, *IsoP* Isoprostane, *PhytoP* phytoprostane, *PhytoF* phytofuran, *Q1 mass* mass selected at the first quadrupole, *Q3* mass chosen at the third quadrupole

3.5 Experiment Checkpoints

An analytical method is applied after validating the method. In addition to the method validation criteria, some points have to be verified to ensure good progress of the experiment:

1. Cross-over. The acquisition batch starts and ends generally with a blank. A blank is passed for every five to ten acquired samples. Verify the absence of contaminants in blanks.
2. Repeatability. In addition to the repeatability method studied during method validation, quality control including two IS verifies the stability of the machine at the end of each acquisition batch. Moreover, the variability of preparation procedures is informed by the coefficient of variation (CV) of internal standards. The value of CVs must be below 15% for inter-day experiments.
3. Quality control sample of extraction. A sample constituted by standards and ISTD at known concentration is subjected to the same extraction steps as matrix samples. This sample can not only verify the presence or absence of contaminants originating from the extraction procedure but also give information on extraction quality, for example, the extraction recovery and extracted analytes.
4. Matrix effect and extraction yield. These two parameters depend on the matrix type and will be studied to improve the precision of quantification if enough sample quantity is provided.

4 Notes

1. Animal tissues and vegetable samples have to be stored at −80 °C. Avoid freeze-thawing cycles that can generate oxidized lipids after sampling. Oil can be stored at 4 °C.
2. BHT is added to samples before extraction to quench radical-catalyzed reactions and to prevent the formation of oxylipins during sample preparation.
3. Do not hold the tube caps during shaking because of possible leakage.
4. Hydrolysis and drying temperature are controlled at 40 °C; if not, analytes may be denatured. Ensure the caps are inserted correctly before rotating the centrifuge tubes to avoid solvent leakage.
5. Filtration can be applied before SPE to avoid sorbent plugging.
6. Drying the sorbent after the washing step eliminates contaminants and leads to a cleaner elution.
7. Samples can be vortexed continuously for 15 min for better resolubilization.
8. Filtration before injection allows eliminating possible precipitates and all particles bigger than the column diameter to avoid column plugging.
9. Molecules are ionized by an electrospray (ESI) operated in the negative mode due to their common carboxylic acid function.
10. Normalization of the calculated concentration to the weight of the matrix is necessary, and the final concentration is expressed as an amount of analyte per gram of matrix.
11. Actually, parametric tests may also be done when the number of samples is smaller than 30 if two conditions are met: symmetric distribution of values of each group and homogeneous variances between groups. The smaller the sample number, the worse the statistical quality.

Acknowledgement

We warmly thank our colleagues for sharing their data.

Annexes

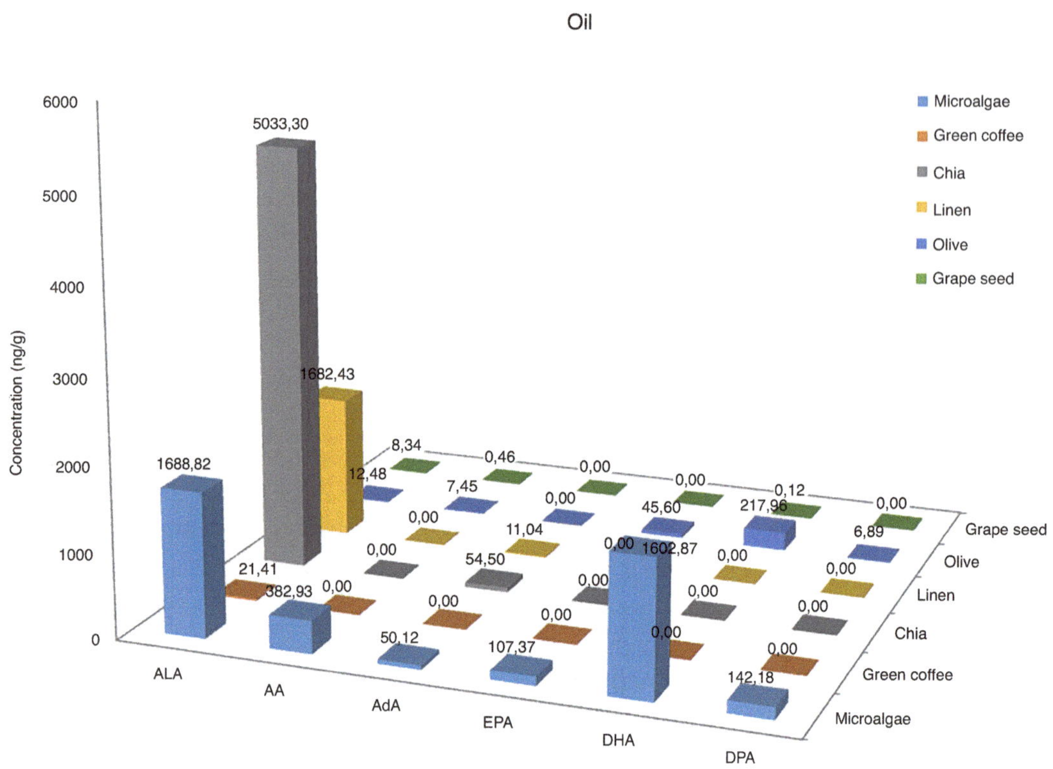

Fig. 3 Sum of amounts of metabolites depending on their fatty acid precursors in different oils (ng/g oil)

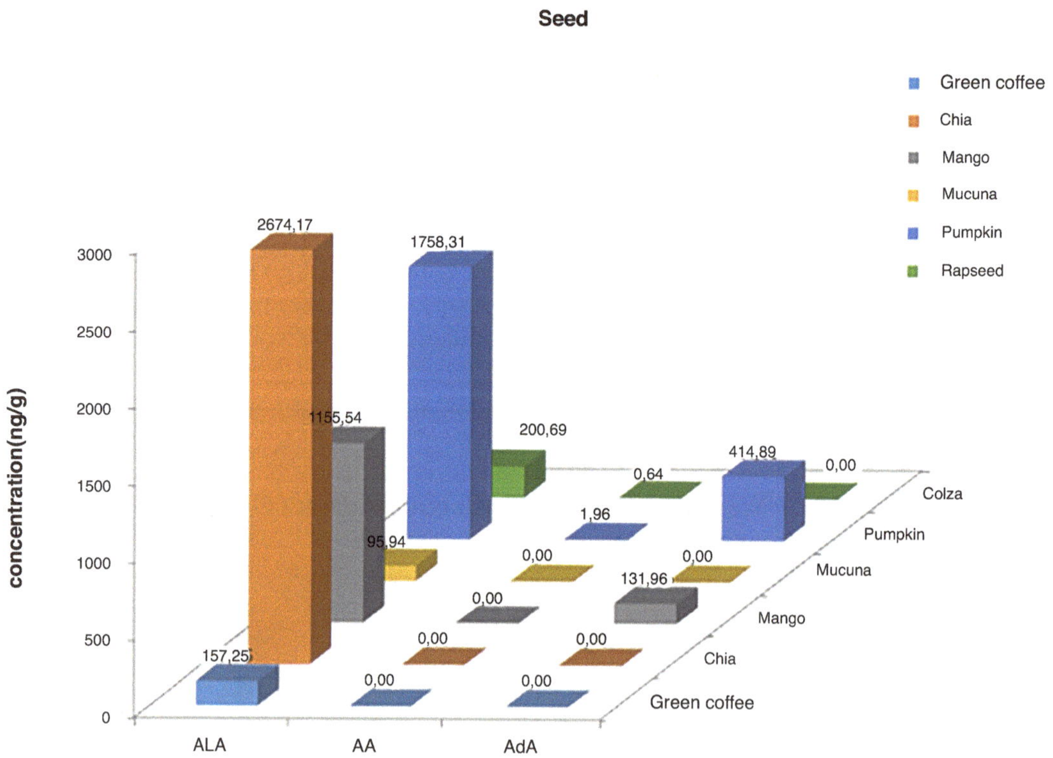

Fig. 4 Amounts of metabolites depending on their precursors in different seeds (ng/g seed). Mucuna seed was studied without integument

References

1. Halliwell B (2020) Reflections of an aging free radical. Free Radic Biol Med 161:234–245
2. Galano J-M, Lee YY, Oger C, Vigor C, Vercauteren J, Durand T, Giera M, Lee JC-L (2017) Isoprostanes, neuroprostanes, phytoprostanes. An Overview of 25 years of research in chemistry and biology. Prog Lipid Res 68:83–108
3. Ahmed OS, Galano J-M, Pavlickova T, Révol-Cavalier J, Vigor C, Lee JC-Y, Oger C, Durand T (2020) Moving forward with isoprostanes, neuroprostanes and phytoprostanes – where are we know? Essays Biochem 64:463–484
4. Collado-Gonzalez J, Perez-Lopez D, Memmi H, Gijon MC, Medina S, Durand T, Guy A, Galano JM, Ferreres F, Torrecillas A, Gil-Izquierdo A (2015) Water deficit during pit hardening enhances phytoprostanes content, a plant biomarker of oxidative stress, in extra virgin olive oil. J Agric Food Chem 63(14)
5. Yonny ME, Torresi AR, Cuyamendous C, Reversat G, Oger C, Galano JM, Durand T, Vigor C, Nazareno MA (2016) Thermal stress in melon plants: phytoprostanes and phytofurans as oxidative stress biomarkers and the effect of antioxidant supplementation. J Agric Food Chem 64(44):8296–8304
6. Morrow JD, Hill KE, Burk RF, Nammour TM, Badr KF, Roberts LJ (1990) A series of prostaglandin F2-like compounds are produced in vivo in humans by a non-cyclooxygenase, free radical-catalyzed mechanism. Proc Natl Acad Sci U S A 87:9383–9387
7. Milne GL, Yin H, Hardy KD, Davies SS, Roberts LJ (2011) Isoprostane generation and function. Chem Rev 11:5973–5996
8. Vigor C, Bertrand-Michel J, Pinot E, Oger C, Vercauteren J, Le Faouder P, Galano J-M, Durand T (2014) Non-enzymatic lipid oxidation products in biological systems: assessment of the metabolites from polyunsaturated fatty acids. J Chromatograph B 964:65–78
9. Galano J-M, Roy J, Durand T, Lee JCY, Le Guennec JY, Oger C, Demion M (2018) Biological activities of non-enzymatic oxygenated metabolites of polyunsaturated fatty acids (NEO-PUFAs) derived from EPA and DHA: New anti-arrhythmic compounds? Mol Aspect Medicine 64:164–168

Part IV

Preparation and Characterization of Emulsions

Chapter 20

Preparation of Dietary Lipid Emulsions Using Laboratory Methods

Valérie Beaumal and Sébastien Marze

Abstract

Emulsification is a common unit operation in the production of different types of food, used to control the size of dispersed droplets, in turn making food homogeneous, stable, and pleasant. Similarly, in the food lab, the composition and structural elements of dietary emulsions must be controlled to enable studying their behavior in a variety of environments and contexts. For this matter, many emulsification methods exist to produce emulsions at the laboratory scale. However, good practice protocols are not widely available to choose from and use in appropriate conditions. In this chapter, we highlight the main criteria that have to be considered to choose a particular emulsification method. Then, we give detailed protocols for three widely used methods, namely, rotor-stator emulsification, high-pressure homogenization, and sonication.

Key words Triglycerides, Rotor-stator emulsification, High-pressure homogenization, Sonication, Emulsifiers

1 Introduction

Triglycerides are the main dietary lipids. As triglycerides are immiscible with water in most foods, they are dispersed in the form of droplets that are stabilized by amphiphilic molecules at the oil–water interface, such as other lipids or proteins. This type of dispersion is known as an oil-in-water (o/w) emulsion (water continuous phase). Exceptionally, some foods are based on an inverse emulsion (oil continuous phase), that is, a water-in-oil (w/o) emulsion (butter, margarine, chocolate). In any case, for the sake of food homogeneity, stability, and edibility, any undispersed lipid is usually emulsified in the kitchen or in the factory. Thus, studying lipids in foods often involves emulsification in the laboratory.

Many technologies exist to choose from, depending on the targeted droplet size distribution (some measurement protocols are given in the next chapter), the required emulsion amount, and the food type and composition. In food science, the main

Christelle Lopez et al. (eds.), *Multidimensional Characterization of Dietary Lipids*, Methods and Protocols in Food Science, https://doi.org/10.1007/978-1-0716-3758-6_20,

emulsification methods used are based on rotor-stator emulsification, high-pressure homogenization, sonication, membrane homogenization, colloid mill shearing, droplet microfluidics, and microfluidization [3, 4, chapter 6, 6].

In this chapter, detailed protocols for the most used first three methods are given, together with guidelines to select the most suitable ones for different applications based on various criteria.

2 Materials and Chemicals

2.1 Materials

- Analytical balance 0.1g.
- Rotor-stator apparatus.
- High-pressure homogenizer.
- Sonication device.
- Thermal water bath or ice bath.
- Orbital shaker or magnetic stirrer.
- pH meter.
- Timer.
- Water.
- Glass vessel (beaker, homogenization beaker).
- Stir bar.
- Micropipette or pipette.

2.2 Chemicals

- HCl or NaOH 1M.
- Salts (NaCl, phosphate, etc.)
- Emulsifiers: protein or lipid isolates from animals or plants

Do not forget to wear individual safety equipment: gloves, ear plugs, and safety glasses.

3 Methods

Before an emulsification method can be selected, emulsion characteristics should first be defined carefully. The main features are the properties of the dietary lipid (fatty acid profile, density, viscosity, etc.), its content in the emulsion (all methods presented here allow high dispersed phase contents to be reached, see, for example, [5]), the targeted droplet size distribution, and the nature of the emulsifier. First, it is important to note that a high vs. a low HLB (hydrophile–lipophile balance) of the emulsifier will favor an o/w emulsion vs. a w/o emulsion, respectively [2, chapter 15]. Then, the emulsifier concentration in the continuous phase (C_{cont} in kg

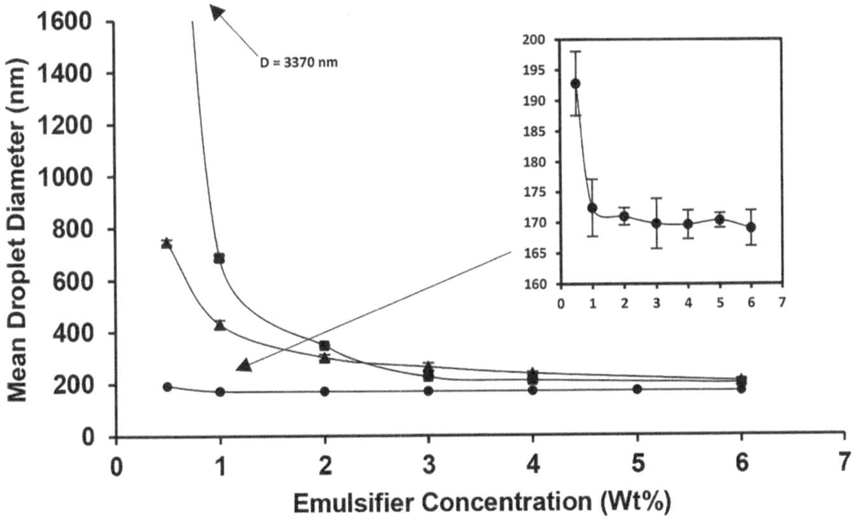

Fig. 1 Mean droplet diameter as a function of emulsifier concentration for 10 wt% sunflower oil-in-water emulsions prepared by sonication, stabilized by saponin (circle), whey protein isolate (triangle), or lecithin (square). (Reproduced from Nejatian and Abbasi [5] under a Creative Commons Attribution-NonCommercial 3.0 Unported Licence)

m^{-3}, equivalent to g dm^{-3}) required to stabilize the emulsion can be estimated using

$$C_{\text{cont}} = \frac{6\varnothing\Gamma_{\text{sat}}}{(1-\varnothing)d_m} \tag{1}$$

where $\varnothing$ is the dispersed phase volume fraction, d_m is the mean droplet diameter in m, and Γ_{sat} is the saturation emulsifier concentration at the interface, usually close to $2{\cdot}10^{-6}$ kg m^{-2} [4, chapter 6].

In a typical optimization work, emulsions with various emulsifier concentrations are prepared and their droplet size distribution is measured (for protocols, see next chapter). The mean droplet size is then plotted as a function of emulsifier concentration to choose an optimal concentration (for an example, see Fig. 1).

Equation (1) is only valid if enough energy is provided to break large droplets into small ones, the maximum diameter of unbroken droplets being given by the law [1]:

$$d_{\max} = k\left(\frac{\gamma + \upsilon'\mu_d}{\rho_c}\right)^{3/5} / P_M^{2/5} \tag{2}$$

where k is a dimensionless constant close to unity, P_M is the power output per unit mass of emulsion in J s^{-1} kg^{-1}, γ is the interfacial tension in J m^{-2}, ρ_c is the density of the continuous phase in kg m^{-3}, υ' is the turbulent fluctuation velocity in m s^{-1}, and μ_d is the viscosity of the dispersed phase in Pa.s, d_{max} being in m.

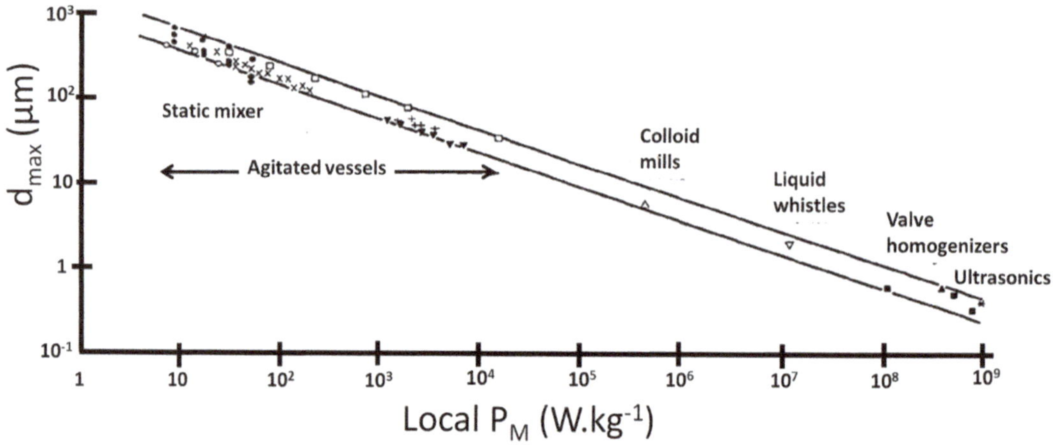

Fig. 2 The Davies curve, maximum diameter of unbroken droplets as a function of power output per unit mass of emulsion, for various emulsification methods. (Adapted from [1])

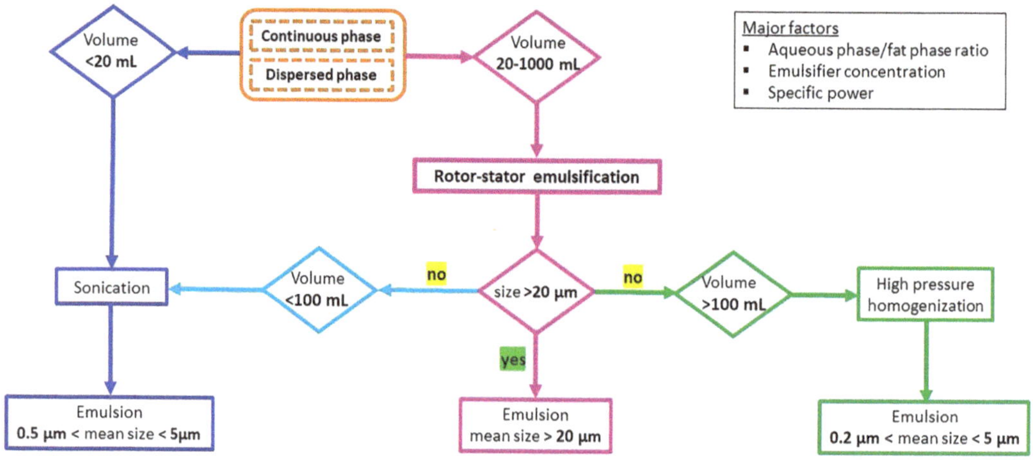

Fig. 3 Decision tree to select emulsification methods

Equation 2 or the Davies curve, presented in Fig. 2, can be used to estimate the power needed to obtain a required droplet size and choose the emulsification method accordingly. This choice will also depend on the mass (or volume) of the emulsion to be produced. Note that droplet diameters below 1 µm are often targeted, because this is about the upper limit of colloidal stability. A decision tree is presented in Fig. 3 to help select suitable methods based on these criteria.

In practice, we advise to produce a coarse emulsion first using a rotor-stator apparatus, then to perform further emulsification using sonication or high-pressure homogenization when smaller droplets are required (Fig. 4) (*see* **Notes 1** and **2**).

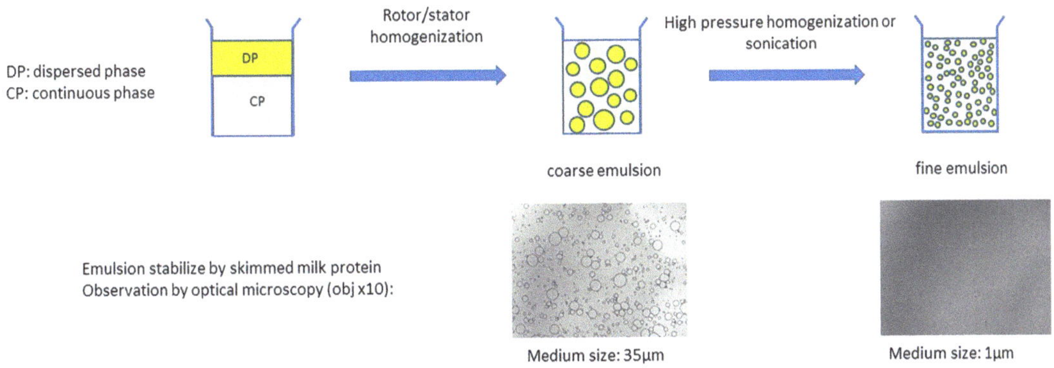

Fig. 4 Schematic representation of the emulsification process

3.1 Fluid Preparation

3.1.1 Continuous Phase

Emulsifier: Protein Extract

- If necessary, prepare the buffer.
- Weight the emulsifier in a beaker or in a tube (*see* **Note 3**).
- Add water or buffer.
- Mix slowly using a magnetic stirrer or an orbital shaker. The mixing time depends on the nature of the emulsifier. It can range from one hour to one night, at room temperature or at 10 °C. In general, to solubilize proteins, let one night at 10 °C.

Emulsifier: Phospholipids

- If necessary, prepare the buffer.
- The phospholipids are stocked in a solvent. Before using them, you must evaporate the solvent.
- The solvent is completely evaporated under nitrogen flow.
- Add water or buffer to rehydrate phospholipids during one night.

3.1.2 Dispersed Phase

The fat phase is stocked at 4 °C. Before using, it must be at ambient temperature. When this phase is solid at ambient temperature (such as fat), it should be maintained above the melting point using a water bath.

3.1.3 Mix Continuous and Dispersed Phases

Both phases are used at the same temperature:

- Weight the continuous phase in a homogenization beaker (*see* **Note 4**)
- Add fat or oil on the continuous phase

If addition of a progressive dispersed phase during rotor-stator emulsification is required (typically when $\varnothing \geq 0.5$), use another vessel suitable to minimize material loss or use a micropipette or a pipette up to 10 mL or 100 mL, respectively.

3.2 Rotor-Stator Emulsification

1. The usual setup of a rotor-stator apparatus is shown in Fig. 5. Choose the most suitable drive unit and dispersing element based on the volume of emulsion to be produced, as recommended by the manufacturers (see [8, 9, 14] for examples). Typical volumes that are treated range from 10 mL to 1 L. In the laboratory, a single drive unit is generally sufficient, with various dispersing elements with specific geometrical features used to treat different volumes and formulations (*see* **Note 5**).
2. Immerge the dispersing element in the fluids contained in the vessel. Vessel width and height should enable the minimal immersion depth to be covered (*see* **Note 5**).
3. Run the rotor-stator apparatus until a homogenous (usually white) emulsion is obtained. Rotor velocity should be set so as to create a small vortex while avoiding air bubble formation. Typical rotor velocity ranges from 5000 rpm to 25000 rpm, and 15000 rpm is often used as a first guess (*see* **Note 6**). When the dispersed phase needs to be added progressively, pour it continuously for a given time with the rotor-stator apparatus running, then allow more emulsification time to obtain a homogenous emulsion. Typical emulsification time ranges from 1 min to 10 min. Rotor velocity and emulsification time should be adjusted to obtain the required droplet size distribution (for measurement protocols, see Chap. 21) and to minimize sample heating, which can also be minimized using a water bath or an ice bath.
4. Remove the dispersing element and clean using a bottle brush with hot water and dishwashing liquid. Rinse with pure hot water and then with pure cold water. Rinse with a solvent (ethanol or acetone) and let dry.

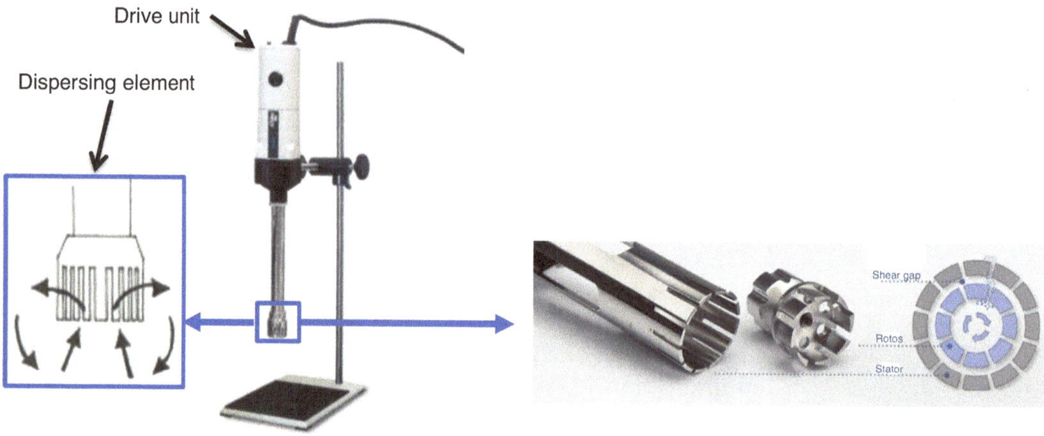

Fig. 5 Usual setup of a rotor-stator apparatus showing the drive unit and an example of a dispersing element geometry. (Adapted from Wikimedia commons and Fullbrook systems)

3.3 Sonication

1. The usual sonication setup is shown in Fig. 6. When small emulsion volumes are prepared, sonication is a suitable technique for further emulsification to decrease the droplet size. Choose the most suitable sonicator and probe based on the volume of emulsion to be produced, as recommended by the manufacturers (see [10, 11] for examples). Typical volumes that are treated range from 1 mL to 100 mL. In the laboratory, a single sonicator is generally sufficient, with various probes used to treat different volumes.
2. Immediately after rotor-stator emulsification, immerge the sonication probe in the coarse emulsion contained in the same vessel as before. The probe tip should be located at the center of the coarse emulsion.
3. Run the sonicator using a pulse (or cycle) mode in which alternate periods of ultrasound generation and pause are set to prevent sample heating. Typical periods lie between 10 s and 100 s. As a first guess, alternate periods of 30 s are often used, for a total sonication time of 5 min (10 min including pauses). Power output should also be set to prevent sample heating, depending on the required droplet size distribution (for measurement protocols, see Chap. 21). Placing the vessel in a water bath or an ice bath is recommended, as sonication works at a very high power output per sample volume, inducing large temperature increase.

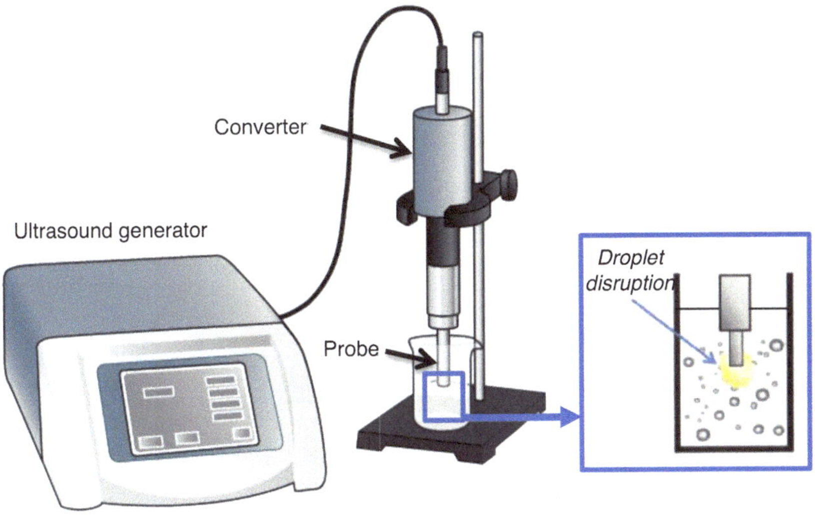

Fig. 6 Usual sonication setup, showing the ultrasound generator, the converter (into mechanical vibration), and the probe (propagating vibration in the sample). (Adapted from Wikimedia commons and from McClements and Rao [14])

4. Clean the probe using a bottle brush with hot water and dish-washing liquid. Rinse with pure hot water and then with pure cold water. Rinse with a solvent (ethanol or acetone) and let dry.

3.4 High-Pressure Homogenization

1. The emulsification section of a high-pressure valve homogenizer is shown in Fig. 7. It can be used in open or close configurations, and pressure can be applied if required (Figs. 8 and 9). There are two types of homogenizers, with one or two high-pressure valves. The most common one has two high-pressure valves. When large emulsion volumes are prepared, high-pressure homogenization is a suitable technique for further emulsification to decrease the droplet size. Choose the most suitable homogenizer based on the maximum pressure required to obtain the targeted droplet size distribution, and on the volume of emulsion to be produced, as recommended by the manufacturers (see [12, 13] for examples). Typical pressure ranges from $5 \cdot 10^7$ to $20 \cdot 10^7$ Pa. Typical volumes that are treated at the laboratory scale range from 100 mL to 1 L.
2. Immediately after rotor-stator emulsification, transfer the coarse emulsion into the homogenizer hopper. Depending on the nature of the fat or the emulsifier, use a bath to control the temperature.
3. Before homogenization, rinse the system with pure water. Run the homogenizer and remove the emulsion (a few milliliters). Close the system and increase the pressure to achieve the selected pressure. The entire emulsion volume should recirculate for at least five times at the selected pressure to obtain an

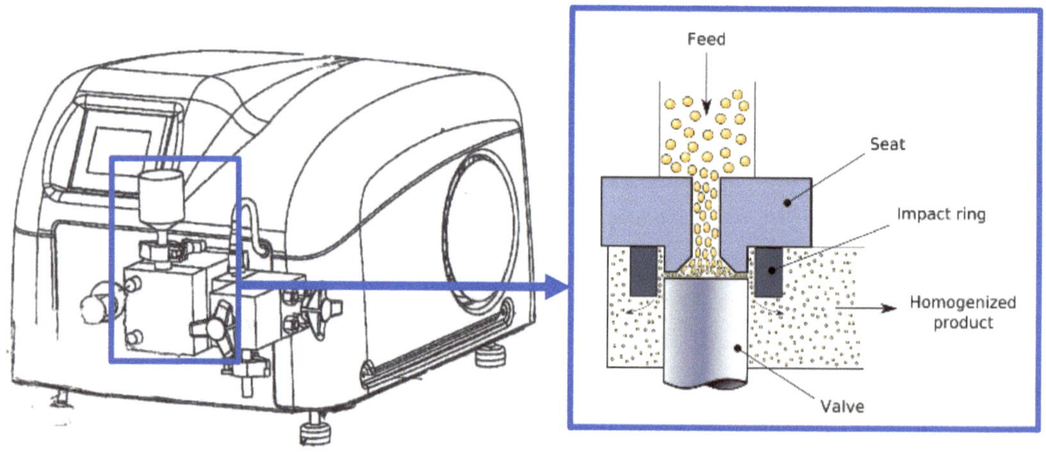

Fig. 7 Emulsification section of a high-pressure valve homogenizer. Note that the parts are static, and the feed is maintained under high pressure. (Adapted from Wikimedia commons and GEA)

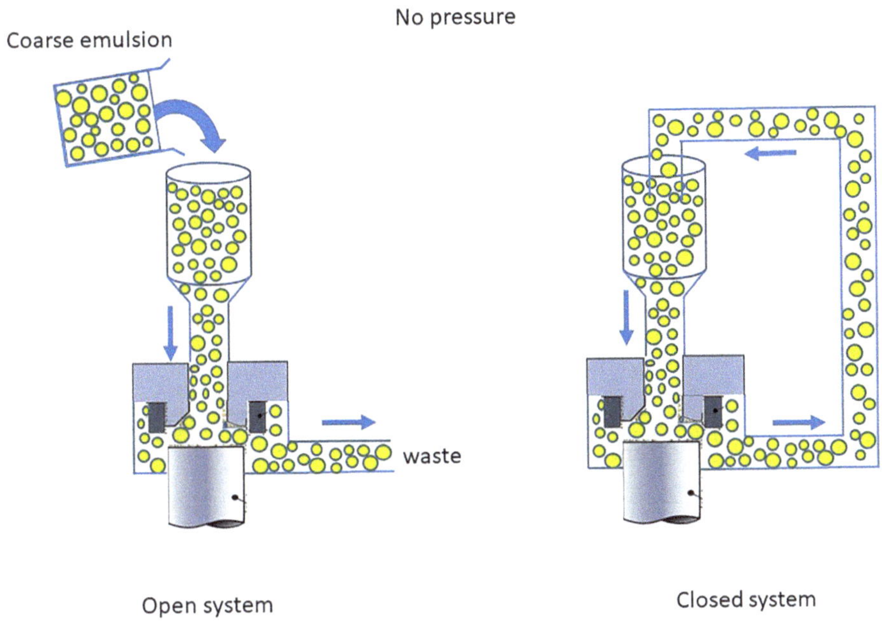

Fig. 8 Homogenization system without pressure

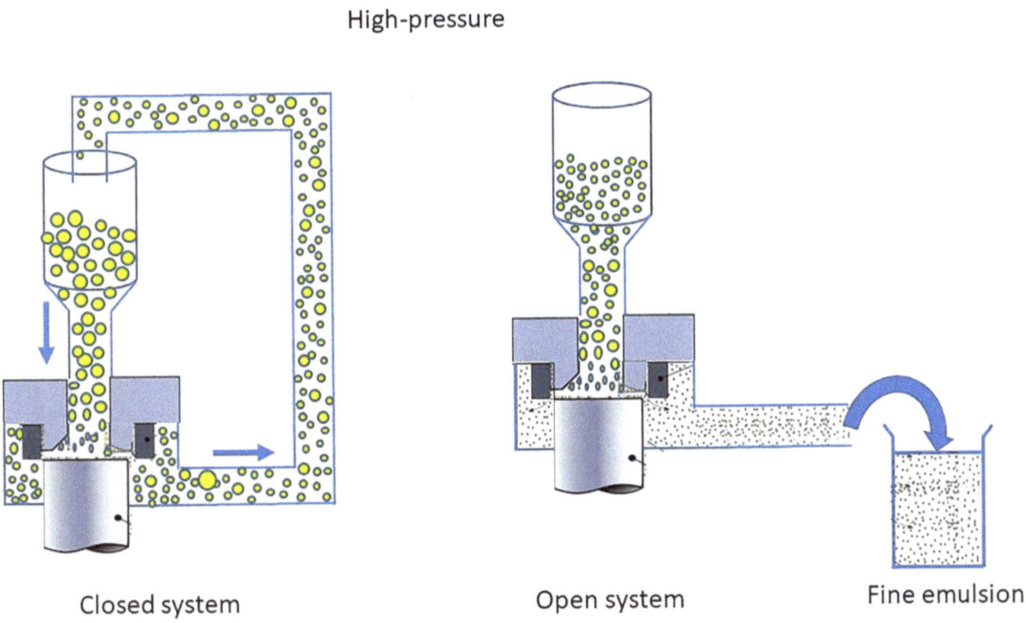

Fig. 9 Homogenization system with high pressure

efficient homogenization that decreases droplet size. The recirculation time thus depends on the volume of emulsion to be homogenized and its flow rate. Typically, five circulation cycles take less than 10 min, yet this depends on the flow rate of the homogenizer. When the selected pressure is above $10 \cdot 10^7$ Pa, the temperature of the sample increases significantly. It is then recommended to use a water bath connected to the homogenizer to control the temperature.

4. When homogenization is finished, stop the recirculation and recover the emulsion in a vessel (*see* **Note 7**). Control the droplet size distribution (for measurement protocols, see Chap. 21) and store the emulsion for further analysis.

4 Notes

1. When the volume is less than 20 mL, it is possible to prepare an emulsion directly by sonication.
2. Ultrasound sonication allows the preparation of fine emulsions.
3. Solubilize the emulsifier in the beaker when the volume is greater than 50 mL. Under this volume, solubilize the emulsifier in a tube. A magnetic stirrer and an orbital shaker are, respectively, adapted to solubilize the emulsifier in a beaker and in a tube. If necessary, the pH value may be adjusted before emulsification using HCl or NaOH 1M solution.
4. In our laboratory, we have two dispersing elements: the smallest 12 F and the largest 18F. The stator diameter and the width of slot are different. The speed range and the volume range are recorded in Table 1.
5. Immerge the rotor-stator dispersing element in the fluid. You must respect the minimum and maximum immersion depths (Fig. 10).
6. Generally, with the dispersing element 12F, the rotor velocity used is 20000 rpm during a short time (30–60 s). With the dispersing element 18F, the velocity used is lower than 20000 rpm and depending on sample volume the time may be longer (maximum 5 min).
7. Decrease the pressure only when the emulsion is out of the homogenizer.

Table 1
Rotor-stator dispersion element characteristics

	Rotor/Stator system	
Dispersion element	12F	18F
Speed (rpm)	5000/20000	5000/20000
Volume (mL)	25/250	100/1500

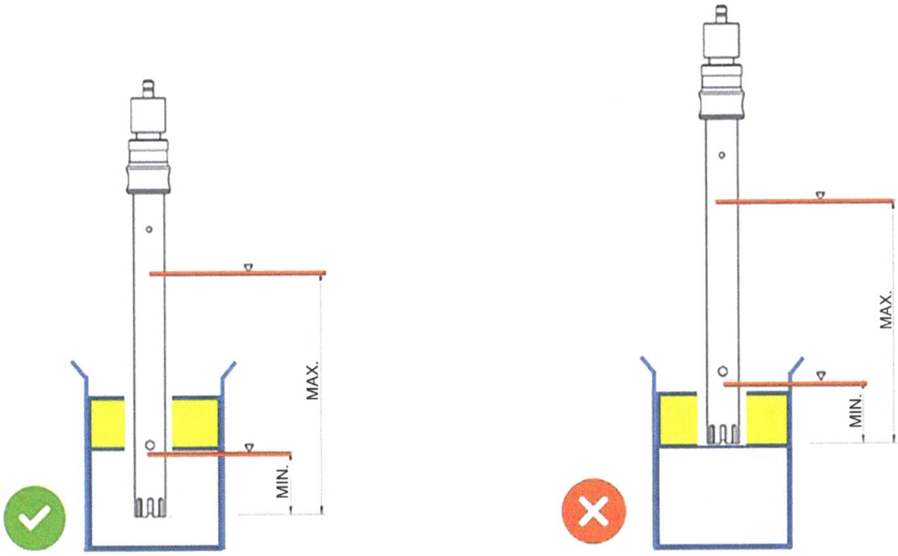

Fig. 10 Schematic representation of dispersion element immersion

References

1. Davies JT (1987) A physical interpretation of drop sizes in homogenizers and agitated tanks, including the dispersion of viscous oils. Chem Eng Sci 42:1671–1676
2. Hasenhuettl GL, Hartel RW (2019) Food emulsifiers and their applications. Springer, Cham
3. Maan AAL, Nazir A, Khan MKL, Boom R, Schroen K (2015) Microfluidic emulsification in food processing. J Food Eng 147:1–7
4. McClements DJ (2016) Food emulsions – principles, practices, and techniques. CRC Press, Boca Raton
5. McClements DJ, Rao J (2011) Food-grade nanoemulsions: Formulation fabrication properties performance biological fate and potential toxicity. Crit Rev Food Sci Nutr 51(4): 285–330. https://doi.org/10.1080/10408398.2011.559558
6. Nejatian M, Abbasi S (2019) Formation of concentrated triglyceride nanoemulsions and nanogels: natural emulsifiers and high power ultrasound. RSC Adv 9:28330–28344
7. Vladisavljevic GT, Kobayashi I, Nakajima M (2012) Production of uniform droplets using membrane, microchannel and microfluidic emulsification devices. Microfluidics Nanofluidics 13:151–178
8. https://www.ika.com/en/Products-Lab-Eq/Dispersers-Homogenizer-csp-177/
9. https://kinematica.ch/en/products/productfinder/category/23
10. https://www.sonicator.com/collections/sonicators
11. https://www.hielscher.com/lab.htm
12. https://www.gea.com/en/products/homogenizers/laboratory-homogenizers/index.jsp
13. http://www.avestin.com/products.htm
14. https://www.fullbrooksystems.co.uk/rotor-stator1.html

Chapter 21

Particle Size Distribution of Dietary Lipid Assemblies

Valérie Beaumal and Sébastien Marze

Abstract

Many foods can be described as dispersed systems whose microstructure consists of various types of objects usually dispersed in a continuous medium. The size distributions of these objects influence several food properties such as rheology, physical and chemical stabilities, sensorial perception, and digestion. In this chapter, we focus on the characterization of lipid-based assemblies, which are mainly droplets, liposomes/ vesicles, and micelles in the context of food and digestion. We present the principle of and give detailed protocols for the laser diffraction (LD) and dynamic light scattering (DLS) methods for measuring particle size distributions and mean sizes of lipid assemblies.

Key words Dynamic light scattering, Laser diffraction, Particle size distribution, Emulsion, Liposome, Vesicle, Micelle, Assemblies

1 Introduction

Dietary lipids, mainly composed of triglycerides, organize in various supramolecular assemblies with other components (e.g., phospholipids and proteins) to form natural or processed emulsions. In natural foods, lipids are commonly found in the form of triglyceride-rich droplets, essentially as fat globules in milk, lipoproteins in egg yolk, lipid droplets (also called oil bodies or oleosomes) in plants and seeds, or fat cells (also called adipocytes) in animal bodies. In processed foods, triglyceride droplets are also the main engineered assemblies (except for butter, margarine, and chocolate which contain aqueous droplets). Yet, a growing number of processed foods now include other lipid structures, like micelles or liposomes, often based on phospholipids. Such lipid structures are often mimicked after biological objects, like phospholipid-based cellular vesicles, or mixed bile assemblies that form upon intestinal digestion of triglyceride droplets.

Controlling lipid assemblies in terms of size and morphology is of prime importance for the physical and chemical stabilities of food products as well as other functional aspects. Particle size

Christelle Lopez et al. (eds.), *Multidimensional Characterization of Dietary Lipids*, Methods and Protocols in Food Science, https://doi.org/10.1007/978-1-0716-3758-6_21,

distributions (PSDs) not exceeding 1 μm are often targeted because this is about the upper limit of colloidal stability. Moreover, small droplets enhance sensory perception and satiety [14, 15], lipid digestion [23], and lipid absorption [24]. However, chemical stability (e.g., against oxidation) may be lower for small droplets in certain foods [2]. In some cases, intermediate particle size distributions may thus be targeted (see previous chapter on emulsion preparation).

In food science, studying dietary lipids in terms of assembly functionalities thus requires the knowledge of particle size distribution (PSD). It is generally characterized using dynamic light scattering (DLS), laser diffraction (LD), and, more recently, nanoparticle tracking (NTA). Typical measurement ranges are 1 nm to 5 μm for DLS, 50 nm to 5000 μm for LD, and 10 nm to 1 μm for NTA.

In this chapter, protocols for LD and DLS characterizations of micelle, liposome, or emulsion PSD are described. These two techniques are indeed complementary, and together they cover the whole size range of lipid assemblies in foods. As NTA covers a narrow size range, it is only suitable for specific assemblies, such as nanoemulsions [13].

2 Materials

- Pipettes, tips.
- Water.
- Salts (NaCl, phosphate, etc.).
- Sodium dodecyl sulfate solution 1%.
- UV-disposable cells.
- Individual safety equipment: safety glasses and gloves.

3 Methods

LD analysis is based on the static measurement of the diffraction of a laser light by a sample in which objects induce different diffraction angles depending on their size. The intensities of the diffracted light are collected on at least one ring-shaped detector array, a modern LD apparatus using two lasers, and several detector arrays to extend the measurable size range. Figure 1 shows two optical setups that are typically used. The measured angular diffraction pattern(s) is then converted to a volume-based PSD using the Mie theory assuming spherical particles [19, chapter 10]. To apply this theory, the refractive index and the absorption values for both phases must be known at the laser wavelength to account

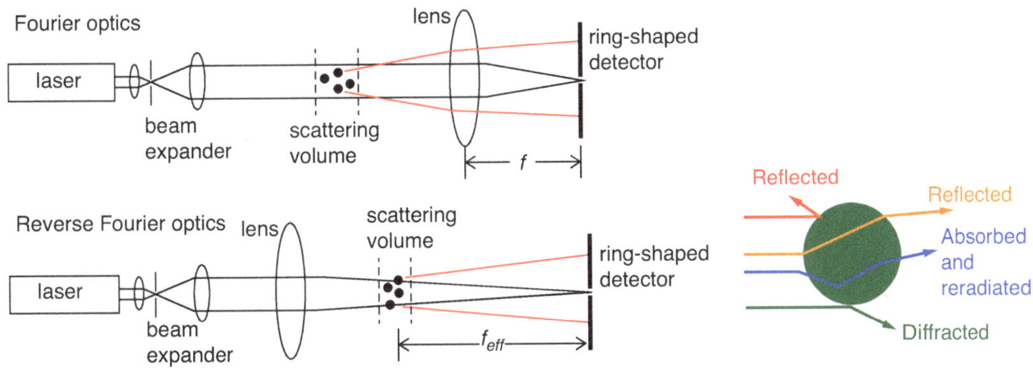

Fig. 1 Main optical setups used for laser diffraction analysis (From [4]), and detail of the light-particle interactions. (After Horiba)

for these light–particle interactions (see Fig. 1). Absorption is low for water and translucent lipids (0.00) but may be significant for colored materials (0.30). The refractive index could be difficult to know precisely for specific dietary lipids, unless measured with a refractometer. Table 1 lists the most common refractive index values reported in the literature for water and some dietary lipids (pure triglycerides, pure fatty acids, and natural dietary fats and oils).

DLS analysis is based on the dynamic measurement of light scattered by a sample collected on a single detector at numerous time intervals. The measured temporal intensity fluctuation due to particle Brownian motion is then used to calculate a temporal autocorrelation function that is fitted to obtain an intensity-based PSD. The most common analysis is based on cumulants fitting to determine an average size and a polydispersity, suitable for unimodal and narrow distributions. Other analyses such as the Laplace transform-based CONTIN algorithm or a recent statistics-based algorithm [17] are used for multimodal distributions. In any case, to convert time to length, the Stokes–Einstein equation is used (Eq. 1):

$$D = \frac{1}{q^2 \tau} = \frac{k_B T_K}{6\pi \eta_s R_H} \tag{1}$$

with $q = (4\pi n_s/\lambda)\sin(\theta/2)$, where D is the diffusion coefficient, τ is the characteristic decay time, k_B is the Boltzmann constant, T_K is the temperature in Kelvin, η_s is the viscosity of the solvent, R_H is the hydrodynamic radius of the particle, n_s is the refractive index of the solvent, λ is the wavelength of the laser, and θ is the angle between the incident light and the detector. To perform this analysis, two properties must be known: the refractive index and the viscosity of the solvent (continuous phase). Values of the refractive index are found in Table 1 for water and dietary lipids, and viscosity values for oils, when used as the continuous phase, can be found in Coupland

Table 1
Refractive index values for water and some dietary lipids. The readers are referred to the original articles for dependence on temperature and wavelength

	20 °C	40 °C	Measurement wavelength, other remarks	Ref.
Water	1.335	1.332	Averaged over visible wavelengths	[10]
Triacetin	1.435	1.426	Averaged over visible wavelengths	[9]
Tributyrin	1.439	1.431		[9]
Tricaproin	1.446	1.438		[9]
Tricaprylin	1.451	1.444		[9]
Tricaprin		1.448		[9]
Trilaurin		1.451		[9]
Triolein	1.473	1.466		[9]
Trilinolein	1.483	1.475		[9]
Trielaidin		1.465		[9]
Caproic acid	1.423	1.416	Averaged over visible wavelengths	[22]
Caprylic acid	1.432	1.425		[22]
Peanut oil	1.470		589 nm	[7]
High oleic peanut oil	1.468			[7]
Peanut oil	1.465	1.455	600 nm	[16]
Soybean oil	1.475	1.465		[16]
Sunflower oil	1.475	1.465		[16]
Safflower oil	1.478		589 nm, 25 °C	[1]
Peanut oil	1.471			[1]
Sesame oil	1.473			[1]
Butter oil	1.460			[1]
Fish oils	1.478		589 nm, 25 °C, depends on EPA and DHA contents	[8]
Cocoa butter	1.457		Unknown wavelength and temperature	[18]
Coconut oil	1.449			[18]
Cod-liver oil	1.481			[18]
Corn oil	1.473			[18]
Cottonseed oil	1.474			[18]
Linseed oil	1.478			[18]
Mustard oil	1.475			[18]
Olive oil	1.468			[18]
Palm oil	1.458			[18]

(continued)

Table 1 (continued)

	20 °C	40 °C	Measurement wavelength, other remarks	Ref.
Palm kernel oil	1.457			[18]
Poppy seed oil	1.469			[18]
Rapeseed oil	1.471			[18]

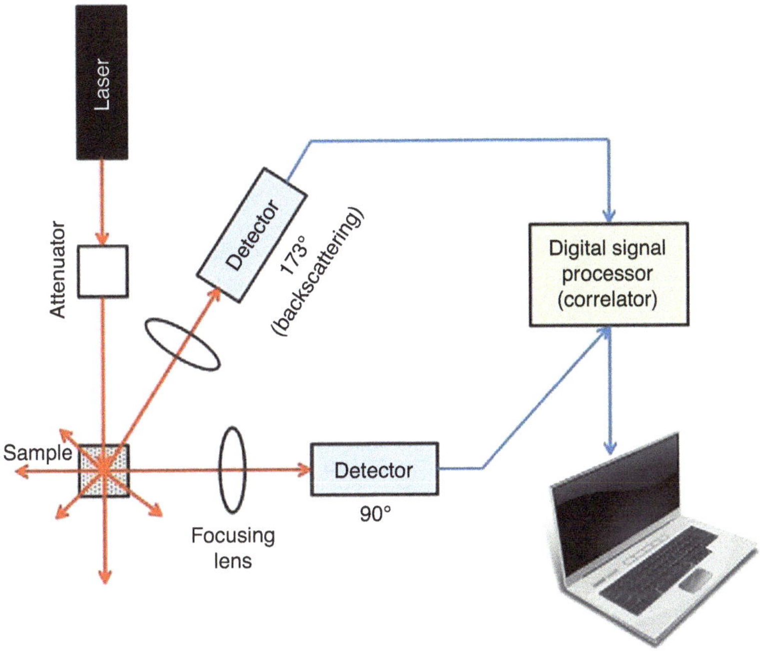

Fig. 2 DLS setup, showing the side scattering and backscattering geometries. (From [3])

and McClements [6]. Note that depending on the scattering angle θ, the DLS technique will be more sensitive to either small or large particles. Side scattering ($\theta \approx 90°$) and especially backscattering ($\theta \approx 173°$) favor the detection of small particles, whereas forward scattering ($\theta \approx 13°$) favors the detection of large particles. Figure 2 shows the optical setups used for side scattering and backscattering.

Both LD and DLS are based on a single scattering condition, meaning the light beam must be scattered by only one object at a time. To fulfill this condition, the sample must be diluted for measurement so that the concentration of assemblies does not exceed 0.01 vol% for LD and 0.1 vol% for DLS [21, chapter 14].

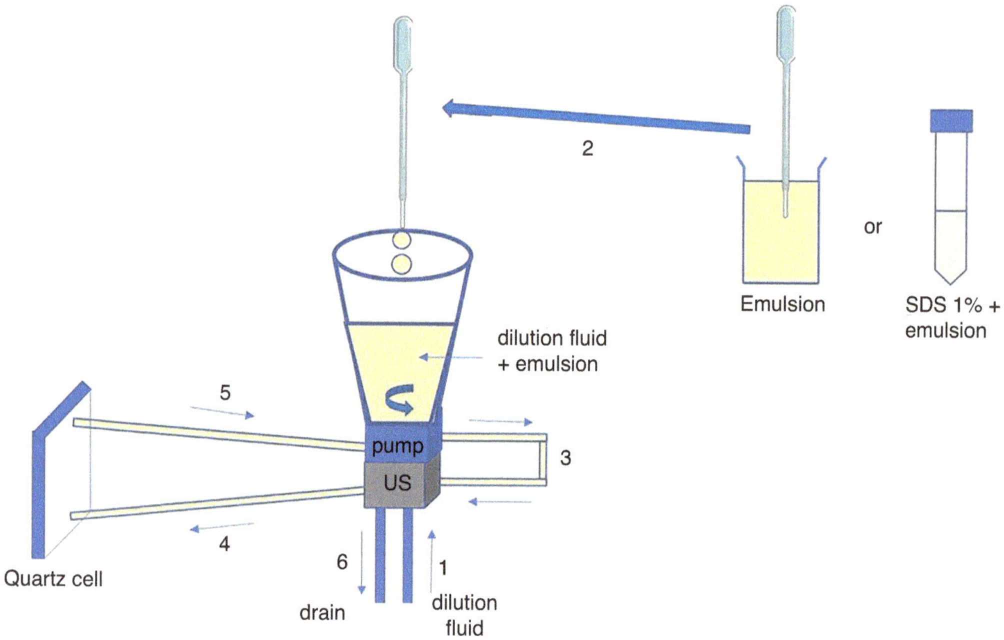

Fig. 3 Schematic representation of sample handling in a laser diffraction sizing system. (Adapted from Horiba)

3.1 Laser Diffraction

Most commercial apparatus come with a dispersion unit that performs the dilution of lipid assemblies in water or inversely of water assemblies in oil (Fig. 3). Dispersion is often obtained by mechanical stirring assisted by ultrasound. The diluted sample is flowed through the measurement cell continuously and light transmission is usually monitored so as to estimate the suitability of the concentration reached. Before measurement, one should indicate the refractive index and the absorption values for both phases.

1. Prepare a dilution fluid that is based on the continuous phase of the assembly dispersion (buffer solution for an aqueous phase or dietary oil for an oil phase). Standard dispersion units require about 100 mL of dilution fluid for each measurement, which can be too much for certain applications. Most manufacturers now offer small-volume units requiring about 10 mL of fluid or lower.
2. Fill the dispersion unit with the dilution fluid and circulate it through the measurement cell. At this point, optical background and alignment are usually performed by the apparatus. This allows one to control the cleanliness of the measurement cell based on the light intensity profile on the detectors, constituting a blank reference.
3. Add an appropriate volume of sample into the dispersion unit so as to reach the recommended concentration range, which is indicated by the apparatus. A highly concentrated sample can

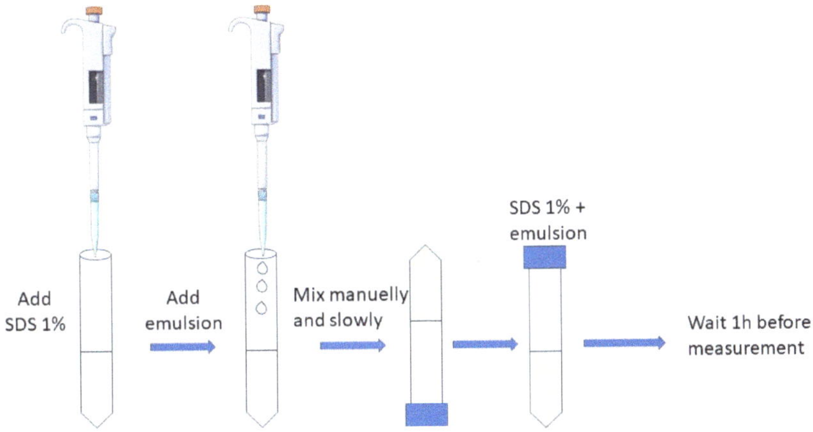

Fig. 4 Sample preparation with SDS 1%

be pre-diluted with the continuous phase fluid before addition in the dispersion unit. When the sample concentration is low, a small-volume dispersion unit may be used, in which less than 1 mL is typically required, but this also depends on the size of the assemblies. Concentrated large liposomes and medium-sized emulsion droplets (in the range of 0.1–100 μm) are generally added directly. Smaller assemblies could require sample volumes much larger than 1 mL. Depending on the nature of the sample, more or less stirring and ultrasounds will be needed to obtain a homogeneous dilution.

4. Record the measurement when a stable intensity profile on the detectors is obtained, indicating a homogeneous dilution.
5. Empty the dispersion unit and run the automatic clean procedure of the apparatus (or clean the apparatus manually) until a satisfactory light intensity profile on the detectors is restored. A new measurement can be made starting at **step 3**.
6. A variant of this protocol is used when there is a suspicion of lipid assembly flocculation/aggregation in an aqueous continuous phase. In this case, the PSD indeed represents the distribution of aggregates, not of individual assemblies. To measure individual assemblies, an additional measurement following **steps 1–5** is required, the only difference being that a 1% SDS (sodium dodecyl sulfate) solution is used to pre-dilute the sample, with a dilution factor between 10 and 50 depending on the lipid concentration (Fig. 4).

   SDS is an anionic surfactant that is used above its critical micelle concentration to deflocculate lipid assemblies such as lipid droplets. If flocculation was present, the PSD with SDS will be shifted toward smaller sizes, corresponding to the size distribution of individual particles (lipid droplets). If there was no flocculation, the PSD with SDS will be identical to the one

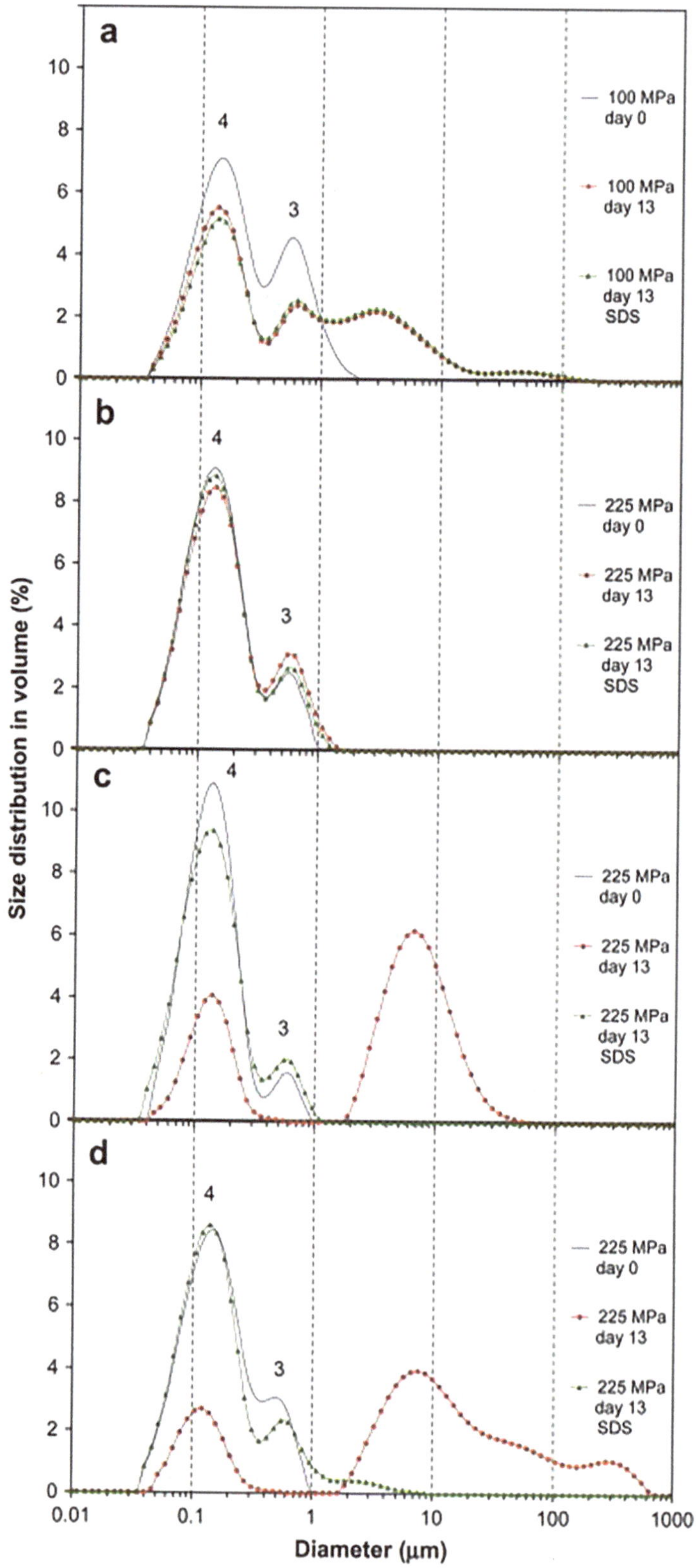

Fig. 5 Droplet size distribution of peanut oil-in-water (o/w) emulsions prepared by high-pressure homogenization and stabilized by whey protein isolate,

without SDS. Note that when investigating the PSD as a function of sample storage time, assembly coalescence leading to larger sizes may also occur, in which case the PSD with SDS will be identical to the one without SDS. These droplet interactions are illustrated in Fig. 5.

7. Expression of the Results. As specified above, LD analysis uses Mie theory to provide a volume-based PSD of spheres. This consists in a histogram of the volumetric proportion of the particles in different size classes. These size classes are usually plotted in log scale. From this volume-based PSD, other PSDs can be calculated, principally the surface-based PSD and the number-based PSD (Fig. 6) and rarely the length-based PSD.

   Mean particle sizes are calculated from any of these PSDs using (ISO 9276-2:2014):

$$\overline{D}_{p,q} = \left[\frac{\sum_i q_r(D_i) D_i^{p-r}}{\sum_i q_r(D_i) D_i^{q-r}} \right]^{1/(p-q)} \quad (2)$$

   where $q_r(D_i)$ is the particle quantity in the i-th size class, D_i is the midpoint of the i-th size class interval, and r represents the type of quantity, equal to 0, 1, 2, or 3 for the number, length, surface, or volume, respectively.

   The most used mean values are $\overline{D}_{4,3}$, $\overline{D}_{3,2}$, and $\overline{D}_{1,0}$, which are the volume-weighted, surface-weighted, and number-weighted mean diameters, respectively (Fig. 6). Another useful parameter is the average specific surface area, which has different expressions, such as $6/\overline{D}_{3,2}$ in surface area per unit dispersed volume, $6\varnothing/\overline{D}_{3,2}$ in surface area per unit total volume, or $6/\rho_d\overline{D}_{3,2}$ in surface area per unit dispersed mass, where $\varnothing$ is the volume fraction of the dispersed phase, and ρ_d is the density of the dispersed phase.

   The broadness of the number-based PSD is usually indicated by the standard deviation s_o using (ISO 9276-2:2014)

Fig. 5 (continued) measured by LD just after preparation (day 0) and after freezing for 13 days. The latter was measured with or without a 0.1% SDS pre-dilution. (**a**) 15 wt% peanut oil at 100 MPa: same enlarged PSD with or without SDS = droplet coalescence, (**b**) 15 wt% peanut oil at 225 MPa: all PSDs are similar to the initial one = no droplet interaction, (**c**) 30 wt% peanut oil at 225 MPa: enlarged PSD only without SDS = droplet flocculation, and (**d**) 45 wt% peanut oil at 225 MPa: enlarged PSD without SDS >> with SDS = mainly droplet flocculation with slight coalescence. (From [5])

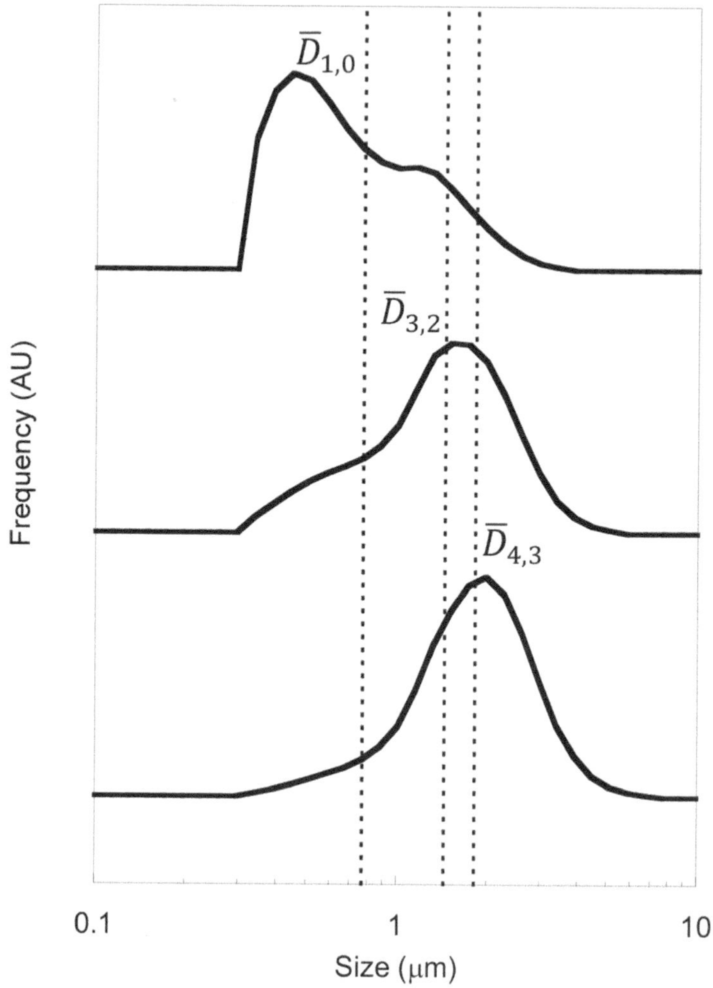

Fig. 6 PSD measured by LD for an o/w emulsion composed of sunflower oil and stabilized by skimmed milk proteins, produced by high-pressure homogenization. From bottom to top: volume-based, surface-based, and number-based PSDs. Corresponding mean diameter values are indicated by the dashed lines

$$s_0 = \sqrt{v_0} = \sqrt{\frac{\sum_i n_i \left(D_i - \overline{D}_{1,0}\right)^2}{-1 + \sum_i n_i}} \tag{3}$$

where v_0 is the variance and n_i is the number of particles in the i-th size class (with $r = 0$).

3.2 Dynamic Light Scattering

Most commercial apparatus use disposable measurement cells of volume ranging from 0.1 to 1 mL. Sample dilution is not automatic and has to be done before cell loading. Given the dilution factor, this technique typically requires a 10–100 μL sample, but this depends on the assembly size and concentration of the sample.

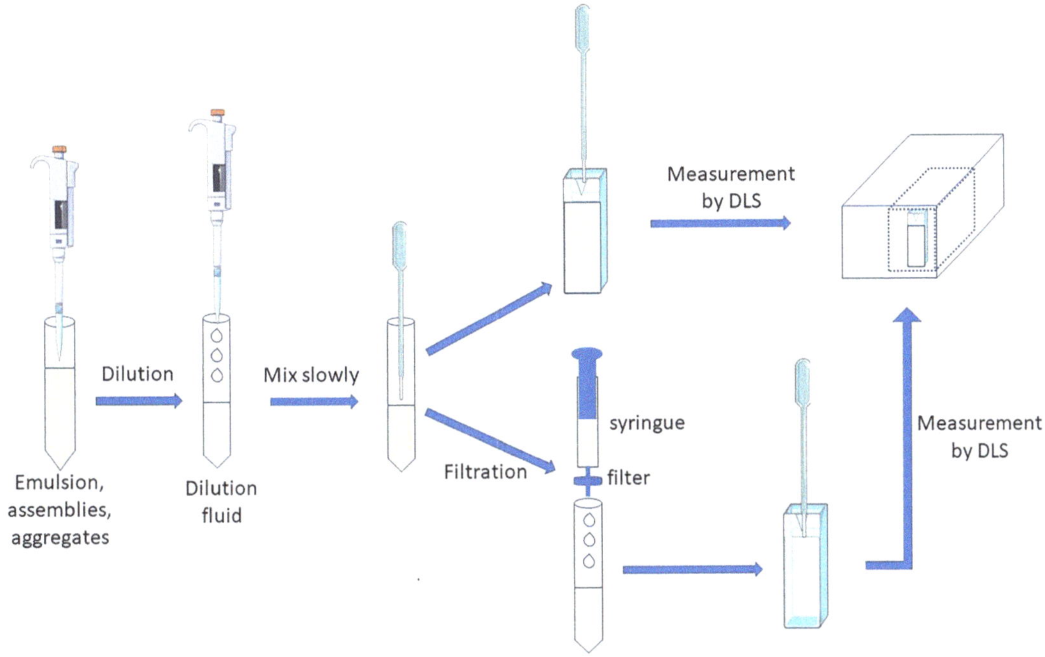

Fig. 7 Sample preparation to measure the particle size distribution by DLS

When Mie theory is applied, one should indicate the refractive index (and absorption) and the viscosity values for the continuous phase before the measurement, optionally the refractive index (and absorption) for the dispersed phase. *See* **Note 1**.

1. Prepare a dilution fluid that is based on the continuous phase of the assembly dispersion (buffer solution for an aqueous phase or dietary oil for an oil phase) (Fig. 7).
2. Dilute the sample to have a concentration of assemblies lower than 0.1 vol%. The exact concentration depends on the assembly size. Micelles, small liposomes, and small emulsion droplets (in the range 1–1000 nm) are usually suitable (Figs. 8 and 9).
3. Fill the cell with the diluted sample and remove any visible air bubbles by tapping or gently vortexing the cell. DLS measurement is also sensitive to small dusts. Sample filtration using cellulosic filters of 0.2–2 μm porosity might be required to obtain a clean and stable signal. As an indicator of stability, the average intensity count rate should be constant as a function of measurement time (no decreasing and/or increasing trend).
4. Place the cell in the apparatus and run a measurement of 30 s (duration is typically 10–50 s). The average intensity count rate (expressed in cps, count per second) of the detected light should be constant and typically lie between 200 and

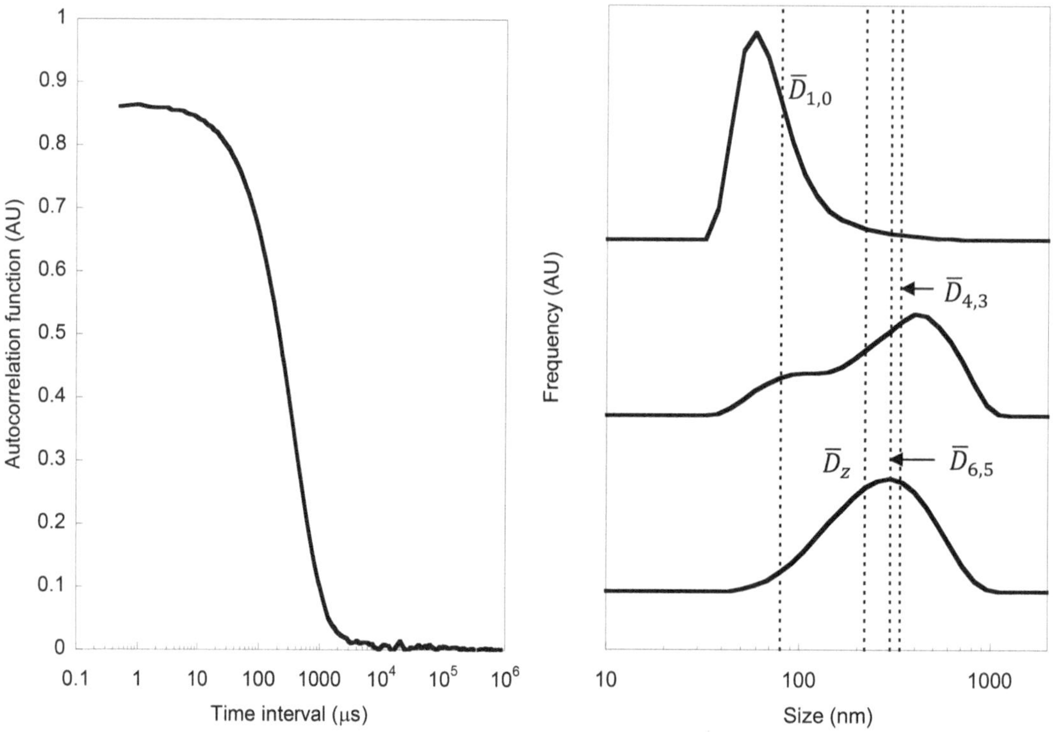

Fig. 8 Autocorrelation function and PSD measured by DLS for an o/w emulsion composed of 0.5% sunflower oil and stabilized by beta-lactoglobulin, produced by rotor–stator emulsification followed by sonication. From bottom to top: intensity-based, volume-based, and number-based PSDs. Corresponding mean diameter values are indicated by the dashed lines

300 kcps. If higher values are obtained, the sample should be diluted further. If lower values are obtained, sample dilution was too high. Another measurement validation is obtained by examining the (auto)correlation function, which should be smooth and have an intercept close to 1 at zero interval time (Figs. 8 and 9).

5. When all conditions mentioned above are fulfilled, run at least 10 measurements of 30 s on the same sample. This will enable us to check the repeatability of the measurement.
6. The dilution fluid variant with SDS to investigate flocculation/ coalescence may also be done as explained in Subheading 3.1, **step 6**.
7. Expression of the Results. As specified above, DLS analysis provides an intensity-based PSD. The characteristic mean size of this PSD is a hydrodynamic diameter called the z-average ($\overline{D}_z$). For unimodal distributions, the z-average is close to $\overline{D}_{6,5}$ as determined using Eq. (2) [20, chapter 1]. Other PSDs (mainly volume-based and number-based) can be deduced from the intensity-based PSD using Mie theory, requiring the

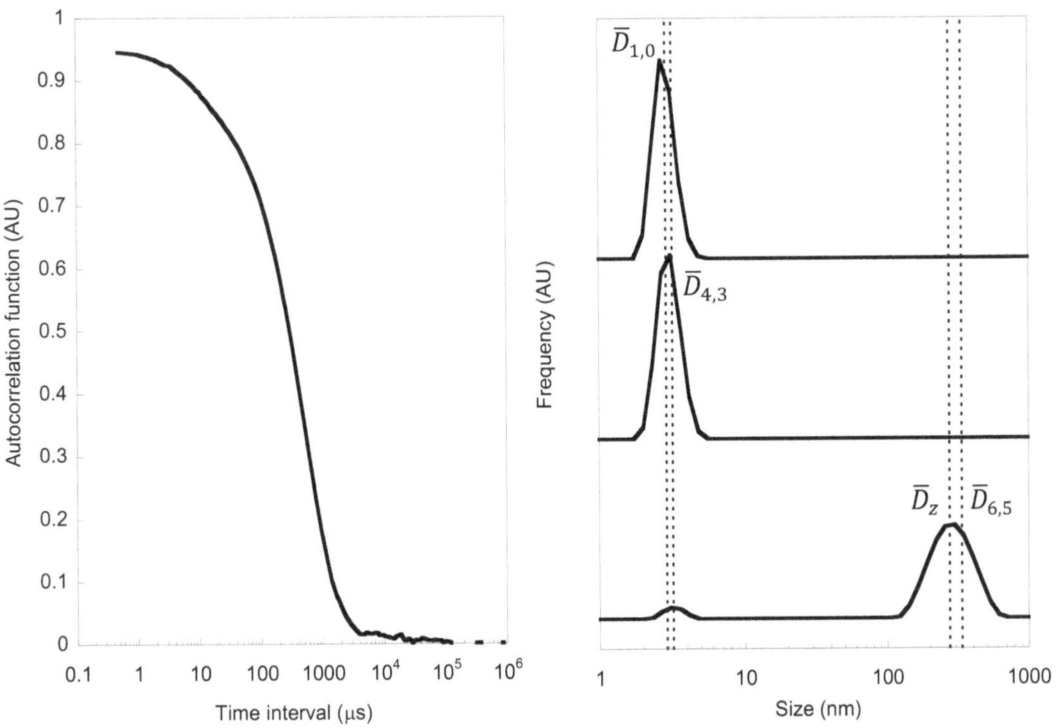

Fig. 9 Autocorrelation function and PSD measured by DLS for mixed bile/lipid micelles, produced by sonication. From bottom to top: intensity-based, volume-based, and number-based PSDs. Corresponding mean diameter values are indicated by the dashed lines

refractive index and absorption values for both phases (Figs. 8 and 9).

The broadness of the intensity-based PSD is usually indicated by the polydispersity index (*PI*) using (ISO 22412: 2017):

$$PI = 2\overline{D}_z^2 \frac{\sum_i q_{\text{int}}(D_i)\left(1/D_i^2 - 1/\overline{D}_z^2\right)}{\sum_i q_{\text{int}}(D_i)} \tag{4}$$

where $q_{\text{int}}(D_i)$ is the intensity-weighted amount of particles in the *i*-th size class. When comparing Eqs. (3) and (4), it is obvious that *PI* is equivalent to a relative variance.

4 Notes

1. Some DLS apparatus are designed to allow sample temperature control. Then, the viscosity values for the continuous phase must be known and indicated for the required temperatures.

Most commercial DLS apparatuses are also designed to enable the determination of particle zeta-potential (electric potential difference between the particle double layer and the continuous phase, usually expressed in mV), through the measurement of the electrophoretic mobility using a special disposable cell equipped with opposite electrodes. The zeta-potential calculation is based on Henry's equation, requiring the knowledge of the dielectric constant and the viscosity values for the continuous phase [3].

References

1. Arya SS, Ramanuja S, Pk V (1969) Refractive index as an objective method for evaluation of rancidity in edible oils and fats. J Am Oil Chem Soc 46:28–30
2. Berton-Carabin CC, Ropers MH, Genot C (2014) Lipid oxidation in oil-in-water emulsions: involvement of the interfacial layer. Compr Rev Food Sci Food Safety 13:945–977
3. Bhattacharjee S (2016) DLS and zeta potential – what they are and what they are not? J Control Release 235:337–351
4. Chen J, Wang HR, Shen JQ (2012) Light scattering of particles illuminated by a divergent beam. Optics Lasers Eng 50:1410–1415
5. Cortés-Muñoz M, Chevalier-Lucia D, Dumay E (2009) Characteristics of submicron emulsions prepared by ultra-high pressure homogenisation: effect of chilled or frozen storage. Food Hydrocoll 23:640–654
6. Coupland JN, McClements DJ (1997) Physical properties of liquid edible oils. J Am Oil Chem Soc 74:1559–1564
7. Davis JP, Sweigart DS, Price KM, Dean LL, Sanders TH (2013) Refractive index and density measurements of peanut oil for determining oleic and linoleic acid contents. J Am Oil Chem Soc 90:199–206
8. de Koning AJ (1994) The relationship between the refractive-index of fish oils and their content of eicosapentaenoic acid (EPA), docosahexaenoic acid (DHA) and total polyunsaturated fatty-acids (PUFA). Fett Wissenschaft Technologie-Fat Sci Technol 96: 352–356
9. Gouw TH, Vlugter JC (1966) Physical properties of triglycerides. I. Density and refractive index. Fette Seifen Anstrichmittel 68:544–549
10. Harvey AH, Gallagher JS, Sengers J (1998) Revised formulation for the refractive index of water and steam as a function of wavelength, temperature and density. J Phys Chem Ref Data 27:761–774
11. ISO 9276-2:2014. Representation of results of particle size analysis – Part 2: calculation of average particle sizes/diameters and moments from particle size distributions
12. ISO 22412:2017. Particle size analysis – dynamic light scattering (DLS)
13. Ji J, Zhang JP, Chen JS, Wang YL, Dong N et al (2015) Preparation and stabilization of emulsions stabilized by mixed sodium caseinate and soy protein isolate. Food Hydrocoll 51:156–165
14. Lett AM, Norton JE, Yeomans MR (2016a) Emulsion oil droplet size significantly affects satiety: a pre-ingestive approach. Appetite 96: 18–24
15. Lett AM, Yeomans MR, Norton IT, Norton JE (2016b) Enhancing expected food intake behaviour, hedonics and sensory characteristics of oil-in-water emulsion systems through microstructural properties, oil droplet size and flavour. Food Qual Prefer 47:148–155
16. Li XC, Zhao JM, Liu LH, Tan JY (2015) Optical properties of edible oils within spectral range from 300 to 2500 nm determined by double optical pathlength transmission method. Appl Optics 54:3886–3893
17. Malm AV, Corbett JCW (2019) Improved Dynamic Light Scattering using an adaptive and statistically driven time resolved treatment of correlation data. Sci Rep 9:13519
18. Malvern Instruments Ltd (2007) Sample dispersion and refractive index guide. Available at: https://www.malvernpanalytical.com/fr/learn/knowledge-center/user-manuals/MAN0396EN
19. Merkus HG (2009) Particle size measurements – fundamentals, practice, quality. Springer, Dordrecht
20. Merkus HG, Meesters GMH (2015) Production, handling and characterization of particulate materials. Springer, Cham

21. McClements DJ (2016) Food emulsions – principles, practices, and techniques. CRC Press, Boca Raton
22. Rubio JEF, Arsuaga JM, Taravillo M, Caceres M (2005) Refractive index temperature and wavelength dependencies of normal saturated fatty acids in liquid state. Exp Thermal Fluid Sci 29:681–684
23. Salvia-Trujillo L, Qian C, Martin-Belloso O, McClements DJ (2013) Influence of particle size on lipid digestion and beta-carotene bioaccessibility in emulsions and nanoemulsions. Food Chem 141:1472–1480
24. Tan KWJ, Sun LJ, Goh KKT, Henry CJ (2016) Lipid droplet size and emulsification on postprandial glycemia, insulinemia and lipidemia. Food Funct 7:4278–4284
25. https://smart.servier.com/
26. https://www.horiba.com/en_en/products/by-segment/scientific/particle-characterization

Chapter 22

Surface Load and Interfacial Composition of Protein-Stabilized O/W Emulsions

Emma B. A. Hinderink, Karin Schroën, Leonard M. C. Sagis, and Claire C. Berton-Carabin

Abstract

Oil-in-water (O/W) emulsions, i.e., the dispersion of oil droplets in a continuous water phase, are widely present in foods. In such systems, oil droplets are often physically stabilized by proteins—amphiphilic biopolymers that can act as emulsifiers. Despite the fact that the interface is very thin (typically 1–15 nm for protein-stabilized interfaces), it controls to a large extent the physical stability of the droplets. Therefore, characterizing the adsorption behavior of proteins is of high importance. Measurements of the surface load (i.e., mass of emulsifier adsorbed per unit of oil–water interfacial area) and interface composition are useful in that respect. This chapter provides a protocol and guidelines for the determination of the surface load and interfacial composition of fluid O/W emulsions. Although centered on protein-stabilized emulsions, the procedure described may be easily adapted to emulsions prepared with other water-soluble emulsifiers.

Key words Emulsions, Oil–water interface, Proteins, Droplet surface, Adsorbed emulsifiers, Surface load, Surface coverage, Interfacial composition

1 Introduction

Proteins are commonly used to stabilize food emulsions, in which the oil–water interface is a concentrated region of proteins. Surface load is a measure of the amount of proteins needed to stabilize the oil–water interface, usually denoted by Γ and expressed in mg/m^2. Furthermore, it provides information about the structure and thickness of the layer. Protein ingredients are typically blends of proteins, e.g., whey proteins contain mainly β-lactoglobulin, α-lactalbumin, and other proteins in minor amounts. When using a mixture of proteins, they may compete for adsorption during emulsification [3–5, 9, 10] and possibly displace each other in time [4, 7, 8]. The proteins thus partition between the interface and the water phase, meaning that their overall composition does not reflect the interfacial composition [1]. Therefore, it is of

Christelle Lopez et al. (eds.), *Multidimensional Characterization of Dietary Lipids*, Methods and Protocols in Food Science, https://doi.org/10.1007/978-1-0716-3758-6_22,

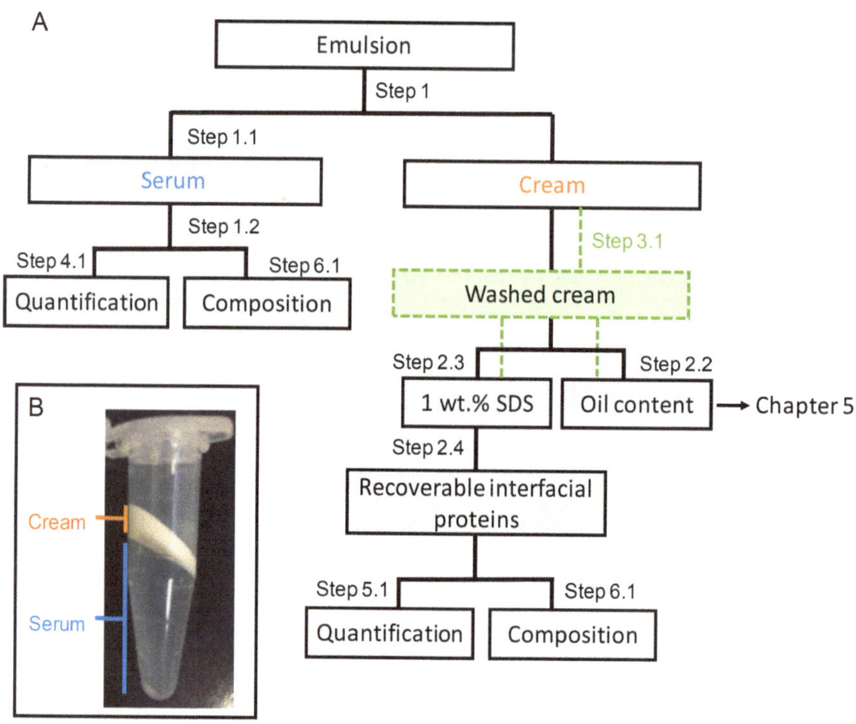

Fig. 1 (**a**) Overview of the procedure to determine the surface load of protein-stabilized emulsions via the serum, creamed or washed cream phases. (**b**) Picture of an Eppendorf tube containing a 10 wt.% O/W emulsion after centrifugation

importance to characterize not only the surface load but also the interfacial composition.

Protein partitioning can be measured in situ using front surface fluorescence [6], which relies on the intrinsic fluorescence of the tryptophanyl residues in aqueous or hydrophobic environments. Unfortunately, this method is not suitable for all proteins due to the small shifts in the fluorescence spectra between adsorbed and non-adsorbed proteins [11]. Therefore, invasive methods, including a centrifugation step to separate the emulsion's phases, are normally used [2]. The surface load and compositions can be quantified indirectly in the water phase (also often called "serum") or directly in the creamed phase (Fig. 1). Using the serum, the amount of adsorbed proteins is calculated by the difference between the total protein amount used to make the emulsion and the measured amount in the serum phase (the non-adsorbed proteins). Measurements via the serum are relatively simple and contain less steps compared to the direct measurements via the creamed phase. However, for emulsions with small droplet sizes ($d_{3,2}$ <0.4 μm), the serum phase still contains many small droplets that cannot be removed by extensive centrifugation or filtration [9]. Proteins adsorbed at the interface of these small droplets falsely contribute to the non-adsorbed protein fraction, which leads to underestimation of the surface load. On the other hand, possible sedimentation of a fraction of non-adsorbed proteins upon centrifugation may lead to overestimation of the surface load, and, therefore, direct measurements via the creamed phase are more appropriate [8, 9]. A washing step of the cream can be applied to remove loosely bound proteins at the interface or those captured between the oil droplets in the creamed phase, which would otherwise be considered adsorbed [8].

The interfacial composition is mostly quantified using sodium dodecyl sulfate polyacrylamide gel electrophoresis (SDS-PAGE) in combination with a calibrated densitometer using the same samples as for the surface load measurements [8, 9, 12]. In case of proteins with similar molecular weights (MWs), more advanced techniques could be considered such as proteomics [15]. This chapter provides a protocol and some practical tips for the determination of the surface load and interfacial composition of fluid O/W emulsions indirectly via the serum and directly via the (washed) cream [8].

2 Materials and Methods

2.1 Materials

Protein-stabilized emulsion (measure the individual droplet size ($d_{3,2}$) of the emulsion by diluting it twofold in 1 wt.% SDS solution prior to droplet size analysis—a higher dilution factor may be considered for emulsions with higher oil fractions), Eppendorf tubes, Eppendorf centrifuge (adapted for 15,000× g), small spatula, pushpin, Stuart™ Rotator Disks (Cole-Parmer™, or any other rotator disk), and a scanner for SDS-PAGE gels connected to software for densitometry analysis.

NOTE: The SDS molecules displace the adsorbed proteins from the interface, giving the droplets a highly negative charge, which results in non-flocculated droplets. This method works well at pH values above the isoelectric point (pI) of the proteins, i.e., proteins are negatively charged. At acidic pH values (pH<pI), the (adsorbed) proteins are positively charged and the addition of SDS may lead to physical destabilization. The use of cationic (e.g., cetrimonium bromide (CTAB)) or nonionic (e.g., Tween20) surfactants is therefore suggested instead of SDS.

2.2 Sample Collection and Preparation

1. Add aliquots of 1.5 mL of emulsion in Eppendorf tubes (2 mL) and centrifuge at 15,000× g for 1 h.

NOTE: These centrifugation settings are suitable for droplets with a $d_{3,2}$ of 400–600 nm. Reduce the time and speed for larger droplets or increase for smaller droplets. The centrifugation settings should not induce droplet coalescence (i.e., the formation of a separate oil layer on top of the sample is a clear indication that the

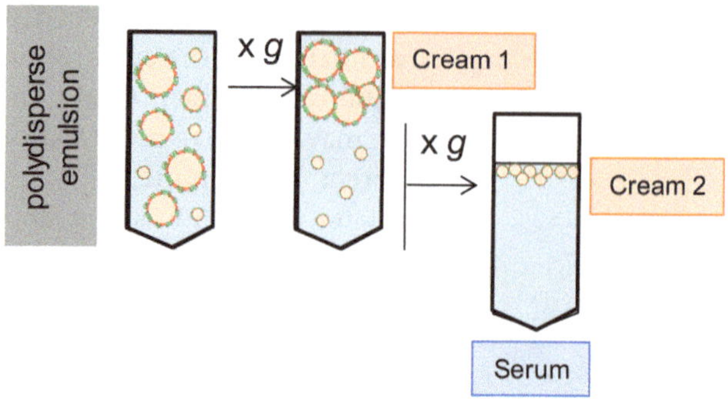

Fig. 2 Schematic representation of the separation procedure via centrifugation for polydisperse emulsions

centrifugation conditions are not appropriate. When no such layer can be seen, it can still be useful to check the droplet size distribution after re-dispersing the creamed phase in 1 wt.% SDS solution to verify that no droplet coalescence occurred).

NOTE: In case of a polydisperse emulsion, two centrifugation steps may be required: (1) at low centrifugation speed to collect the large droplets and (2) at high centrifugation speed of the remaining subnatant to collect the small droplets (Fig. 2).

2.2.1 Serum

1.1 Use a centrifuged Eppendorf tube. Collect the supernatant by carefully making a hole at the bottom of the tube with a pushpin without disturbing the creamed layer.

1.2 Filter the collected supernatant (0.2 μm).

Final sample for the surface load and interface composition via the serum phase. See ***step 4.1*** *for further analysis.*

2.2.2 Cream

NOTE: To determine the surface load via the creamed phase, two parameters have to be measured:

- The oil content of the (washed) cream
- The protein content of the (washed) cream

2.1 Continue with the centrifuged emulsions (obtained after **step 1**). After removal of the serum phase, the cream can be carefully collected using a spatula.

2.2 Carefully collect a known mass of cream (in grams, record 4 decimals) and determine the oil content according to the protocol described in Chap. 5 of this book.

2.3 Carefully collect a known mass of cream (record 4 decimals) and re-disperse in a recorded mass of 1 wt.% SDS (4 decimals) and rotate vertically at 20 rpm for 1 h using a rotator disk, after vigorous vortexing. SDS displaces the adsorbed proteins from the oil–water interface into the serum phase. Subsequently centrifuge the sample at 15,000× *g* for 1 h.

2.4 Collect the final subnatant containing the proteins that were initially adsorbed at the interface.

Final sample for the surface load and interface composition via the creamed phase. See ***step 5.1*** *for further analysis.*

2.2.3 Washed Cream

3.1 Continue with the centrifuged emulsions (**step 1**) and carefully collect the cream. Re-disperse the cream in buffer or ultrapure water in the same weight ratio as the starting O/W emulsion. Rotate the sample at 20 rpm for 1 h, subsequently centrifuge at 15,000× *g* for 1 h. Treat the centrifuged samples further as described for the cream in **steps 2.2–2.4**.

2.3 Protein Analysis and Calculations

2.3.1 Surface Load via the Serum Phase

4.1 Measure the protein concentration of the filtered serum phase (**step 1.2**) using a protein colorimetric assay (e.g., BCA assay [13]) or total nitrogen determination (e.g., Dumas method [14]).

4.2 The surface load (mg/m^2) can be calculated as follows: $\Gamma = \frac{C_s . d_{3,2}}{6\varphi}$, where C_s (mg/L) is the adsorbed protein concentration (calculated by subtracting the protein concentration in the serum phase from the initial protein concentration of the solution used for emulsion preparation), $d_{3,2}$ is the surface-weighed mean droplet diameter after twofold dilution of the emulsion in 1 wt.% SDS, and φ is the dispersed phase volume fraction in the emulsion.

2.3.2 Surface Load via the (Washed) Cream

5.1 Measure the protein concentration in the subnatants containing the initially adsorbed proteins (**step 2.4**) using a protein assay as in **step 4.1** and calculate the total amount adsorbed (mg) per gram cream using the mass of the SDS solution added and the initial mass of the cream.

5.2 Calculate the total surface area ($A_{\text{oil,total}}$, m^2) in the emulsion based on the oil content and the emulsion's average droplet size ($d_{3,2}$ after twofold dilution of the emulsion in 1 wt.% SDS):

- Calculate the number of oil droplets (N_{drop}) in the cream using the total oil volume ($V_{\text{oil,total}}$, m^3) per gram cream (**step 2.2**) and the volume of one droplet ($V_{\text{drop}} = 4/3\ \pi\ (d_{3,2}/2)^3$): $N_{\text{drop}} = V_{\text{oil,total}}/V_{\text{drop}}$.
- Calculate the total area of the oil droplets ($A_{\text{oil,total}}$) in the cream using the N_{drop} and the area of one droplet ($A_{\text{drop}} = 4\pi\ (d_{3,2}/2)^2$): $A_{\text{oil,total}} = N_{\text{drop}}\ A_{\text{drop}}$.

5.3 Finally, the surface load can be calculated by dividing the total amount of proteins adsorbed (mg) by the total surface area ($A_{oil,total}$).

2.3.3 Protein Partitioning

6.1 Take the sample containing the adsorbed proteins (**step 2.4**). Determine the composition of the sample using SDS-PAGE under reducing conditions in combination with a calibrated densitometer (Fig. 3). In case of the serum proteins, the interfacial composition can be estimated by subtracting the serum protein composition from the initial protein composition. The starting solution is used as a reference for all measurements.

NOTE: when estimating the interfacial composition by difference with the serum composition, a substantial fraction of the proteins must be adsorbed for the difference to be significant. If this condition is not met, the washed cream method should be preferred:

- Load the diluted protein samples (~0.5 μg of protein per band) on the SDS-PAGE, using a molecular weight (MW) marker (preferably twice, on both edges of the gel, as illustrated in Fig. 3 (left)).
- After staining the gel, the MW and optical density of the bands are determined using a calibrated densitometer.

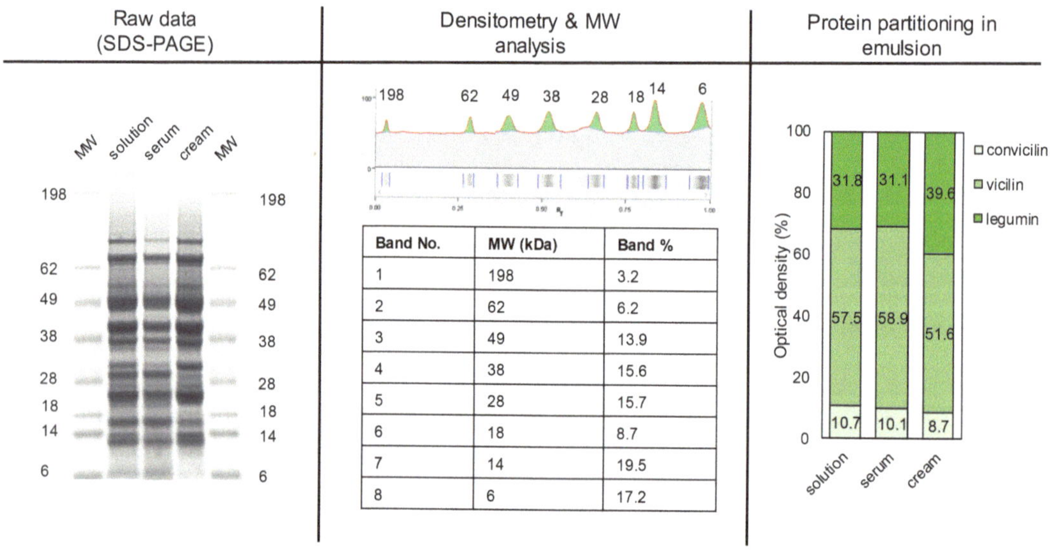

Band No.	MW (kDa)	Band %
1	198	3.2
2	62	6.2
3	49	13.9
4	38	15.6
5	28	15.7
6	18	8.7
7	14	19.5
8	6	17.2

Fig. 3 Schematic representation of the experimental steps to determine protein partitioning in protein-stabilized emulsions. In the example shown, a pea protein-stabilized emulsion was prepared and analyzed, and the sum of the subunits of the different protein constituents (i.e., convicilin, vicilin, and legumin) is reported

- It is recommended to report the average of at least two independently prepared samples.

NOTE: under reducing conditions, the proteins dissociate into their subunits leading to multiple bands in the SDS-PAGE. In the final results, the sum of the subunits or the individual subunits can be reported.

References

1. Berton-Carabin CC, Ropers M-H, Genot C (2014) Lipid oxidation in oil-in-water emulsions: involvement of the interfacial layer. Compr Rev Food Sci Food Safety 13(5): 945–977. https://doi.org/10.1111/1541-4337.12097
2. Berton C, Genot C, Ropers MH (2011) Quantification of unadsorbed protein and surfactant emulsifiers in oil-in-water emulsions. J Colloid Interface Sci 354(2):739–748. https://doi.org/10.1016/j.jcis.2010.11.055
3. Damodaran S (2004) Adsorbed layers formed from mixtures of proteins. Curr Opin Colloid Interface Sci 9(5):328–339. https://doi.org/10.1016/j.cocis.2004.09.008
4. Dickinson E (2011) Mixed biopolymers at interfaces: Competitive adsorption and multilayer structures. Food Hydrocoll 25(8): 1966–1983. https://doi.org/10.1016/j.foodhyd.2010.12.001
5. Euston SE, Singh H, Munro P, a., & Dalgleish, D. G. (1996) Oil-in-water emulsions stabilized by sodium caseinate or whey protein isolate as influenced by glycerol monostearate. J Food Sci 61(5):916–920. https://doi.org/10.1111/j.1365-2621.1996.tb10901.x
6. Granger C, Barey P, Toutain J, Cansell M (2005) Direct quantification of protein partitioning in oil-in-water emulsion by front-face fluorescence: avoiding the need for centrifugation. Colloids Surf B Biointerfaces 43(3–4): 158–162. https://doi.org/10.1016/j.colsurfb.2005.04.002
7. Hinderink EBA, Münch K, Sagis L, Schroën K, Berton-Carabin CC (2019) Synergistic stabilisation of emulsions by blends of dairy and soluble pea proteins: contribution of the interfacial composition. Food Hydrocoll 97. https://doi.org/10.1016/j.foodhyd.2019.105206
8. Hinderink EBA, Sagis L, Schroën K, Berton-Carabin CC (2021) Sequential adsorption and interfacial displacement in emulsions stabilized with plant-dairy protein blends. J Colloid Interface Sci 583:704–713. https://doi.org/10.1016/j.jcis.2020.09.066
9. Hunt JA, Dalgleish DG (1994) Adsorption behaviour of whey protein isolate and caseinate in soya oil-in-water emulsions. Topics Catal 8(2):175–187. https://doi.org/10.1016/S0268-005X(09)80042-8
10. McClements DJ, Jafari SM (2018) Improving emulsion formation, stability and performance using mixed emulsifiers: a review. Adv Colloid Interface Sci 251:55–79. https://doi.org/10.1016/j.cis.2017.12.001
11. Rampon V, Genot C, Riaublanc A, Anton M, Axelos MAV, McClements DJ (2003) Front-face fluorescence spectroscopy study of globular proteins in emulsions: displacement of BSA by a nonionic surfactant. J Agric Food Chem 51(9):2482–2489. https://doi.org/10.1021/jf026168g
12. Schröder A, Berton-Carabin CC, Venema P, Cornacchia L (2017) Interfacial properties of whey protein and whey protein hydrolysates and their influence on O/W emulsion stability. Food Hydrocoll 73:129–140. https://doi.org/10.1016/j.foodhyd.2017.06.001
13. Smith PK, Krohn RI, Hermanson GT, Mallia AK, Gartner FH, Provenzano MD et al (1985) Measurement of protein using bicinchoninic acid. Anal Biochem 150(1):76–85. https://doi.org/10.1016/0003-2697(85)90442-7
14. Wiles PG, Gray IK, Kissling RC (1998) Routine analysis of proteins by Kjeldahl and Dumas methods: review and interlaboratory study using dairy products. J AOAC Int 81(3): 620–632. https://doi.org/10.1093/jaoac/81.3.620
15. Xiong T, Ye X, Su YT, Chen X, Sun H, Li B, Chen Y (2018) Identification and quantification of proteins at adsorption layer of emulsion stabilized by pea protein isolates. Colloids Surf B Biointerfaces 171(April):1–9. https://doi.org/10.1016/j.colsurfb.2018.05.068

Part V

Characterization of the Physical Properties of Dietary Lipids and Lipid Structures

Chapter 23

Thermal Properties of Dietary Lipids Investigated Using Differential Scanning Calorimetry

Christelle Lopez

Abstract

Dietary lipids can change their physical state as a function of temperature. Particularly, dietary fats containing saturated fatty acids exhibit crystallization properties. Differential scanning calorimetry (DSC) is the technique the most frequently used to determine the thermal properties of dietary fats, i.e., their crystallization and melting properties as a function of temperature, and to determine the characteristic temperatures and enthalpies. This chapter presents the DSC technique applied to dietary fats.

Key words Crystallization, Melting, Temperature, Enthalpy

1 Introduction

Dietary lipids are mainly composed of triacylglycerols (TAGs) that are neutral lipids in which 3 fatty acids are esterified on a glycerol backbone (Fig. 1). Each TAG can contain various fatty acids (mixed TAG). Each fatty acid exhibits a melting point that depends on the saturation of the acyl chain and on the number of carbon atoms (Fig. 1).

The thermal properties of TAG molecules depend on their fatty acid composition (combination of the 3 fatty acids), on the position of the fatty acids in the 3 positions of glycerol (*sn-1*, *sn-2*, and *sn-3* positions), and on the polymorphism of TAGs. TAG molecules are able to pack differently in the solid state and the α, β', and β polymorphic forms have different melting temperatures: $\alpha < \beta' < \beta$ (Fig. 2).

Dietary fats that contain a wide diversity of TAGs can exhibit melting and crystallization behaviors in a large range of temperatures (e.g., from −40 °C to 40 °C for milk fat), contrary to pure TAG molecules.

Crystallization and melting of dietary fats not only occur in the anhydrous or bulk states (e.g., palm oil and anhydrous milk fat) but

Christelle Lopez et al. (eds.), *Multidimensional Characterization of Dietary Lipids*, Methods and Protocols in Food Science,
https://doi.org/10.1007/978-1-0716-3758-6_23,

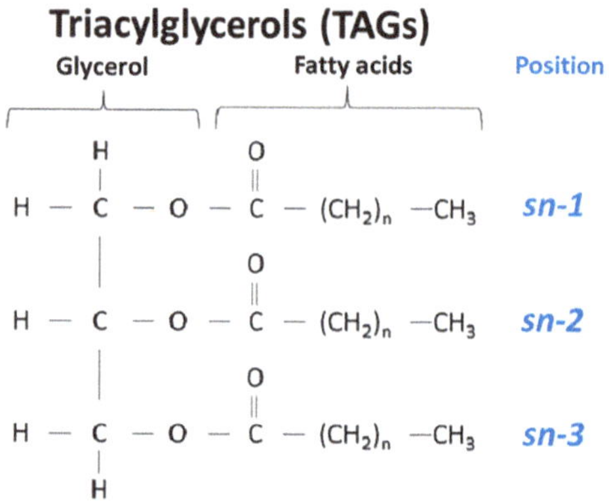

Fatty acids		
Formula	Name	Melting point (°C)
C4:0	Butyric acid	-8
C6:0	Caproic acid	-4
C8:0	Caprylic acid	16
C10:0	Capric acid	31
C12:0	Lauric acid	44
C14:0	Myristic acid	54
C16:0	Palmitic acid	63
C18:0	Stearic acid	70
C18:1	Oleic acid	16
C18:2	Linoleic acid	-5
C18:3	Linolenic acid	-14

↗ melting point with the increase in the number of atoms of carbon

↘ Melting point with the number of unsaturation

Fig. 1 Structure of a triacylglycerol (TAG) and melting points of fatty acids depending on their chain length and saturation

TAG molecular species	Melting temperature (°C) α	β′	β
Tri-saturated			
SSS	54·7	64·3	72·5
PPP	44·7	55·7	65·9
SPS	50·7	64·0	68·0
SSP	50·1	61·8	64·4
PSP	47·2	67·7	65·3
SPP	46·4	58·7	62·6
Monounsaturated			
SOS	37·0	41·5	44·0
SSO	30·3	41·9	NA
POP	26·5	33·5	37·2
PPO	18·4	31·0	34·5
SOP	19·6	33·0	37·1
SPO	17·9	NA	40·2
PSO	25·9	40·0	41·0
PPL	NA	26·5	36·0
SSL	NA	NA	35·8
Di-unsaturated			
OPO	NA	NA	19·6
OOP	NA	NA	18·5
OSO	NA	20·5	23·9
OOS	NA	NA	23·5
Tri-unsaturated			
OOO	−33·7	−10·0	4·8
LLL	NA	−25·3	−12·3

S, stearic acid; P, palmitic acid; O, oleic acid; NA, not available; L, linoleic acid.

Fig. 2 Melting temperature of triacylglycerol (TAG) as a function of their fatty acid composition and polymorphic form (α, β′, or β)

also in dispersed systems such as oil-in-water emulsions and in food products (crystallization of milk fat in butter, dairy creams and cheeses, and crystallization of cocoa butter in chocolate).

Differential scanning calorimetry (DSC) is the technique most frequently used for studying the thermal properties of dietary fats.

2 Method: Differential Scanning Calorimetry

2.1 Principle of Differential Scanning Calorimetry

Differential scanning calorimetry (DSC) is a technique used for measuring the energy necessary to establish a nearly zero temperature difference between a substance and an inert reference material, as the two specimens are subjected to identical temperature regimes in an environment heated or cooled at a controlled rate.

A calorimeter is a commercially available instrument used for DSC, or the process of measuring the heat of chemical reactions or physical changes as well as heat capacity.

DSC measures the differential heat flows associated with transitions in substance such as dietary lipids (crystallization, melting, and polymorphic evolutions) and reference materials. Both the sample and the reference (e.g., an empty pan) are linearly heated or cooled at a predetermined rate as a function of time, dT/dt (T = Temperature, t = time), in a controlled environment. During a thermal event in the sample, the system will transfer heat to or from the sample pan to maintain the same temperature in the reference and sample pans. DSC measurements provide quantitative and qualitative information about physical and chemical changes that involve endothermic or exothermic processes, or changes in heat capacity.

DSC is the generic term for the following two measurement methods: power compensation DSC and heat flux DSC.

- *Power Compensation DSC*

It is a technique in which thermal energy difference that is applied to the sample and the reference material per unit of time is measured as a function of the temperature to equalize their temperature, while temperature of the sample unit, formed by the sample and reference material, is varied in a specified program. The sample and the reference are placed into two different ovens but in the same calorific heat chamber (Fig. 3).

- *Heat Flux DSC*

It is a technique in which the temperature of the sample unit, formed by a sample and reference material, is varied in a specified program, and the temperature difference between the sample and the reference material is measured as a function of temperature (Fig. 4). The sample and the reference are placed into the same oven.

Figure 4 shows the block diagram of the heat flux DSC. Heat flux DSC comprises the sample and reference holder, the heat resistor, the heat sink, and the heater. Heat generated from the heater is supplied into the sample and the reference through the heat sink and the heat resistor. Heat flow is proportional to the heat difference between the heat sink and holders. The heat sink has

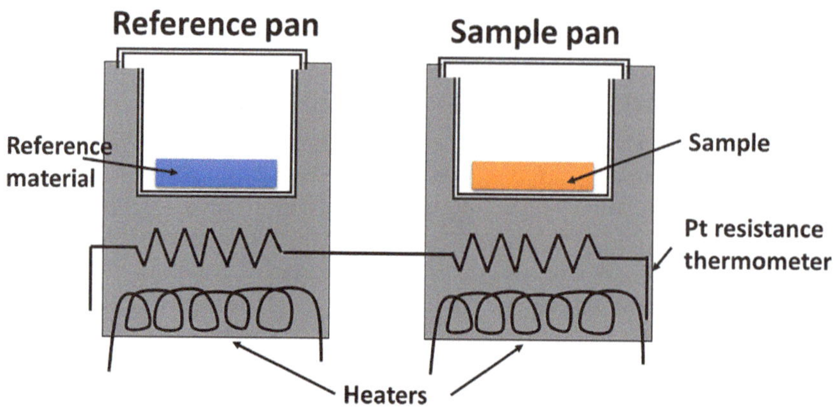

Fig. 3 Block diagram of power compensation DSC

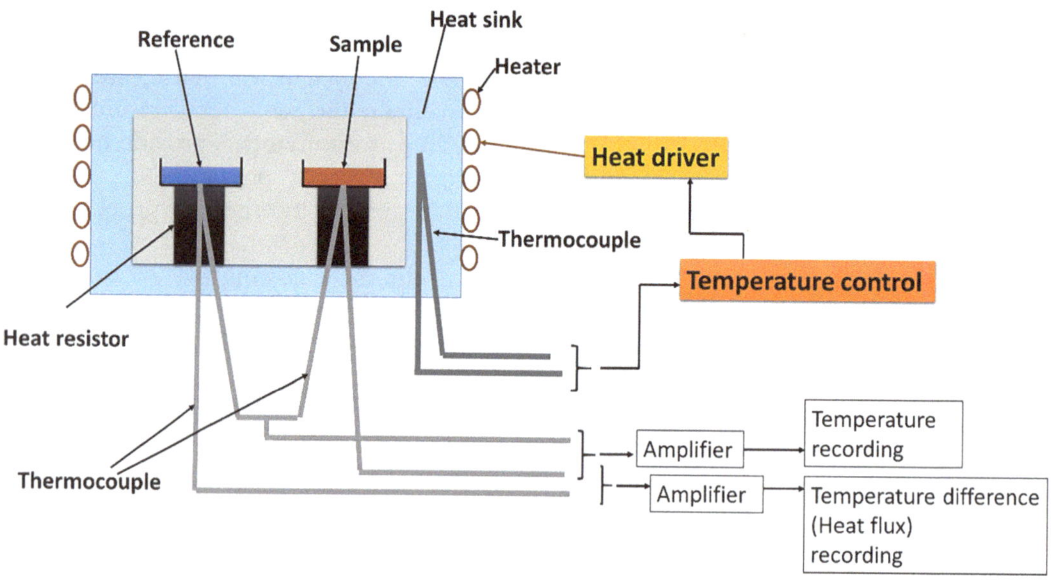

Fig. 4 Block diagram of heat flux DSC

enough heat capacity compared to the sample. In case the sample occurs endothermic or exothermic phenomena such as transition and reaction, this endothermic or exothermic phenomena is compensated by heat sink. Thus, the temperature difference between the sample and the reference is kept constant. The difference in the amount of heat supplied to the sample and the reference is proportional to the temperature difference of both holders. By calibrating the standard material, the unknown sample quantitative measurement is achievable.

NOTE: Modulated DSC can also be used to investigate the thermal properties of materials. Heating and cell arrangement are the same as in the heat flux DSC. With the help of the programmed Fourier transform, the overall signal is mathematically deconvoluted into two parts: (1) a reversing heat flow signal, and (2) a non-reversing heat flow signal. Usually step transitions such as glass transition appear only in the reversing heat flow signal, and exothermic or endothermic events may appear in both signals.

2.2 Sample Conditioning

Samples can be weighed in hermetic aluminum pans (volume about 50 μL) or other kinds of pans adapted for DSC. An empty pan is used as a reference.

- Use a tong to manipulate the pan and the lid.
- Weigh accurately the pan and the lid.
- Fill the pan with the sample.
- Crimp the lid on the top of the filled pan.
- Weigh accurately the crimped pan to determine the weight of the sample.
- The crimped pan filled with the sample is ready for DSC experiments.
- A reference pan (an empty pan or a pan containing a reference material such as the buffer contained in the sample, e.g., in the case of oil-in-water emulsions) is also prepared for the DSC experiments.
- The pans are loaded into the calorimeter.

2.3 Calibration

Calibration of the calorimeter is essential to obtain results with correct enthalpies and temperatures. Calibration is generally performed by using molecules with well-known thermal properties (e.g., indium and pure fatty acids or triacylglycerols)

2.4 Examples of DSC Curves

2.4.1 Pure TAG

In Fig. 5, the DSC curve recorded on cooling of the TAG tripalmitin (PPP; C16:0-C16:0-C16:0) at 10 °C/min shows a single exotherm with $T_{onset} = 41.15$ °C. The DSC curve recorded on subsequent heating at 10 °C/min shows a complex thermotropic behavior with an endotherm, an exotherm, and a final endotherm with a melting temperature $T_{offset} = 67.3$ °C. This complex DSC curve reveals the polymorphic transitions occurring on heating of PPP: melting of α form (first endotherm) → formation of β′ form (exotherm) → melting of β′ form and formation of β form → melting of the β form (last endotherm) → liquid state of PPP above T_{offset}. Only the combination of DSC with X-ray diffraction permits the identification of the crystalline polymorphic forms of TAGs [2].

2.4.2 Complex Mixture of TAGs in Dietary Fats

DSC studies have given an insight into the thermodynamics of fat phase transition of dietary fats in bulk and in emulsions. For example, authors who studied anhydrous milk fat by DSC observed that it crystallizes and melts in several steps (Fig. 6). This is due to its complex fatty acid and TAG composition (more than 400 fatty

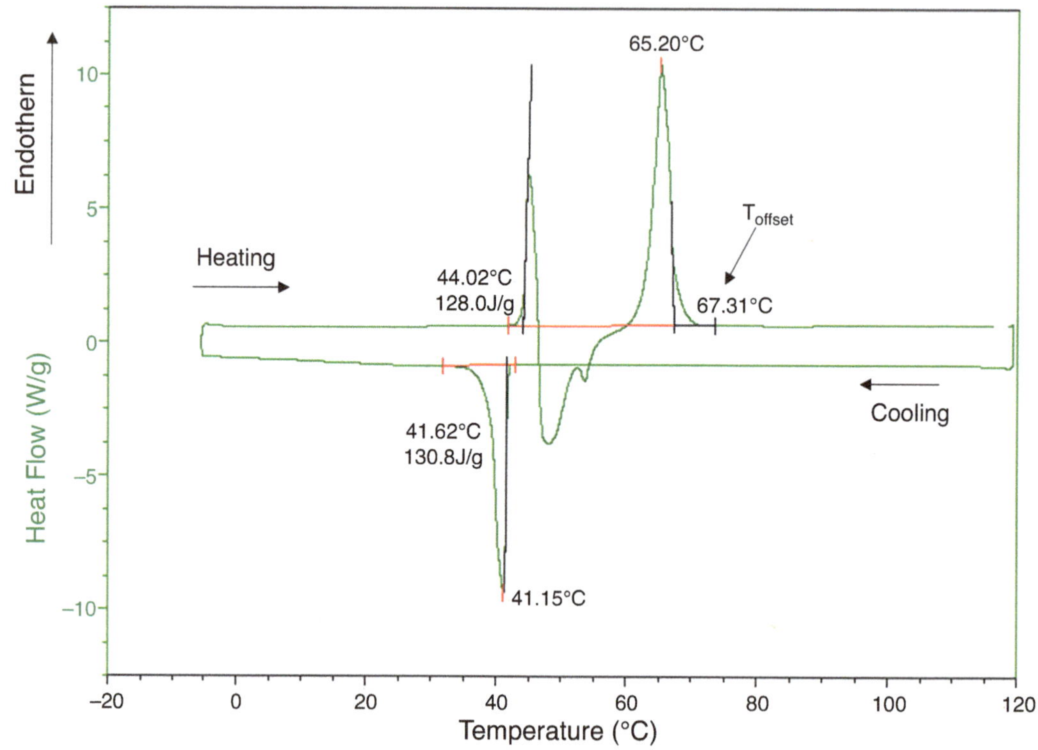

Fig. 5 DSC curves recorded on cooling and subsequent heating of the triacylglycerol tripalmitin (PPP; C16:0-C16:0-C16:0)

acids and 100 TAGs identified in milk fat) and also to the complex polymorphism of TAG molecules.

A typical crystallization curve of anhydrous milk fat recorded from the melt (from the liquid state of milk fat) shows two exothermic peaks on cooling corresponding to the successive crystallization of the high crystallization temperature fraction of TAG that is mainly composed of TAGs with saturated long-chain fatty acids and the low crystallization temperature fraction that contains TAGs with unsaturated fatty acids (oleic acid; C18:1 n-9) and/or short-chain fatty acids (butyric acid; C4:0). On subsequent heating, a typical melting curve of anhydrous milk fat shows three endothermic peaks, corresponding to the successive melting of the low melting point (LMP), medium melting point (MMP), and high melting point (HMP) fractions of TAGs. These peaks correspond to large groups of TAGs that melt separately and behave like solid solutions. The main TAGs that melt in the LMP, MMP, and HMP fractions are indicated in Fig. 6. Only the combination of DSC with X-ray diffraction permits the identification of the crystalline polymorphic forms of TAGs [1].

Figure 7 compares the DSC curves recorded on heating of various dietary fats. The melting profiles of the dietary fats results

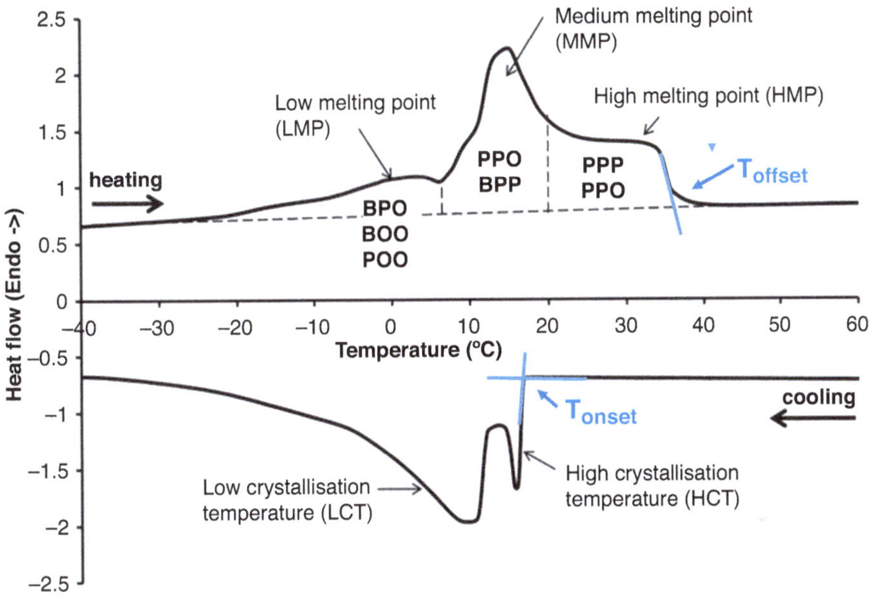

Fig. 6 DSC curves recorded for anhydrous milk fat, on cooling from the melt and on subsequent heating at 2 °C/min. The initial temperature of crystallization (T_{onset}) and the final temperature of melting (T_{offset}) are indicated in the figure. The main triacylglycerols that melt in the LMP, MMP, and HMP fractions are indicated in the figure. Abbreviations: B butyric acid (C4:0), P palmitic acid (C16:0), O oleic acid (C18:1 n-9)

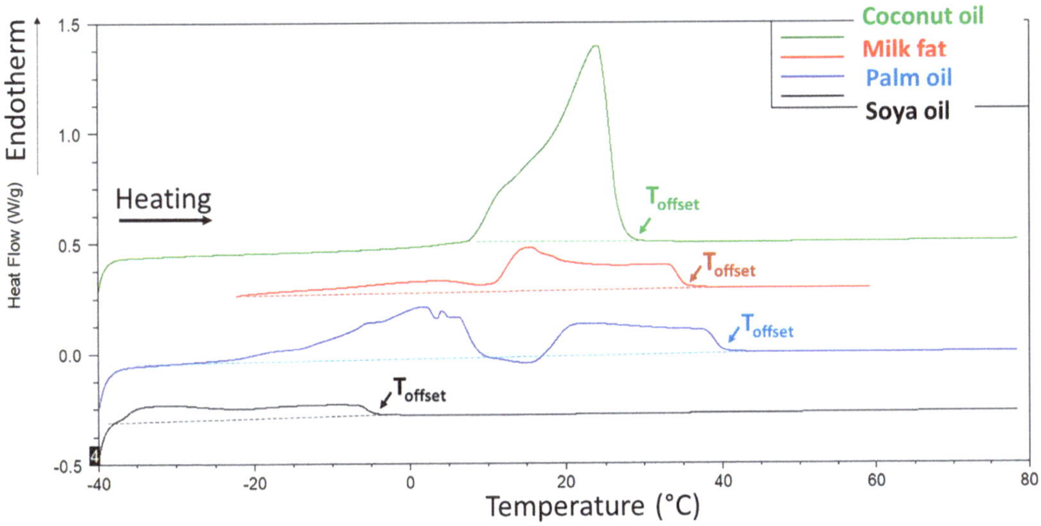

Fig. 7 DSC curves recorded on heating of various dietary fats: coconut oil, milk fat, palm oil, and soya oil

from their fatty acid and TAG composition. Some dietary fats are liquid at room temperature (e.g., soya oil), and other dietary fats are partially solid at room temperature and melt above 20 °C (e.g., coconut oil, milk fat, and palm oil).

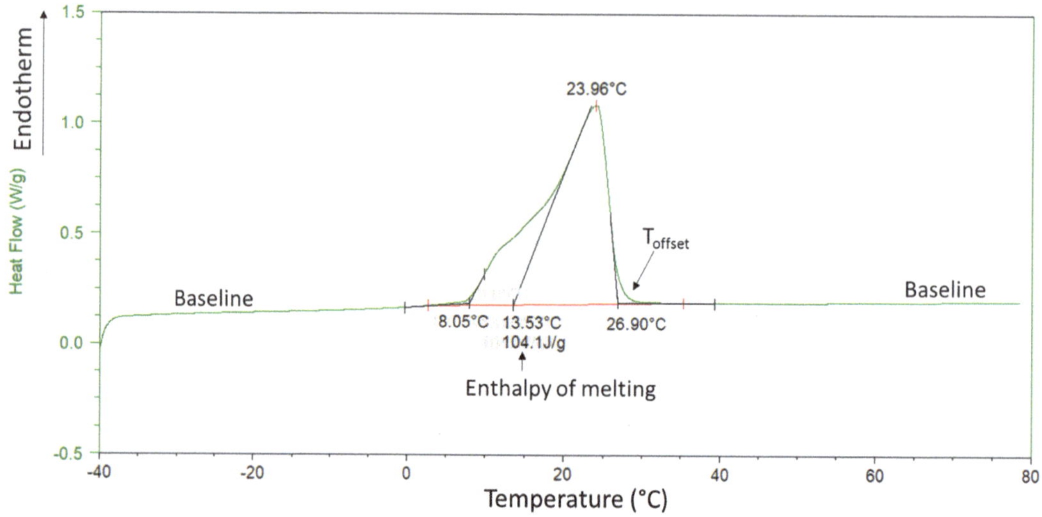

Fig. 8 DSC curve recorded on heating of coconut oil. Determination of characteristic parameters such as T_{offset} (the temperature of final melting) and the enthalpy of melting

The occurrence of numerous thermal transitions during DSC experiments, the partial overlapping of the melting peaks, and their respective enthalpies and temperatures depend strongly on the thermal treatments (e.g., heating and cooling rates, and tempering) and on the entire thermal history of the sample. The complex DSC recordings are difficult to interpret. Thus, DSC experiments need to be coupled with other techniques, such as X-ray diffraction.

2.5 Treatment of the DSC Curves

It is essential to determine the baseline and to differentiate the endothermic events from the exothermic events. The endothermic signal, up or down from the baseline, must be indicated in the DSC figures.

Several parameters can be extracted from the DSC curves, such as the temperature of the beginning of crystallization (T_{onset}), the temperature of final melting (T_{offset}), and the enthalpy of melting (ΔH).

These parameters are generally calculated by the software. An example is given Fig. 8.

3 Specific Care

1. *Presence of water in the samples*: The thermal properties of anhydrous fats can be investigated in a wide range of temperatures, e.g., from −60 to 80 °C. In case of the presence of water in the samples, e.g., in oil-in-water emulsions, decreasing the temperature below 0 °C can induce ice formation with consequences on the structure of the sample.

2. *Keeping the DSC cell clean*: One of the first steps to ensure good DSC data is to keep the DSC cell clean. The sample must be loaded into the pans, and the external part of the pans must be clean. DSC cells can get dirty after decomposition of the samples or sample spilling out of the pan during DSC runs. It is absolutely necessary to be careful about possible pollution of the oven by a sample. In case of pollution, the oven must be cleaned. Brush gently both sensors and the cell if necessary.

References

1. Lopez C (2020) Crystallization and melting properties of milk fat. In: Truong T, Lopez C, Bhandari B, Prakash S (eds) Dairy fat products and functionality: fundamental science and technology. Springer International Publishing, Cham, pp 205–243. https://doi.org/10.1007/978-3-030-41661-4_9
2. Ollivon M, Keller G, Bourgaux C, Kalnin D, Villeneuve P, Lesieur P (2006) DSC and high resolution X-ray diffraction coupling. J Therm Anal Calorim 85(1):219–224. https://doi.org/10.1007/s10973-005-7351-y

Chapter 24

Crystallization Properties of Dietary Lipids Characterized by X-Ray Diffraction

Christelle Lopez

Abstract

X-ray diffraction (XRD) is a technique widely used to elucidate the organization of dietary lipids, mainly triacylglycerols (TAGs), in their solid state. XRD can be performed at the laboratory beamline or using high-flux synchrotron radiation. XRD patterns can be recorded at a single temperature or as a function of temperature to follow the crystallization properties of TAG on cooling, their melting behavior on heating, and polymorphic changes that can occur as a function of temperature and time.

Key words Crystallization, Solid lipid phase, Polymorph

1 Introduction

Dietary lipids (also called fats or oils) are mainly composed of triacylglycerols (TAGs; 95% of dietary lipids) that are esters of fatty acids (FAs) and glycerol. As a result of their fatty acid composition, dietary lipids can be (i) in a liquid state at room temperature (e.g., vegetable oils rich in unsaturated fatty acids such as linseed oil, sunflower oil, and rapeseed oil) or (ii) in a semi-solid state corresponding to a mixture of TAG crystals and liquid oil over a wide range of temperatures, including the temperature of storage (4–7 °C), consumption, and digestion (37 °C) of food products. Dietary fats in a semi-solid state are, for example, milk fat, cocoa butter, and palm oil that are rich in high melting temperature long-chain saturated fatty acids.

TAG crystals are involved in the rheological and sensorial properties of food products (e.g., chocolate containing cocoa butter, and butter, cream, and cheeses containing milk fat). TAG crystals can be responsible for quality defaults, e.g., the blooming in chocolate that is due to crystallization of cocoa butter in a specific polymorphic form of TAGs. The crystallization properties of dietary fats can be affected by many parameters such as the cooling rate

Christelle Lopez et al. (eds.), *Multidimensional Characterization of Dietary Lipids*, Methods and Protocols in Food Science,
https://doi.org/10.1007/978-1-0716-3758-6_24,

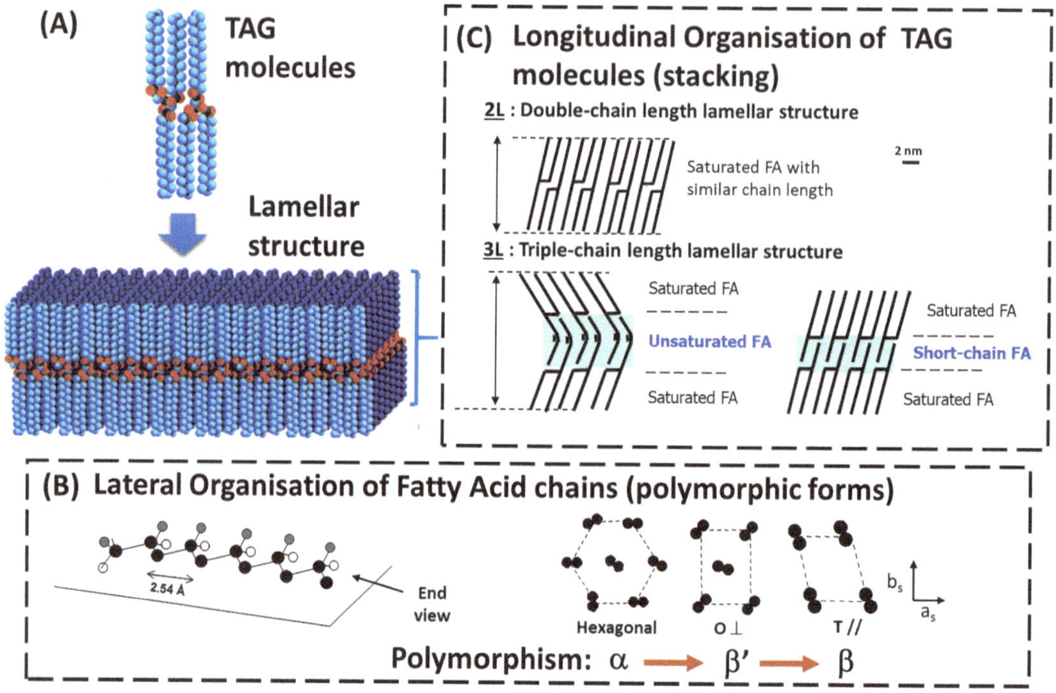

Fig. 1 Main types of TAG packings. (**A**) Lamellar structure formed by triacylglycerol (TAG) molecules in the solid state: example of the β form of trilaurin. (**B**) Left: Stable conformation of the hydrocarbon chains of saturated fatty acid (FA) is a planar zigzag shown here as a 3D view along its main axis. Right: Three main types of chain lateral packings (only carbon atoms are drawn, hydrogen atoms are not drawn): hexagonal, orthorhombic perpendicular (O⊥), and triclinic parallel (T//) subcells to which α, β', and β polymorphic forms correspond. (**C**) Two main types of TAG chain longitudinal stacking (fatty acids are drawn as straight lines). (Adapted from [1])

and thermal history, the shear, the presence of minor lipid components, the dispersion state (anhydrous *versus* emulsified), and changes in the fatty acid and TAG compositions. Besides the solid fat content, which is determined by nuclear magnetic resonance (NMR) at a given temperature, the functionality of a fat is also affected by the molecular organisation of TAG (polymorphic forms), the number, the size and the shape of TAG crystals that can form networks.

In the solid state, TAG molecules can organize as lamellar structures (Fig. 1). Each TAG can exhibit several crystalline forms due to polymorphism, the occurrence of which strongly depends on the thermal history of the dietary fat. TAG polymorphism relates to the ability of TAG molecules to arrange themselves within a crystal lattice in a number of different ways such as lateral packing of the fatty acid chains and longitudinal stacking of TAG molecules in lamellar structures (Fig. 1). Thus, pure TAGs and mixtures of TAGs such as those naturally occuring in dietary fats (e.g., cocoa butter, milk fat, palm oil) can adopt several crystalline arrangements.

The three main polymorphic forms frequently observed for the lateral packing of fatty acid chains correspond to different subcells that have been described in detail: hexagonal (α form), orthorhombic perpendicular (β′ form), and triclinic parallel (β form) (Fig. 1B). Transitions between the polymorphic forms are mostly irreversible; monotropism implies that they are only possible via a liquid and when leading to the formation of more stable forms. The density, enthalpy of fusion, melting point, and stability increase in the order of α, β′, and β, according to the monotropic character of the polymorphism. In the α form, the packing of chains is not very tight and the chains have considerable rotational freedom, whereas in the β form, the chains are very densely packed. Each polymorphic form is characterized by its own melting point. The fact that TAG mixtures exhibit multiple melting points, depending on their composition and polymorphism, renders the overall melting behavior of dietary fats even more complex.

TAG crystals are made by the stacking of TAG molecule layers (Fig. 1C), the thickness of which depends on the length and unsaturation of the fatty acid chains and their angle of tilt with respect to the basal planes formed by the methyl end groups of the TAGs. The longitudinal organizations of TAGs in lamellar structures are primarily related to the number of chains stacked one on top of the other in the main crystalline cell. For TAGs, this number frequently takes a value of 2 or 3 and corresponds to the stacking of double (2L) or triple (3L) chain length lamellar structures, leading to distances in the range 35–50 Å and 55–75 Å, respectively. Roughly, 3L forms are usually related to long-chain monounsaturated and/or mixed long- and short-chain TAGs, whereas 2L forms are generated mostly by similar long-chain trisaturated TAG as illustrated Fig. 1C.

Understanding the crystallization and melting behaviors of dietary TAGs as well as the polymorphism of dietary TAGs is of great importance from a scientific point of view and with regard to the economic impact of dietary fats, especially in fat-rich food products.

The identification of the crystalline structures formed by TAG molecules is possible by using X-ray diffraction (XRD) recorded at both small and wide angles to have information on the longitudinal organization of TAG molecules, e.g., double chain length (2L) or triple chain length (3L) and on the packing of the FA chains (polymorphic forms: α, β′, and β), respectively.

2 Method: X-Ray Diffraction

2.1 Principle of X-Ray Diffraction

X-ray diffraction (XRD) is an essential tool for elucidating the molecular packing of TAG molecules in the solid state and polymorphism of pure TAGs and complex fats including dietary fats, since it provides structural information on the organization of TAG molecules, and it complements differential scanning calorimetry (DSC) experiments.

X-rays are the ideal direct probe to determine the internal structure of crystals, since they provide information on the repetitive patterns of the electron density of the array of atoms. The atoms that constitute the TAG molecules inside a perfect crystal are arranged periodically in planes at a repetitive distance, d (Fig. 2).

As a beam of X-rays comes to the crystal at an angle θ, the X-rays of wavelength λ are reflected by the planes at that same angle θ. The optical path length required to produce constructive interference is nλ, with integral values of n, providing specific crystal symmetry allows it. The equation that determines the angle at which constructive interference occurs is known as Bragg's law: $2d\sin\theta = n\lambda$. The actual diffraction measurement during an experiment is the one formed between the incident beam and the reflected beam, which is denoted by 2θ. It is often convenient to use the scattering vector q instead of the scattering angle 2θ since the former is independent of the X-ray wavelength. They are related as follows: $q = (4\pi/n\lambda)\sin\theta$. Thus, $q = 2\pi/d$ (Fig. 2).

The two levels of organization, i.e., the lateral packing of the fatty acid chains and the longitudinal stacking of TAG molecules in lamellae formed in the solid state (Fig. 1), are easily identifiable from XRD patterns recorded at wide and small angles, respectively.

The cross-sectional packings of the aliphatic chains are characterized by specific short spacings, independent of the chain length

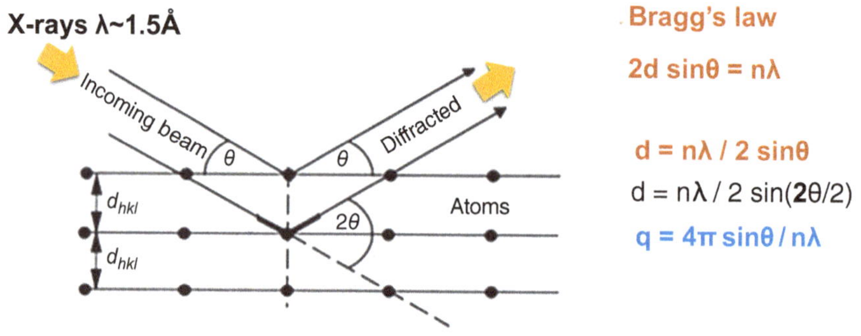

Fig. 2 Demonstration of Bragg's law of diffraction: $2d\sin\theta = n\lambda$ (d in Å is the repetitive distance between the planes arranged periodically; θ in degrees is the angle of incidence of X-ray relative to the crystalline plane; λ in Å is the X-ray wavelength; the integral values of n are required to produce constructive interference nλ)

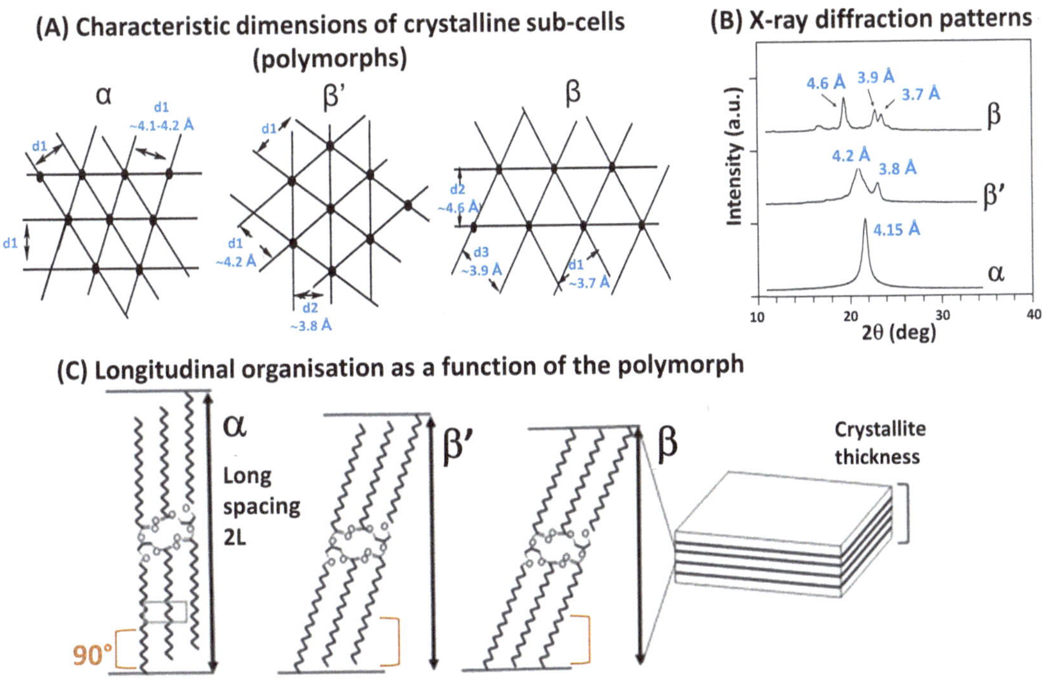

Fig. 3 TAG polymorphs characterized by X-ray diffraction (XRD). (**A**) Characteristic dimensions of crystalline sub-cells (polymorphs). (**B**) XRD patterns recorded at wide angles. (**C**) Longitudinal organization as a function of the polymorphic form

and observable at wide angles in the range of $10° < \theta < 30°$. Short spacings are widely used for identifying the various crystalline subcells characterizing the polymorphic forms (Figs. 1B and 3). A single line around 4.15 Å characterizes the α form (hexagonal subcell) and a strong line at 4.6 Å among other sharp lines identifies the β form (triclinic parallel subcell), while the β' form (orthorhombic perpendicular subcell) frequently shows association of two lines, among which one is about 4.2 Å and the other around 3.8 Å (Fig. 3).

The thickness of the lamellar structures, which can be measured by XRD at small angles ($0° < \theta < 5°$), corresponds to TAG longitudinal stacking called TAG long spacings. As a function of the polymorphic form, the TAG stacking can exhibit an angle of tilt (Fig. 3C). From the measurement of the thickness of the stacking d (in Å) and with knowledge of the fatty acid composition (number of carbon atoms, chain length, and unsaturation), it is possible to deduce if the stackings correspond to 2L or 3L organizations (Fig. 1C).

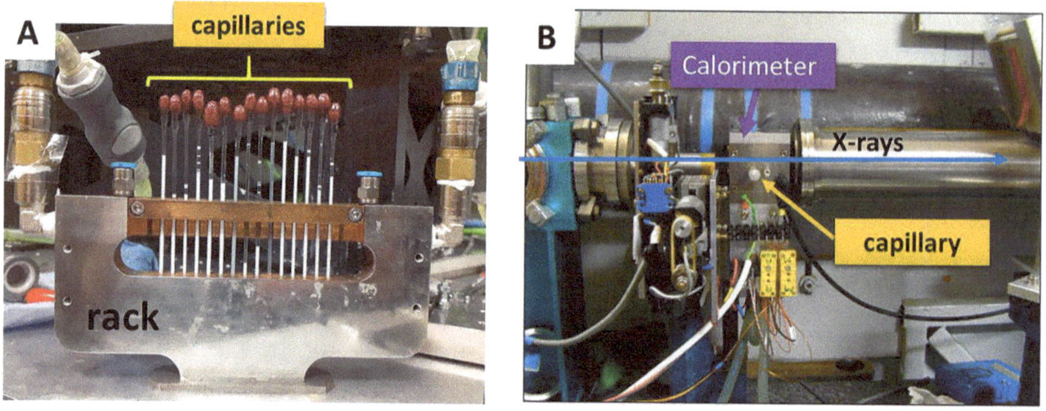

Fig. 4 Capillaries filled with samples positioned in a rack (**A**) or a calorimeter (**B**) [4]. The X-ray beam will pass through the samples and X-rays will be diffracted at defined angles, permitting elucidation of the organization of dietary lipids in the samples

2.2 X-Ray Sources: Laboratory X-Ray Sources Versus Synchrotron Radiation

Recent use of high energy flux of synchrotron radiation XRD, which provides X-ray flux that is 10^3 to 10^6 times more intense than that generated by usual laboratory X-ray sources, permits recordings to be performed within a few seconds or milliseconds. Thus, direct continuous recordings can be performed as a function of time (XRDt) or temperature (XRDT). Moreover, synchrotron radiation permits the study of the organization of TAGs in water-dispersed systems such as emulsions and food products and to quantitatively monitor phase changes within emulsion lipid droplets. This is especially interesting to relate the textural and rheological properties of milk fat, and high-fat food products, to the thermal and crystallographic properties of TAGs investigated thanks to the use of DSC and XRD techniques.

2.3 Sample Conditioning

Samples, e.g., anhydrous fats, emulsions, and pieces of butter or cheese, can be loaded into capillary tubes (quartz, glass, or borosilicate capillaries) that are adapted to record XRD patterns by transmission through these materials. These capillaries are specially designed for X-ray studies since they allow minimum attenuation of the beam and parasitic scattering. In the examples given in Fig. 4, the glass capillaries have 0.01 mm of wall thickness and diameter $\Phi = 1.40 \pm 0.10$ mm. The capillaries are loaded using a syringe and a tube or a needle.

These capillaries can be introduced into various sample environments, such as a capillary rack (operating at room temperature or regulated in temperature, e.g., from 20 to 60 °C), stages regulated in temperature (e.g., Linkam stage; from −200 to 600 °C), or a calorimeter allowing the coupling of XRD with DSC [4] (Fig. 4B).

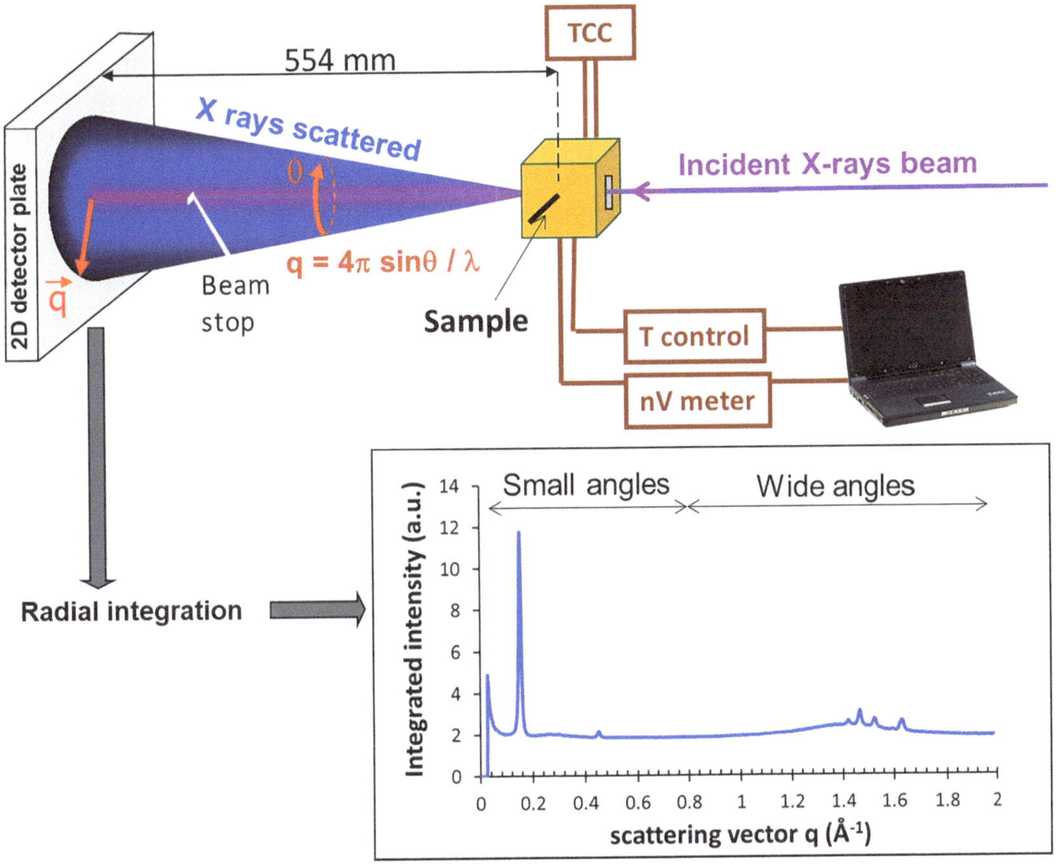

Fig. 5 Set-up showing the sample filled in a capillary introduced in a calorimeter positioned into the X-ray beam allowing diffraction. The X-rays scattered at both small and wide angles, θ, are collected into a detector and radially integrated. (Adapted from [2])

2.4 Detection of the Diffracted X-Rays and Calibration

The diffracted X-rays can be detected by a single detector (Fig. 5) or by two separate detectors permitting XRD at small angles for the characterization of the longitudinal organization of TAG molecules and at wide angles for the identification of the chain packing, respectively (Fig. 1).

The channels of the detectors are calibrated with molecules having strong diffraction and well-known structures to express the XRD data in scattering vector q ($q = 4\pi \sin \theta/\lambda$; q in $Å^{-1}$, θ in degrees is the angle of incidence of the X-ray relative to the crystalline plane and λ in Å is the X-ray wavelength). For example, silver behenate, which is characterized by a long spacing of 58.380 ± 0.001 Å, can be used to calibrate the detector at small angles (Fig. 6A, B). The crystalline 2L β form of high-purity tristearin can be used as a reference for the wide-angle channel to scattering vector q calibration of the detector (d = 4.59, 3.85, 3.70 ± 0.01 Å) (Fig. 6C, D).

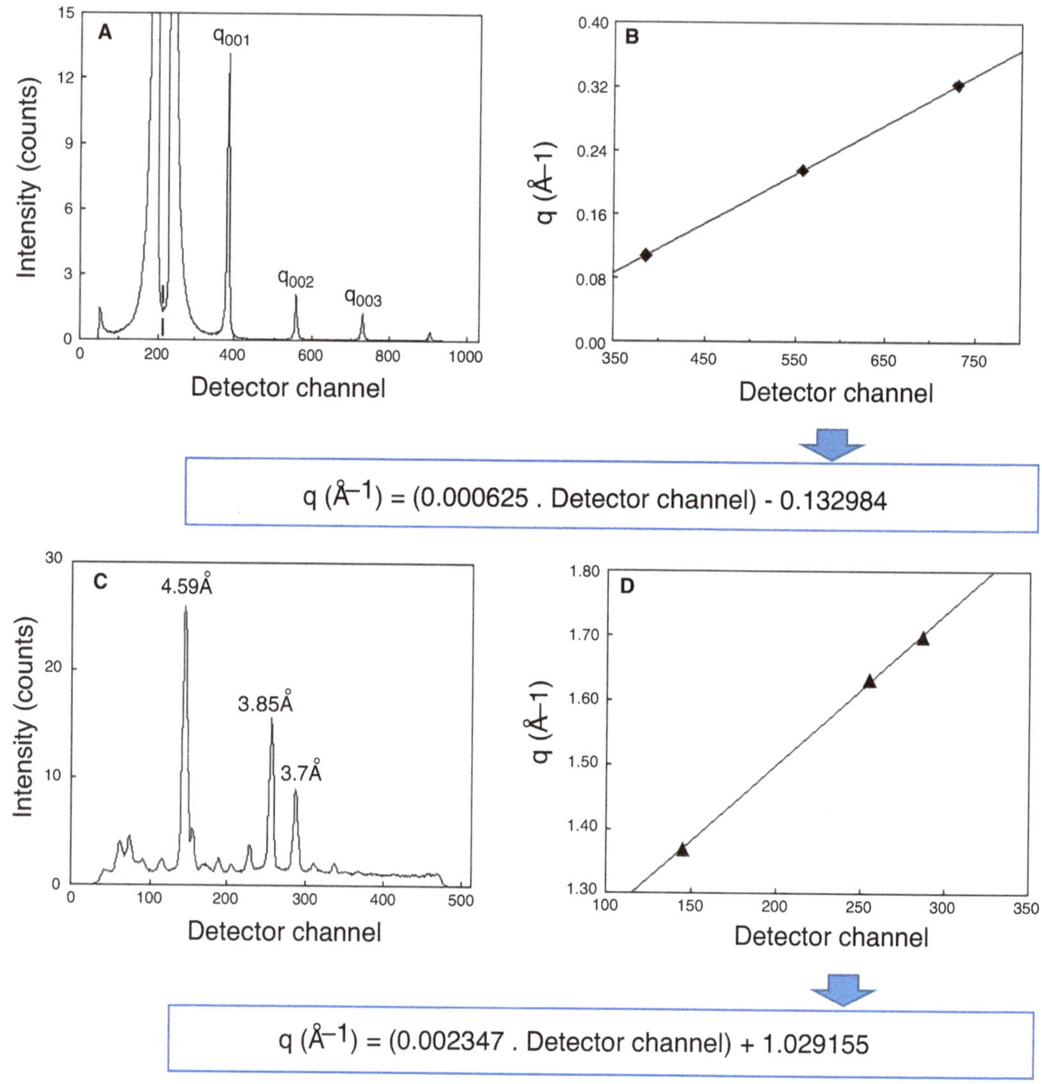

Fig. 6 Example of calibration of detector channels performed using (**A**, **B**) silver behenate (Ag-C24:0) and (**C**, **D**) tristearin (SSS; C18:0-C18:0-C18:0)

2.5 Treatment of XRD Patterns

The XRD patterns can be analyzed using software able to determine structural parameters such as the position of the maximum of the XRD peaks, their maximal intensity, their middle-height width, and the area under each XRD peak (Fig. 7).

The position of the XRD peaks recorded at both small and wide angles permits the identification of crystalline structures.

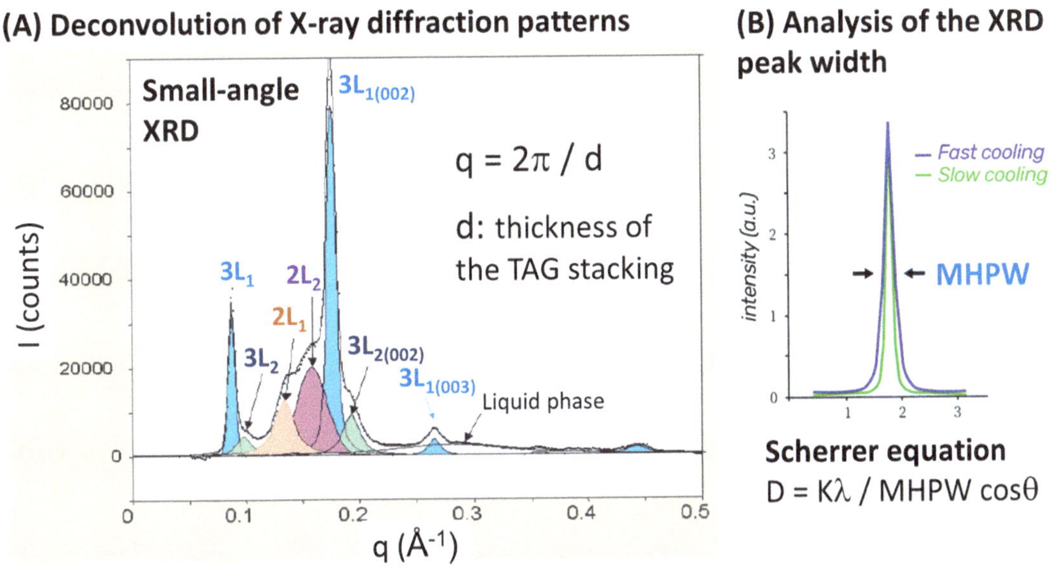

Fig. 7 Analysis of X-ray diffraction (XRD) pattern. (**A**) Example of analysis performed with PeakFit software on a small-angle XRD pattern recorded at −8 °C at the end of cooling of milk fat globules from 60 °C to −8 °C at 0.15 °C·min^{-1}. The area under the peaks corresponding to the same lamellar structure has the same color (Adapted from [3]). (**B**) Analysis of the middle-height peak width (MHPW; Δq) gives information on the thickness of the TAG crystals (stacking of lamellae) thanks to the Scherrer equation

References

1. Lopez C, Bourgaux C, Lesieur P, Ollivon M (2007) Coupling of time-resolved synchrotron X-ray diffraction and DSC to elucidate the crystallisation properties and polymorphism of triglycerides in milk fat globules. Le Lait 87:459–480. https://doi.org/10.1051/lait:2007018
2. Lopez C, Briard-Bion V, Bourgaux C, Pérez J (2013) Solid triacylglycerols within human fat globules: β crystals with a melting point above in-body temperature of infants, formed upon storage of breast milk at low temperature. Food Res Int 54:1541–1552. https://doi.org/10.1016/j.foodres.2013.10.014
3. Lopez C, Lesieur P, Bourgaux C, Keller G, Ollivon M (2001) Thermal and structural behavior of milk fat. J Colloid Interface Sci 240:150–161. https://doi.org/10.1006/jcis.2001.7664
4. Ollivon M, Keller G, Bourgaux C, Kalnin D, Villeneuve P, Lesieur P (2006) DSC and high resolution X-ray diffraction coupling. J Therm Anal Calorim 85:219–224. https://doi.org/10.1007/s10973-005-7351-y

Chapter 25

Rheological Characterization of Butter Samples Using a Lubricated Squeezing Flow

Marie-Hélène Famelart and Coralie Pasquier

Abstract

The functional properties of butter, such as its rheological properties, are of primary importance for its use in many bakery applications. Here, we report on a squeeze flow method to evaluate the rheological properties of butter samples with a test that mimics the mechanical stress encountered by butter in extrusion/laminar processes. This test is performed in the lubricated mode to measure biaxial extension viscosity of small samples. The compression is performed with a constant biaxial strain rate and not with a constant crosshead speed. This allows us to determine the apparent viscosity of butter in the elongation mode. This method is particularly adapted to test butter samples at different temperatures as slippage could be a problem in the measurement of rheological properties of butter by traditional shearing methods, whereas slippage is a prerequisite in squeeze flow tests.

Key words Butter, Rheology, Squeeze flow

1 Introduction

Butter can be used for domestic purposes or as a food ingredient in baking industry. We estimate that, in France, 30% of butter is used for pastry or bakery applications, due to its functional properties, based on its texture. Butter should be firm, or melted, depending on the process parameters, including the used temperature. It can be mixed with many other components or can be laminated with dough like in puff/flaky pastry, without mixing, i.e., by keeping in separated layers. Understanding the texture of butter and its changes with temperature is a critical point for the selection of raw materials in these processes and for the quality of baked food.

The most applied simple tests for measuring rheological properties of butter is inserting a probe, cylinder, cone, or wire at a defined speed into it [1] or compressing the butter sample between two plates [2] and measuring its resistance to strain at a known temperature. Methods based on butter extrusion have been used

Christelle Lopez et al. (eds.), *Multidimensional Characterization of Dietary Lipids*, Methods and Protocols in Food Science, https://doi.org/10.1007/978-1-0716-3758-6_25,

less. Small strain oscillating methods applied on a butter slice placed in a gap between 2 plates can also be found in the literature [3], as methods for the evaluation of butter spreadability [4]. For a description of these methods, see the review by Wright et al. [5].

Here, we describe a method for measuring biaxial extension viscosity of butter by a lubricated squeeze flow test. As reported by Casiraghi et al. [6], using lubricated plates in squeeze flow measurements to reduce frictional effects leads to more reproducible results than using unlubricated plates. This method is widespread in the study of polymer rheology and has been used for the characterization of dough or cheeses [6, 7], but less on butter samples [8, 9]. It consists in compressing a thin butter sample between two lubricated plates and measuring the strength during the extrusion of butter from the gap between the two plates. It could be of particular value for using butter in baking, as it could mimic the mechanical stress encountered by butter in extrusion/laminar processes (*see* **Note 1**).

It can be conducted either at a constant compressing speed (i.e., at a variable biaxial strain rate) or at a constant biaxial strain rate (i.e., at a variable compressing speed). As shown by others [7], using a constant biaxial strain rate is much easier to estimate elongational viscosities, whereas using a constant crosshead speed does not allow us to reach a steady regime, as both stress and biaxial strain rate change during compression. Moreover, as slippage is a requirement in this test, this lubricated test is recommended for butter that easily slips against plates and should melt, depending on the tested temperature.

The Hencky strain, ε, in the vertical direction is calculated as:

$$\varepsilon = -\log_n \left[\frac{H}{H_0}\right] \tag{1}$$

where H_0 and H are the initial height and the height at time t during compression, respectively (in mm or μm). The biaxial strain, ε_b, is equal to ½.ε.

It follows that the strain rate, $\dot{\varepsilon}$ (s^{-1}) is equal to:

$$\dot{\varepsilon} = -\frac{1}{H} \cdot \frac{dH}{dt} \tag{2}$$

and the biaxial strain rate, $\dot{\varepsilon_b}$ (s^{-1}), is equal to:

$$\dot{\varepsilon_b} = -\frac{1}{2} \cdot \dot{\varepsilon} = -\frac{1}{2} \cdot \frac{1}{H} \cdot \frac{dH}{dt} \tag{3}$$

In consequences, maintaining a constant biaxial strain rate leads to a decreasing compression rate, $\frac{dH}{dt}$ (mm.s^{-1} or μm.s^{-1}), during compression.

The stress during compression, σ (Pa), is deduced from normal force measurement, F_N (N), as:

$$\sigma = \frac{F_N}{\pi.R^2}$$

with $\pi.\ R^2$ being the cross-sectional area of the sample (m^2), and the apparent viscosity in the elongational mode, η_b (Pa.s), is deduced as:

$$\eta_{b=} \frac{\sigma}{\dot{\epsilon}_b}$$

The method comprises 3 steps: preparation of butter samples, rheological measurements, and calculations. In addition, some measurements are shown as examples in the "Notes" section.

2 Materials

2.1 Rheometer

- A rheometer or penetrometer that is able to apply constant compression strain rates, measure normal forces, and regulate the temperature can be used. The MCR 301 (Anton Paar) is used for the measurements shown as examples.
- The rheometer is equipped with a Peltier oven that covers the butter sample. The Peltier oven is regulated at the same temperature than the plates. This device avoids any temperature gradient inside the sample, but measurements can probably be performed without it, as the time of measurement is rather short (less than a min at high shear rates, and ca. 7 min at the lowest shear rate shown in the examples).
- The upper crosshead (and plate) is moved into the butter samples with a constant compression strain rate, $\dot{\varepsilon}$ (s^{-1}), i.e., with a decreasing crosshead speed during the time of compression.

2.2 Geometry

- Use two stainless steel plates with smooth surfaces. It is more convenient to use plates (and particularly the upper plate) with diameters equal to the sample diameter as the sample surface is thus considered constant during compression. Nevertheless, the test can be conducted by using plates larger than the sample, provided that the surface of the sample used for the calculation of the stress is corrected during compression by assuming a constant volume of the sample. In the measurements shown as examples, an upper stainless steel plate of 2 cm diameter and a lower one of 5 cm diameter are used.
- Cover copiously the plates with paraffin oil to avoid any friction.

2.3 Devices

- Use a knife to cut the initial block of butter.
- A homemade sampler device (Fig. 1) is used to cut the butter slices. This device has been specifically designed for these experiments, but any other sampler can be used. The sampler device

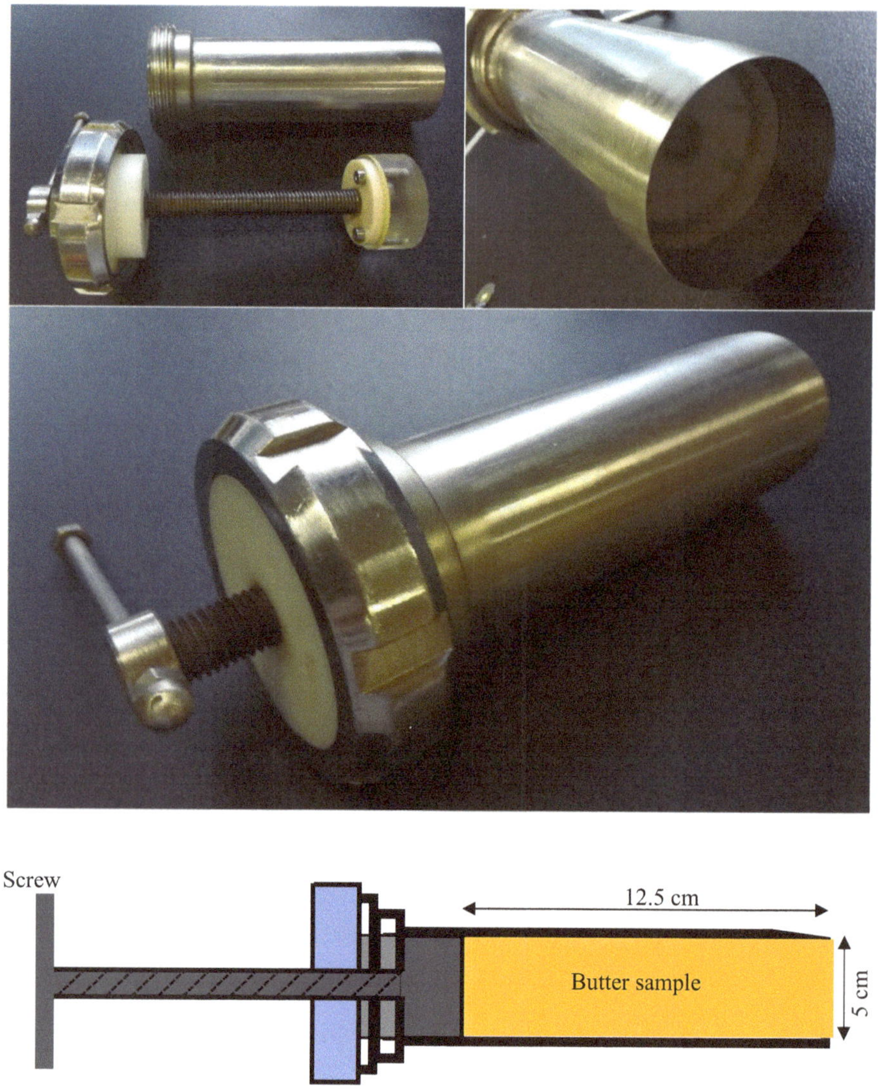

Fig. 1 Homemade sampler device

has an internal diameter of 5 cm and an internal length of 12.5 cm. The outside and inside walls of the sampler are coated with paraffin oil before sampling.

- Use a wire to cut the final butter slice outside of the sampler device.
- A core drill of diameter lower than 5 cm can also be used to recut the slice.
- Store the knife, the sampler device, the drill, and the wire at the temperature of measurement before use, to avoid any change in the sample temperature.

3 Methods

3.1 Preparation of Butter Samples

- Store the initial block of butter for at least 24 h at the temperature of measurement (15 °C in the measurements shown as examples).
- Cut a first block of sample with the knife at the temperature of measurement. Try to reach dimensions close to $3 \times 6 \times 6$ cm^3, the latter being the thickness.
- Then, cut the slices with the homemade sampler device at the target temperature. Sample the butter inside the tube of the sampler device, turn the screw to eject a 3-mm-thick slice of butter, and cut it with a wire (*see* **Note 2**).
- Possibly, recut the sample slice in the center with a core drill. In the measurements shown as examples, a core drill of 2 cm diameter has been used to reduce the butter slice diameter. Place the slice between 2 sheets of anti-adhesive butter special paper and store it at the temperature of measurement for at least 10 min in a closed petri dish. The final dimensions are 2 cm diameter and 3 mm thickness in the measurements shown as examples.
- Place the samples into a thermostated box to avoid any temperature changed sample before measurement.

3.2 Measurements

- Before any measurements, check the temperature of any butter sample of 9 mm thickness placed as further described on the geometry settled at the temperature of measurement by using a temperature sensor and possibly correct the target temperature on the rheometer.
- Use the 2 measurement plates at the target temperature before measurements with the oven in the measured position.
- Move the upper plate to 100 mm from the lower plate and place the sample slice rapidly in the center of the lower plate.
- Lower the upper plate at the highest speed (at 800 μm.s^{-1}, for instance) to a distance 4–5 mm from the lower plate. Then, lower it at a slower speed (at ca. 10 μm.s^{-1}) until the normal force (F_N) reaches 0.1 N, i.e., to be in contact with the sample without crushing it.
- Leave the geometry and the sample at rest for 5 min, during which F_N relaxes from 0.1 N to ~0.01–0.03 N.
- Begin the measurements by moving down the upper plate at a constant biaxial strain rate, $\dot{\varepsilon}_b$. In the data shown as examples, 0.0025, 0.025, and 0.25 s^{-1} are the tested $\dot{\varepsilon}_b$ values. These values correspond to values of $\dot{\varepsilon}$ of 0.005, 0.05, and 0.5 s^{-1}, respectively.

- Stop measurements when normal force exceeds 50 N or the maximal value is reached by the normal force sensor.

3.3 Calculations

- Deduce the biaxial deformation ε_b (no unit), the stress during compression σ (Pa), and the η_b (Pa.s) from measurements and plot the stress and the elongational viscosity against strain and strain rate.

4 Notes

1. Moreover, this test can be used to model the resistance of butter in a puff/flaky pastry before cooking. In these products, butter is sheared and stretched with dough, so that thin slices of butter separate the dough layers. Reducing the thickness of the butter sample in a lubricated squeeze flow could be a good model to understand the behavior of butter in the multilamellar packing of such pastries during processing.
2. One major source of problems can arise from the preparation of butter slices, particularly when low temperatures are used. First, using the sampler device allows us to eject a known thickness of butter. However, the butter can crack during sampling, or the butter slice can be extruded from the sampler device with a tilted angle, due to friction against walls. Pay attention to polishing the walls of the sample device well with a very fine sandpaper to avoid friction and then to coating them well (inside and outside) with oil before each sampling. If you have butter samples that have a highly brittle texture, particularly at low temperatures, and if you still meet problems of cracks, put the sampler device at room temperature for 30 s before sampling. It would allow the outer layer of the butter to slightly melt and could help in the sampling.
3. Fig. 2 shows the increase in stress with biaxial strain for repeated measurements on 3 butter samples at 15 °C tested at 3 different biaxial strain rates (0.0025, 0.025, and 0.25 s^{-1}). At low biaxial strain rates (0.0025 and 0.025 s^{-1}) and at low strain, the slope of stress/strain is high and a maximum can be seen, whereas this slope decreases at larger strain. This has already been reported in previous studies on butter or cheeses [6, 9] and is typical of fracture properties for very brittle material (as for the butter sample A at 15 °C in the example). This sample appears to fracture at a low biaxial strain, say around 0.2, whereas the other two samples do not show clear fractures and present a more continuous behavior. You can also observe the butter slice after the test, as an indicative of such fractures, i.e., if the sample can be removed in one or many pieces.

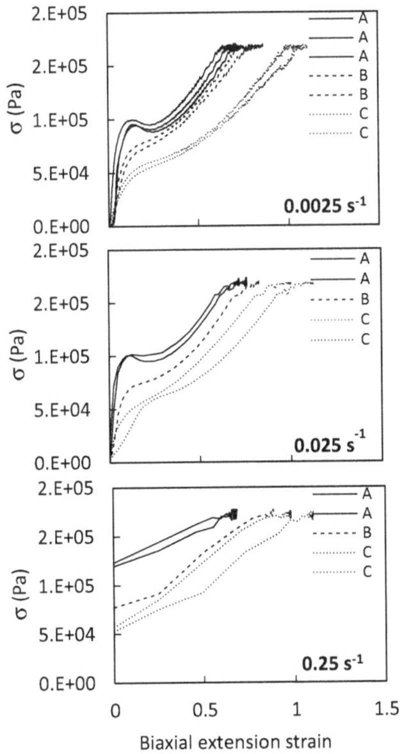

Fig. 2 Measured stress versus strain during lubricated squeeze flow test at 3 biaxial strain rates (0.0025, 0.025, and 0.25 s^{-1}) of 3 butter samples (A, B, and C). Different curves for repeated measurements on different slices (*see* **Notes 3, 4, 5**)

4. When the biaxial strain rate is increased (to 0.25 s^{-1} in the measurements shown as examples), the compression of the butter slice increases so rapidly during the trial that only the part after the fracture can be seen. Adapt the applied strain rates to the resistance and brittleness of samples. The maximal stress has plateaued at ca. $1.6.10^5$ Pa for all samples due to the limited force sensor (50 N in the measurements shown as examples).

5. Also, adapt the thickness and diameter of the sample to the maximal normal force available on the rheometer. A lot of rheometers are equipped with a 50 N normal force sensor, but measurements can also be performed on universal compressing testing machines.

6. Fig. 3 shows the increase in stress with biaxial strain rates on a butter samples at 15 °C calculated at increasing strains (0.12, 0.24, 0.5, 0.6, and 0.7). Stress measured at higher strains increased. This sample at 15 °C shows a plastic behavior with a yield stress at ca. 9.10^4 Pa deduced at low strain values (0.12 and 0.24). Elongational viscosity decreased with biaxial strain rate increases.

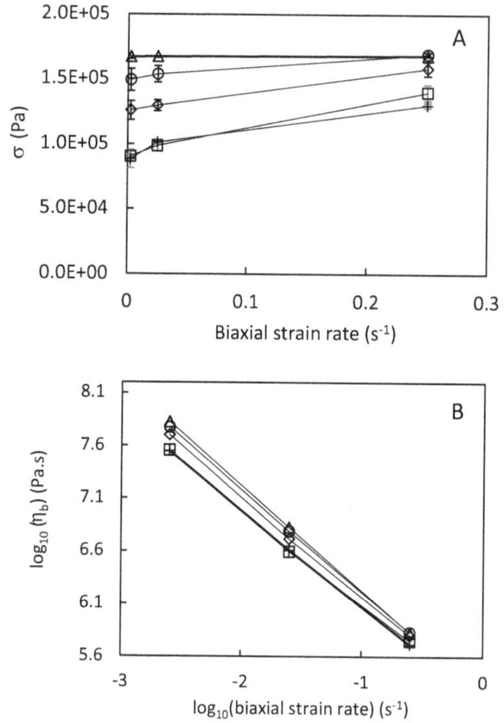

Fig. 3 Stress and elongational viscosity measured during lubricated squeeze flow of butter sample A at different strains (+: 0.12; □: 0.24; ◇: 0.5; ○: 0.6; Δ: 0.7). (**a**) Stress versus biaxial strain rate (0.0025, 0.025, and 0.25 s^{-1}). (**b**) Log_{10} of elongational viscosity versus log_{10} of biaxial strain rate (0.0025, 0.025, and 0.25 s^{-1}) (*see* **Note 6**)

References

1. Narine SS, Humphrey KL (2004) A comparison of lipid shortening functionality as a function of molecular ensemble and shear: microstructure, polymorphism, solid fat content and texture. Food Res Int 37:28–38. https://doi.org/10.1016/j.foodres.2003.09.013
2. Dixon BD (1974) Spreadability of butter: determination. I. Description and comparison of five methods of testing. Aust J Dairy Technol 29:15–20
3. Shukla A, Rizvi SSH (1995) Viscoelastic properties of butter. J Food Sci 60:902–905. https://doi.org/10.1111/j.1365-2621.1995.tb06257.x
4. Huebner VR, Thomsen LC (1957) Spreadability and hardness of butter.1, 2 I. Development of an instrument for measuring spreadability1. J Dairy Sci 40:834–838. https://doi.org/10.3168/jds.S0022-0302(57)94560-5
5. Wright AJ, Scanlon MG, Hartel RW, Marangoni AG (2001) Rheological properties of milkfat and butter. J Food Sci 66:1056–1071. https://doi.org/10.1111/j.1365-2621.2001.tb16082.x
6. Casiraghi EM, Bagley EB, Christianson DD (1985) Behavior of mozzarella, cheddar and processed cheese spread in lubricated and bonded uniaxial compression. J Texture Stud 16:281–301. https://doi.org/10.1111/j.1745-4603.1985.tb00697.x
7. Kouassi-Koffi JD, Launay B, Davidou S, Kouamé LP, Michon C (2010) Lubricated squeezing flow of thin slabs of wheat flour dough: comparison of results at constant plate speed and constant extension rates. Rheol Acta

49:275–283. https://doi.org/10.1007/s00397-009-0414-4

8. Rohm H (1993) Rheological behaviour of butter at large deformations. J Texture Stud 24:139–155. https://doi.org/10.1111/j.1745-4603.1993.tb00040.x

9. Shukla A, Rizvi SSH, Bartsch JA (1995) Rheological characterization of butter using lubricated squeezing flow. J Texture Stud 26:313–323. https://doi.org/10.1111/j.1745-4603.1995.tb00968.x

Chapter 26

Observation of Dietary Lipids Using Microscopy Techniques

Christelle Lopez

Abstract

Dietary lipids can be observed by different microscopy techniques that exhibit complementarity in the given information. This chapter presents light microscopy; polarized light microscopy; fluorescence microscopy including conventional (i.e., wide-field) fluorescence microscopy and confocal laser scanning microscopy; electron microscopy (EM) including transmission EM, cryo-EM, and scanning EM; and atomic force microscopy. The protocols need to be adapted to the samples. In particular, lipid droplets dispersed in oil-in-water emulsions are thermodynamically instable and need special care when preparing the samples for microscopic observations.

Key words Microscopy, Lipid droplet, Emulsion, Crystals, Fluorescent dye

1 Introduction

The microstructure of dietary lipids is involved in many technological and food properties. For example, the size of the lipid droplets in processed oil-in-water emulsions or in natural emulsions such as milk (lipid droplets called milk fat globules) or plant aqueous extracts (lipid droplets called oil bodies or oleosomes) as well as their aggregation and coalescence have consequences on the mechanisms of oxidation, on the texture and on the sensory properties of food products. The organization of the oil–water interface between the lipids dispersed in droplets and the surrounding environment (aqueous as in milk or dense as in gels) is also of interest. The size, morphology, and distribution of fat crystals within food products such as butter, margarine, spreads, or oleogels are of importance to better understand the texture. Examining the microstructure of dietary lipids is therefore of interest and is performed by microscopic observations.

Microscopy is defined as the use of a microscope. Different kinds of microscopy techniques can be used to observe dietary lipids. This chapter presents light microscopy; polarized light microscopy; fluorescence microscopy including conventional (i.e.,

Christelle Lopez et al. (eds.), *Multidimensional Characterization of Dietary Lipids*, Methods and Protocols in Food Science, https://doi.org/10.1007/978-1-0716-3758-6_26,

wide-field) fluorescence microscopy and confocal laser scanning microscopy (CLSM); electron microscopy (EM) including transmission electron microscopy (TEM), cryo-EM, and scanning electron microscopy (SEM); and atomic force microscopy (AFM).

Whatever the microscopy technique that is used, the results of any microscopic observations correspond to two-dimensional images or stacks of two-dimensional images corresponding to three-dimensional representations. The quality of the microscopy images will depend on their resolution. The scale bar (e.g., 10 μm) is an important element of the microscopy images since it gives a reference for the size of the objects observed in the image. The recording of the images in a high resolution and a noncompressed format, for example the .tif format, is recommended. Image analysis, which gives quantitative information, is out of the scope of this chapter.

2 Methods: Microscopy Techniques

2.1 Light Microscopy

Dietary lipids can be organized as lipid droplets, for example the milk fat globules in milk, the oil bodies in plant aqueous extracts, and processed lipid droplets in many food oil-in-water emulsions. These lipid droplets are objects of several micrometers in size that are too small to be seen directly with the naked eye. However, these objects can be observed with an optical instrument called a light microscope.

Light microscopy is a technique based on microscopy that uses visible (white) light to illuminate (light up) the object (sample) being looked at, to focus the light using a system of lenses, to detect and to magnify very small objects, and finally to provide an image of the object. Light microscopes (Fig. 1) can be used only in the presence of light and have lower resolution compared with electron microscopes.

The sample (e.g., about 3 μL for milk) is deposited on a microscope glass slide (dimension 25 × 76 mm; 1–1.2 mm thick. A glass cover slip (dimension 20 × 20 mm; 0.12–0.17 mm thick) is then deposited on the sample (Fig. 1c).

NOTE: Be careful, it is recommended to slowly deposit the cover slip to avoid the introduction of air bubbles in the sample.

Main Parts of a Light Microscope

Most microscopes operate with a LED light as light source. The job of the light source is to illuminate the object evenly. Under a conventional light microscope, the light travels through the whole sample as far into the sample as it can penetrate. The sample should therefore be diluted with a low thickness.

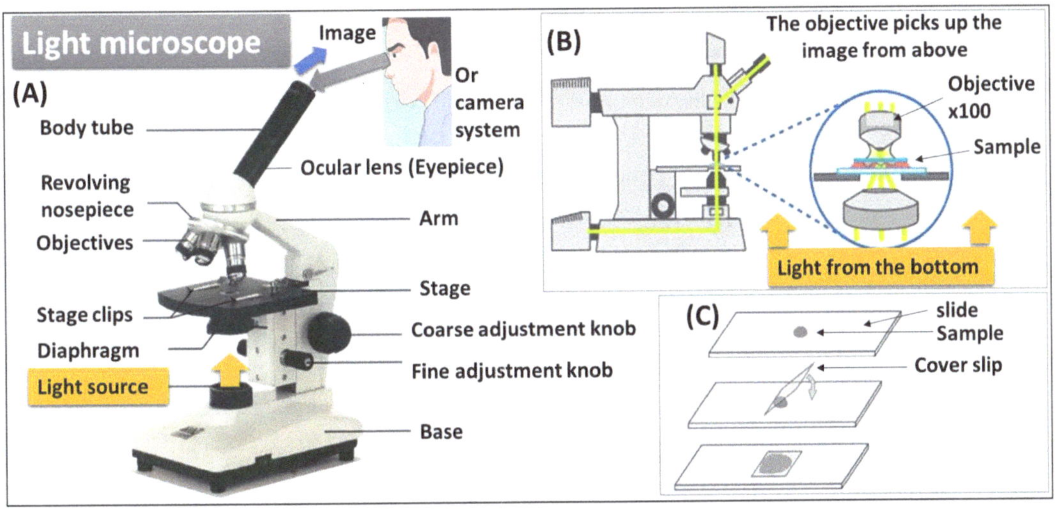

Fig. 1 Light microscopy. (**a**) The different parts of a light microscope, (**b**) the light path, and (**c**) the preparation of a sample between a glass slide and a cover slip

The ocular lens is part of the optical system, which is directed at the user of the microscope. This is the part that has at least one or more lenses. The function of the ocular lens on a microscope is to convert the enlarged real-middle image from the objective into an enlarged virtual image. The size of the incoming light cone will be adjusted to the size of the human eye.

Body tube is part of the microscope that serves to accommodate the lens system that enlarges the specimen.

Objective revolvers are used in microscopes with multiple objective lenses, which have different magnification factors. By rotating the revolver, you can select the lens with the desired level of magnification. Light microscopes have high magnification, e.g., x100.

Arm is a holder connected to all components that function as a support for the microscope so that the microscope can be used properly.

The stage clip acts as a holder for the object plate and ensures that it does not fall out of place accidentally.

The microscope stage is the part that serves to place the object plate with a cover glass on it. By sliding the platen, you can select the part of the object you want to view.

With fine focus, you can adjust the distance between the object and the objective to achieve the required sharpness. Fine focus works by moving the lens smoothly.

Like fine focus, coarse focus also moves the stage to adjust the difference between the object and the objective. The function of the coarse focus is to capture the exact distance roughly and quickly. Optimal sharpness can be adjusted to finer roughness afterwards.

How to Use a Light Microscope

To use a light microscope, you can follow the steps below carefully.

- Start with a low lens (e.g., objective x4) and a clean slide. The microscope stage should be lowered as low as possible.
- Center the slide so that the sample is under the objective lens.
- Use the coarse adjustment knob to get a general focus. Then slowly move up the stage until focus is achieved.
- Use the fine adjustment knob to get clear focus.
- Center the sample on a low plane before moving it to a medium plane.
- Switch the objective lens to medium and focus using a fine adjustment. If you cannot see the sample at this point, return to the low plane and re-center.
- Once you have focused and centered on the medium power, you can move it to even greater magnification (e.g., objectives x20, x40, x100).

2.2 Polarized Light Microscopy

Polarized light microscopy can mean any of a number of optical microscopy techniques involving polarized light. Simple techniques include illumination of the sample with polarized light. Directly transmitted light can, optionally, be blocked with a polarizer orientated at 90 degrees to the illumination. More complex microscopy techniques which take advantage of polarized light include differential interference contrast microscopy and interference reflection microscopy. Scientists will often use a device called a polarizing plate to convert natural light into polarized light.

Polarized light microscopes are conventional microscopes with additional features that work by converting unpolarized light to polarized light and to perform observation under polarized light (Fig. 2). The light source of such an instrument is equipped with a polarizing filter, the polarizer, so that the light it supplies is linearly polarized.

These illumination techniques are most commonly used on birefringent samples where the polarized light interacts strongly with the sample, generating contrast with the background. Polarized light microscopy is used extensively in the case of dietary fats containing a solid or crystallized fat fraction (e.g., milk fat, butter, and margarines but also in oleogels containing crystallized organogelators).

A system permitting regulations in temperature, such as a cooling and/or heating stage, can be positioned on the microscope stage. With this configuration, changes in the temperature of the sample inducing melting or crystallisation of dietary lipids can be investigated.

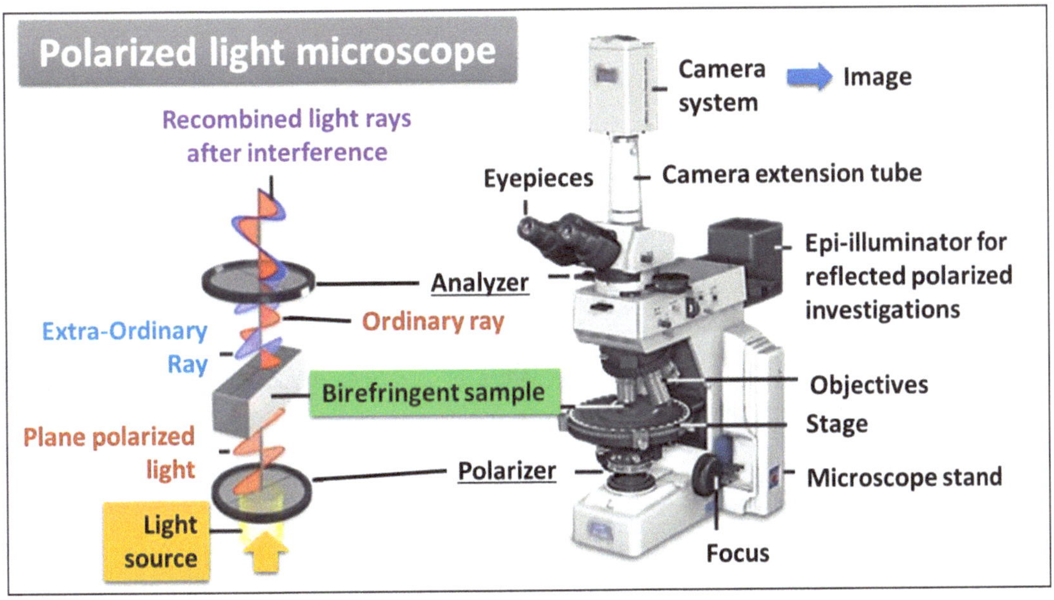

Fig. 2 Polarized light microscopy. Different parts of a polarized light microscope and principle of the polarization of light

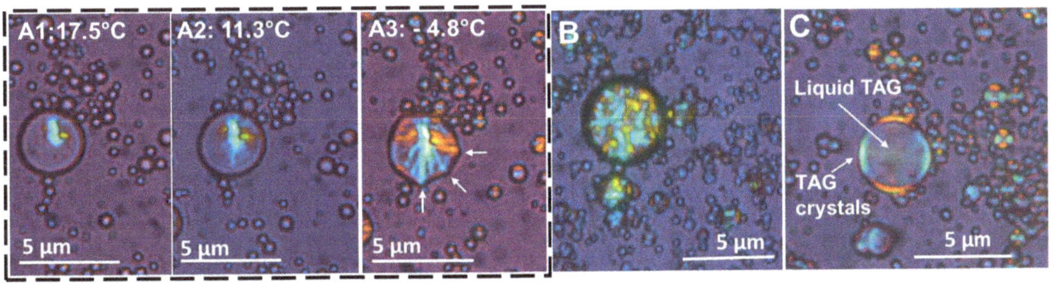

Fig. 3 Polarized light microscopy images of the growth of fat crystals within milk fat globules (**a**) as a function of temperature on cooling at dT/dt = 1 °C/min, (**b**) after rapid cooling, and (**c**) after storage at 4 °C. The birefringence of the fat crystals is clearly observable and the different orientations of the crystals are visualized by different colors. (Adapted from [4])

Figure 3 presents polarized light microscopy images showing the growth of fat crystals within milk fat globules.

2.3 Fluorescence Microscopy

A fluorescence microscope is an optical microscope that uses fluorescence to generate an image, whether it is a simple setup like an epifluorescence microscope where excitation of the fluorophore and detection of the fluorescence are done through the same light path (i.e. through the objective) or a more complicated design such as a confocal microscope (see next section), which uses optical sectioning to get better resolution of the fluorescence image.

For fluorescence microscopy, the sample must be fluorescent. The main technique is labeling of the sample with a fluorescent dye.

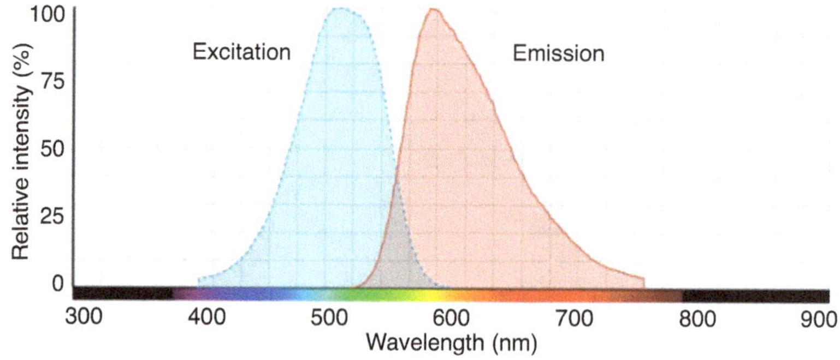

Fig. 4 Fluorescent properties of molecules including excitation spectra and emission spectra

There are many fluorescent molecules called fluorophores or fluorochromes such as fluorescein, Alexa Fluors, or DyLight 488, which are chemically linked to a molecule which binds the target of interest within the sample. The fluorescent molecules are characterized by excitation spectra and emission spectra at higher wavelengths that can change as a function of the environment, e.g. the pH of the sample (Fig. 4).

Fluorophores lose their ability to fluoresce as they are illuminated in a process called photobleaching. Photobleaching occurs as the fluorescent molecules accumulate chemical damage from the electrons excited during fluorescence. Photobleaching can severely limit the time over which a sample can be observed by fluorescence microscopy. Several techniques exist to reduce photobleaching such as the use of more robust fluorophores or by minimizing illumination.

As for light microscopy, the sample (e.g., about 3 μL for fluorescently stained milk) is deposited on a microscope glass slide (dimension 25 × 76 mm; 1–1.2 mm thick). A glass cover slip (dimension 20 × 20 mm; 0.12–0.17 mm thick) is then deposited on the sample.

Figure 5 presents a fluorescence microscope.

In a conventional (i.e., wide-field) fluorescence microscope, the entire sample is flooded evenly in light from a light source. All parts of the sample can be excited at the same time, and the resulting fluorescence is detected by the microscope's photodetector or camera including a large unfocused background part.

The sample is illuminated with light of a specific wavelength (or wavelengths) which is absorbed by the fluorophores, causing them to emit light of longer wavelengths (i.e., of a different color than the absorbed light). The illumination light is separated from the much weaker emitted fluorescence through the use of a spectral emission filter. Typical components of a fluorescence microscope are a light source (xenon arc lamp or mercury-vapor lamp are common; more advanced forms are high-power LEDs and lasers),

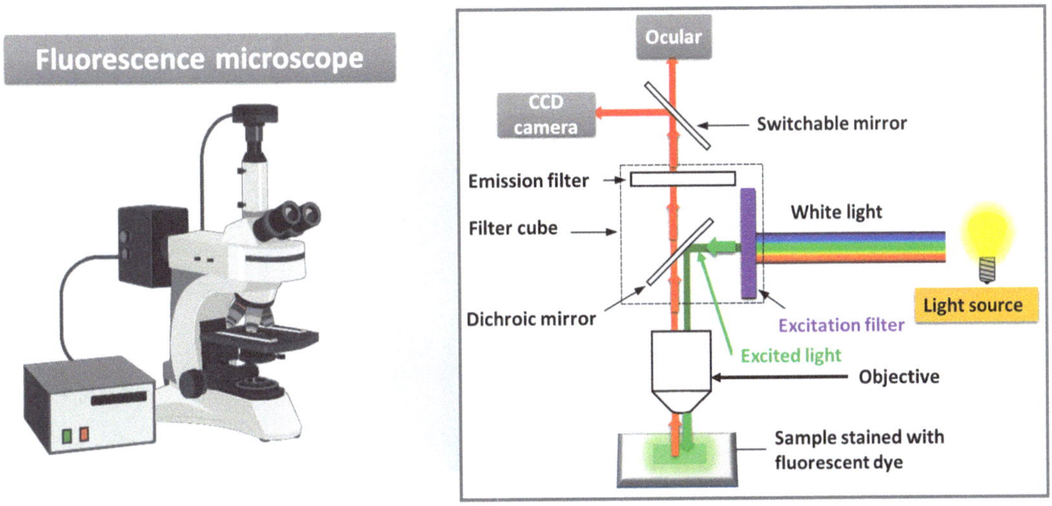

Fig. 5 Fluorescence microscopy: different parts of a fluorescence microscope and principle

the excitation filter, the dichroic mirror (or dichroic beamsplitter), and the emission filter. The filters and the dichroic beamsplitter are chosen to match the spectral excitation and emission characteristics of the fluorophore used to label the specimen. In this manner, the distribution of a single fluorophore (color) is imaged at a time. Multicolor images of several types of fluorophores must be composed by combining several single-color images.

2.4 Confocal Laser Scanning Microscopy

2.4.1 Principle

Confocal laser scanning microscopy (CLSM), most frequently called confocal microscopy or laser confocal scanning microscopy (LCSM), is an optical imaging technique for increasing optical resolution and contrast of a micrograph by means of using a spatial pinhole to block out-of-focus light in image formation (Fig. 6). CLSM allows the observation of objects higher than 0.5 μm in size.

Compared with a conventional light microscope where the light travels through the sample as far into the sample as it can penetrate, a confocal microscope only focuses a smaller beam of light at one narrow depth level at a time. CLSM achieves a controlled and highly limited depth of field. With CLSM, it is therefore possible to examine the structure of optically dense samples such as food products (e.g., a piece of butter or margarine, a piece of cheese) to characterize the microstructure of dietary lipids.

Capturing multiple two-dimensional images at different Z-depths in a sample enables the reconstruction of three-dimensional structures (a process known as optical sectioning) within an object or a sample. Successive slices make up a 'Z-stack', which can either be processed to create a 3D image or it is merged into a 2D stack (predominately the maximum pixel intensity is taken, other common methods include using the standard deviation or summing the pixels).

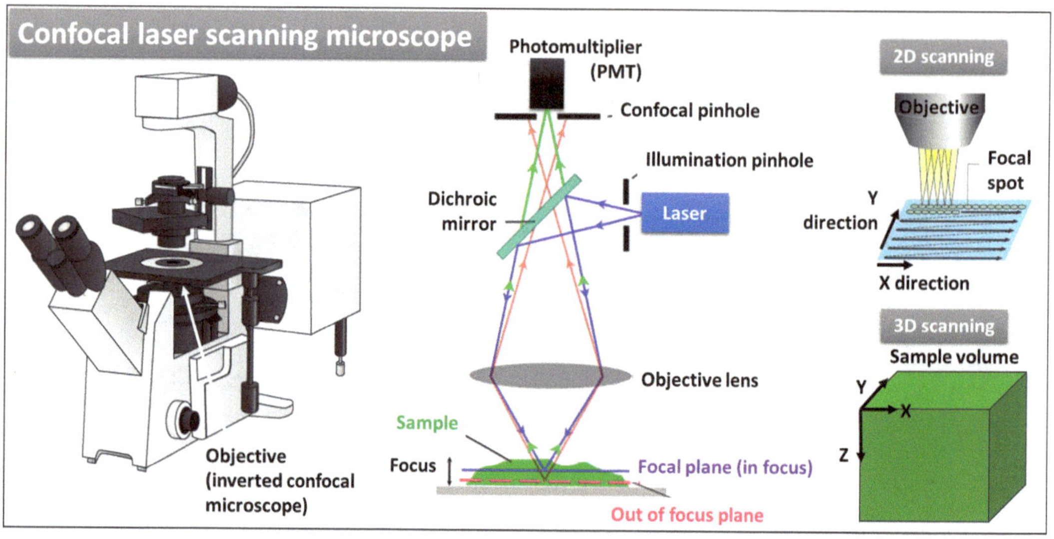

Fig. 6 Confocal laser scanning microscope: different parts of the microscope and principle

This technique is currently used extensively in the scientific and industrial communities.

The confocal microscope can have inverted objectives (observation of the samples from the bottom) or direct objectives.

A confocal microscope uses point illumination and a pinhole in an optically conjugate plane in front of the detector to eliminate out-of-focus signal—the name "confocal" stems from this configuration. As only light produced by fluorescence very close to the focal plane can be detected, the image's optical resolution, particularly in the sample depth direction, is much better than that of the wide-field microscopes. However, as much of the light from sample fluorescence is blocked at the pinhole, this increased resolution is at the cost of decreased signal intensity—so long exposures are often required. To offset this drop in signal after the pinhole, the light intensity is detected by a sensitive detector, usually a photomultiplier tube (PMT) or avalanche photodiode, transforming the light signal into an electrical one.

As only one point in the sample is illuminated at a time, 2D or 3D imaging requires scanning over a regular raster (i.e., a rectangular pattern of parallel scanning lines) in the sample. The beam is scanned across the sample in the horizontal plane by using one or more oscillating mirrors. This scanning method usually has a low reaction latency, and the scan speed can be varied. Slower scans provide a better signal-to-noise ratio, resulting in a better contrast.

The achievable thickness of the focal plane is defined mostly by the wavelength of the used light divided by the numerical aperture of the objective lens but also by the optical properties of the specimen. The thin optical sectioning makes these types of microscopes particularly good at 3D imaging and surface profiling of samples.

2.4.2 Examples of Sample Preparation

Confocal microscopy provides the capacity for direct, noninvasive, serial optical sectioning of intact, thick samples with a minimum sample preparation. The samples are often treated with fluorescent dyes to make selected objects visible. The fluorescent dye concentration should be low to avoid nonspecific staining. Organic solvent (e.g., chloroform) must not be used in order to avoid artifacts in the sample containing dietary lipids. For example, to label the phospholipids, fluorescent dyes are commercially available in chloroform; the fluorescent dye must be inserted in a glass vial (not a plastic tube), and chloroform must be evaporated before the sample is put in contact with the fluorescent dye for staining.

Fluorescent Dyes: Alone or in Combination

Among the diversity of fluorescent dyes that can be used, Nile red is a fluorescent dye widely used to stain the dietary hydrophobic lipids such as the triacylglycerols in the core of lipid droplets and also the hydrophobic part of lipid membranes. To label the phospholipids in the membrane surrounding natural lipid droplets such as the fat globules in milk and the oil bodies in plant seed aqueous extracts, the head-labeled phospholipid rhodamine-DOPE has been successfully used. To label the proteins, Fast green is often used. The chemicals structure of these fluorescent dyes are presented Fig. 7.

Several fluorescent dyes with different fluorescent characteristics (excitation and emission wavelengths) can be used simultaneously to localize the components in the sample.

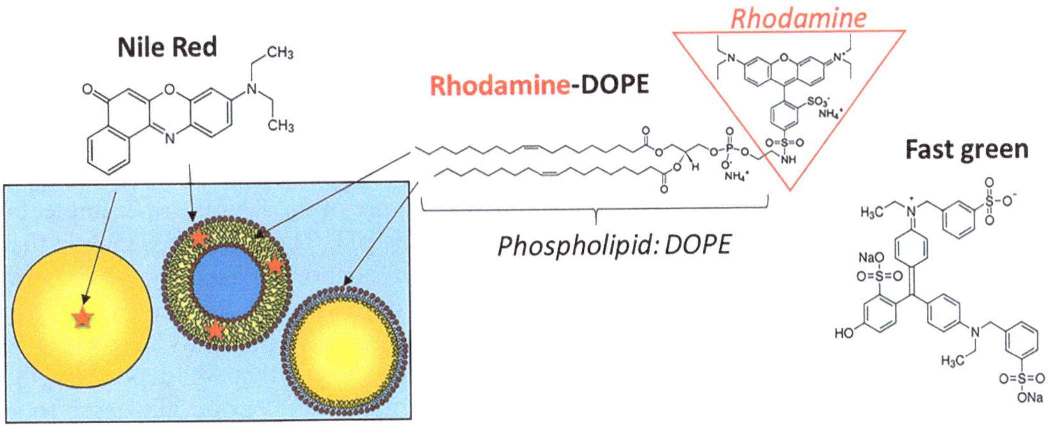

Fig. 7 Examples of fluorescent dyes widely used for the labeling of dietary lipids including the core of lipid droplets rich in triacylglycerols (Nile Red), biological membranes such as the milk fat globule membrane, or the hemi-membrane surrounding plant oil bodies (the exogenous phospholipid Rhodamine-DOPE) that contain polar lipids such as phospholipids and sphingolipids. The fluorescent dye Fast green is widely used for the labeling of proteins, for example, in oil-in-water emulsions

For example, the hydrophobic neutral lipids triacylglycerols (Nile Red; λexc 543 nm), the proteins (Fast Green; λexc 633 nm), and the bacteria (Acridine Orange as a marker of DNA; λexc 488 nm) can be localized simultaneously in a food sample such as cheese.

The fluorescent dye Acridine Orange stains both the DNA of bacteria and the protein network. Acridine orange has been used in combination with Nile Red to stain the lipids and to examine the microstructure of cheese including the localization of colonies of bacteria [8].

Adaptation of the Protocols as a Function of the Viscosity and Physical State of the Samples

- If the sample is liquid, (e.g., milk or oil-in-water emulsions), 0.5 to 1 mL of the sample can be inserted in an Eppendorf tube, and a volume (several microliters) of the solution of fluorescent dye(s) can be added. The sample is kept in the dark for at least 15 min for the labeling step. Then, 3 μL of the stained sample can be deposited on a thin glass slide (in case of the use of an inverted confocal microscope) covered by a cover slip for CLSM observations. To avoid or limit the Brownian motion of lipid droplets in the sample during CLSM observations and image acquisition, the sample can be mixed with high melting point agarose (prepared at 5 g/L and stored at 45 °C until utilization). Examples of oil-in-water dietary emulsions (fat globules in milk, oil bodies in hemp seed aqueous extract) observed by CLSM can be found in Lopez et al. [7, 9].
- If the sample is viscous or solid (e.g., a piece of cheese), a piece of sample (1 cm × 1 cm) is cut with a scalpel to have a flat surface and is then fluorescently stained. The diffusion of the fluorescent dye(s) within the sample is an important step in order to recover CLSM images with a good fluorescent signal. It is important that the solution containing the dye that penetrates the sample does not alter the microstructure of the sample, for example, by changing the hydration or the pH. The side of the sample that has been fluorescently stained is deposited in contact with a thin glass slide (in case of the use of an inverted confocal microscope for CLSM observations). The laser of the confocal microscope will penetrate the sample at a Z-depth that will depend on the optical density of the sample. Examples of optically dense food products in which the microstructure of dietary lipids has been observed by CLSM can be found in Lopez et al. [5, 6, 8].
- If the sample is a powder (e.g., a powdered ingredient or a powdered infant milk formula), the powder must not be hydrated during the staining step. The powder can be dispersed in glycerol or in PEG200 in which the fluorescent dyes (e.g., Nile Red and Fast green) have been previously dispersed. Examples of powders observed by CLSM can be found in Lopez and Schuck [10], Vignolles et al. [13], and Zouari et al. [15].

2.4.3 Acquisition of Individual CLSM Images and Co-localization

The confocal microscope scans the surface of the sample and detects the emission fluorescence of each of the dye used, and the software superimposes the individual images to provide co-localization of the stained elements. For example, Fig. 8 shows the CLSM images of an oil-in-water emulsion in which the lipid droplets were stained by Nile Red (0.125% in propanediol), the proteins by Fast Green (1% in water), and their co-localization showing the proteins around each lipid droplet (adapted from Chevallier et al. [1].

2.5 Electron Microscopy

An electron microscope is a microscope that uses a beam of accelerated electrons as a source of illumination (Fig. 9). As the wavelength of an electron can be up to 100,000 times shorter than that of the visible light photons, electron microscopes have a higher resolving power than light microscopes and can reveal the structure of small objects, from 1 nm to several micrometers.

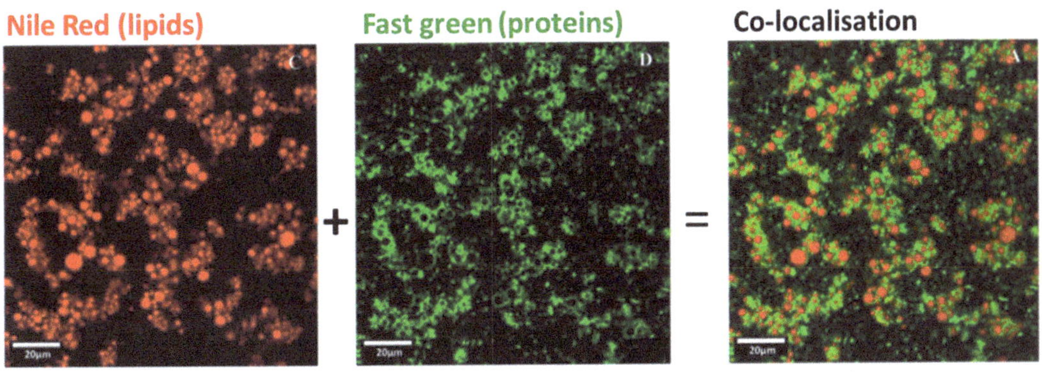

Fig. 8 Confocal laser scanning microscopy images of an oil-in-water emulsion in which the lipid droplets were stained by Nile Red (0.125% in propanediol), the proteins by Fast Green (1% in water), and their co-localization showing the proteins around each lipid droplet. (Adapted from [1])

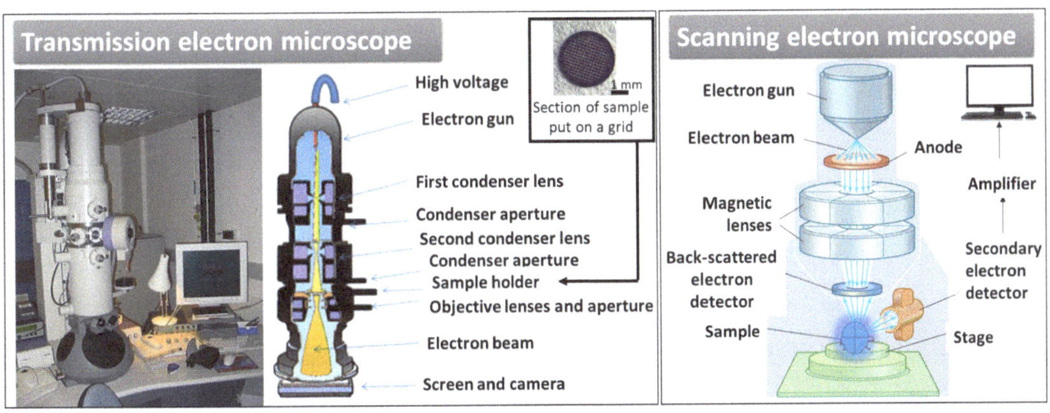

Fig. 9 Schematic representation of an electron microscope including transmission electron microscope and scanning electron microscope

2.5.1 Transmission Electron Microscope (TEM)

The transmission electron microscope (TEM) uses a high voltage electron beam to illuminate the specimen and create an image. The electron beam is produced by an electron gun, commonly fitted with a tungsten filament cathode as the electron source. The electron beam is accelerated by an anode typically at +100 keV (40–400 keV) with respect to the cathode, focused by electrostatic and electromagnetic lenses, and transmitted through the specimen that is in part transparent to electrons and in part scatters them out of the beam. When it emerges from the specimen, the electron beam carries information about the structure of the specimen that is magnified by the objective lens system of the microscope. The spatial variation in this information (the "image") may be viewed by projecting the magnified electron image onto a fluorescent viewing screen coated with a phosphor or scintillator material such as zinc sulfide. Alternatively, the image can be photographically recorded by exposing a photographic film or plate directly to the electron beam, or a high-resolution phosphor may be coupled by means of a lens optical system or a fiber optic light-guide to the sensor of a digital camera. The image detected by the digital camera may be displayed on a monitor or computer.

The samples are chemically fixed, dehydrated, and embedded in a polymer resin to stabilize them sufficiently to allow ultrathin sectioning. TEM requires extremely thin sections of the samples, typically about 100 nm. In addition, sections of samples may require staining with heavy atom labels in order to achieve the required image contrast.

Sample Preparation

Materials to be viewed under an electron microscope may require processing to produce a suitable sample. The technique required varies depending on the specimen and the analysis required.

- *Negative staining*: a rapid method to prepare samples for TEM observations. The samples are briefly mixed with a dilute solution of an electron-opaque solution containing heavy metal compounds, such as ammonium molybdate, uranyl acetate (or formate), or phosphotungstic acid. This mixture is applied to a suitably coated EM grid, blotted, then allowed to dry. Viewing of this preparation in the TEM should be carried out without delay for best results. The TEM observations can only be performed after negative staining. This is quite rapid and efficient to characterize the samples. Examples of negatively stained samples observed by transmission electron microscopy are presented in Fig. 10.

The preparation of the samples can be performed using a procedure most time consuming, by following different steps:

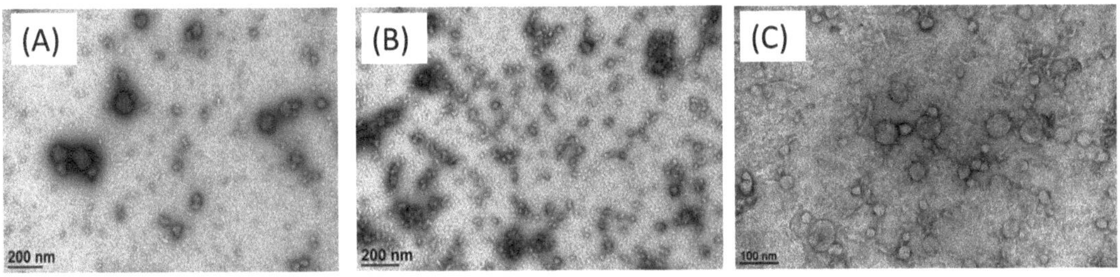

Fig. 10 Transmission electron microscopy (TEM) images of negatively stained (**a**, **b**) whey proteins, and (**c**) lipid vesicles

- *Preparation of the samples: application to oil-in-water emulsions.* A small volume of the oil-in-water emulsion can be placed into a liquid solution of agar 1.5% (w/v) at 40 °C. The sample is then cooled for agar gelation. Then, a part of the agar gel trapping the oil-in-water emulsion is cut with a scalpel in small pieces (1 mm × 1 mm) for the other steps of preparation.
- *Chemical fixation*: Aims to stabilize the sample's mobile macromolecular structure with chemical products. Chemical crosslinking of proteins with aldehydes such as formaldehyde and glutaraldehyde (e.g., 3% glutaraldehyde in 0.1 M phosphate buffer at pH 7.2 for 4 h). The sample is rinsed in one bath of phosphate buffer at pH 7.2 and then with distilled water. Fixation of lipids is performed with osmium tetroxide (OsO_4). The sample is rinsed in one bath of phosphate buffer at pH 7.2 and then with distilled water.
- *Dehydration*: Replacement of water with organic solvents such as ethanol, with increasing concentration till 100%. For example, the pieces of agar gel trapping the oil-in-water emulsion are dehydrated with successive baths containing increasing concentrations of ethanol (30, 50, 70, 85, 95, and 100% v/v). This step is followed by critical point drying or infiltration with embedding resins.
- *Embedding in resin*: After dehydration, sample for observation in the transmission electron microscope is embedded so it can be sectioned ready for viewing. To do this, the sample is passed through a 'transition solvent' such as propylene oxide (epoxypropane) or acetone and then infiltrated with an epoxy resin such as Araldite, Epon, or Durcupan. The samples may also be embedded directly in water-miscible acrylic resin. For the emulsions, the impregnation can be first made in propylene oxide (1 bath 50% ethanol/50% propylene oxide and 1 bath 100% propylene oxide) and secondly in epon resin (1 bath 50% propylene oxide/50% epon, and 3 baths 100% epon). Finally, the sample is included in epon, and the resin be left 3 days at 55 °C for polymerization. After the resin has been polymerized (hardened), the sample is thin sectioned (ultrathin sections) and stained. It is then ready for viewing by TEM.

- *Sectioning*: Produces thin slices of the sample, semitransparent to electrons. These can be cut on an ultramicrotome with a glass or diamond knife to produce ultrathin sections about 60–90 nm thick of the resin containing the sample inclusion. Disposable glass knives are also used because they can be made in the laboratory and are much cheaper. These sections are placed on a TEM grid.
- *Staining*: Uses heavy metals such as lead, uranium (4% (w/v) uranyl acetate), or tungsten to scatter imaging electrons and thus give contrast between different structures, since many (especially biological) materials are nearly "transparent" to electrons (weak phase objects). In biology, specimens can be stained "en bloc" before embedding and also later after sectioning. Typically thin sections are stained for several minutes with an aqueous or alcoholic solution of uranyl acetate followed by aqueous lead citrate.

Examples of transmission electron microscopy images of oil-in-water emulsions containing lipid droplets covered by protein particles are presented in Fig. 11.

For examining lipid membranes and their incorporated proteins in “face on” view, freeze-fracture or freeze-etch is a preparation method particularly useful. Freeze-fracturing helps to peel open lipid membranes to allow visualization of what is inside.

2.5.2 Cryo-EM

Cryogenic electron microscopy (cryo-EM) is a cryomicroscopy technique applied on samples cooled to cryogenic temperatures. With the development of cryo-electron microscopy of vitreous sections (CEMOVIS), it is now possible to observe samples from virtually any biological specimen close to its native state.

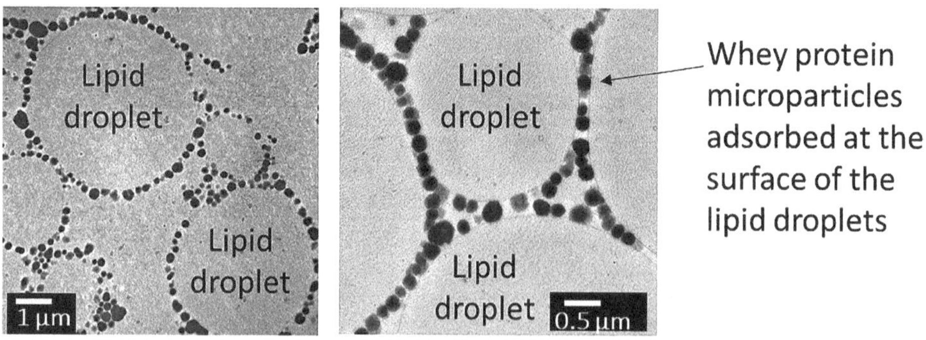

Fig. 11 Transmission electron microscopy (TEM) images of oil-in-water emulsions prepared with whey protein microparticles

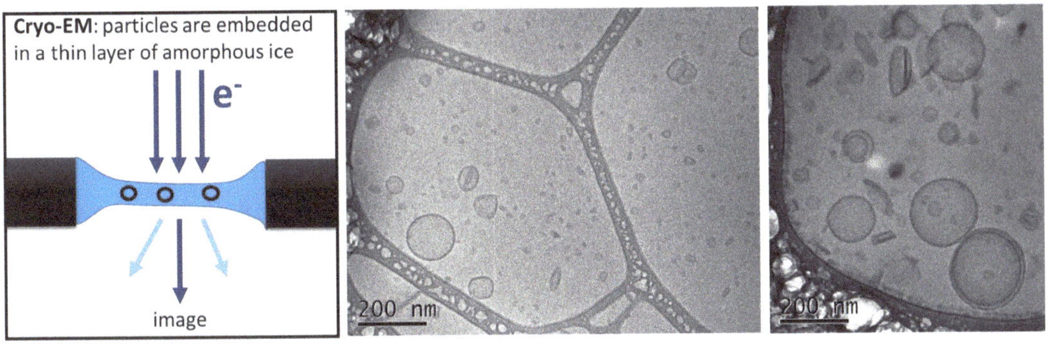

Fig. 12 Cryogenic electron microscopy (cryo-EM) images of lipid vesicles

For cryo-EM observations, cryo-fixation is required; it consists in freezing a sample rapidly. An aqueous sample solution is applied to a grid-mesh and plunge-frozen in liquid ethane or a mixture of liquid ethane and propane. The water forms vitreous (noncrystalline) ice. For biological specimens, the structure of the objects contained in the sample is preserved by embedding in a snapshot of its solution state, in an environment of vitreous ice. An entire field called cryo-electron microscopy has branched from this technique.

Cryo-EM is used, for example, for the observation of lipid vesicles as illustrated in Fig. 12.

2.5.3 Scanning Electron Microscope (SEM)

A scanning electron microscope (SEM) is a type of electron microscope, designed for directly studying the surfaces of objects, that utilizes a beam of focused electrons of relatively low energy as an electron probe that is scanned in a regular manner over the sample. The electron source and electromagnetic lenses that generate and focus the beam are similar to those described for the transmission electron microscope (TEM). The action of the electron beam stimulates emission of high-energy backscattered electrons and low-energy secondary electrons from the surface of the specimen.

Generally, the image resolution of a SEM is lower than that of a TEM. However, because the SEM images the surface of a sample rather than its interior, the electrons do not have to travel through the sample. This reduces the need for extensive sample preparation to thin the specimen to electron transparency. The SEM also has a great depth of field and so can produce images that are good representations of the three-dimensional surface shape of the sample.

Another advantage of SEMs comes with environmental scanning electron microscopes (ESEM) that can produce images of good quality and resolution with hydrated samples or in low, rather than high, vacuum, or under chamber gases. This facilitates imaging unfixed biological samples that are unstable in the high vacuum of conventional electron microscopes.

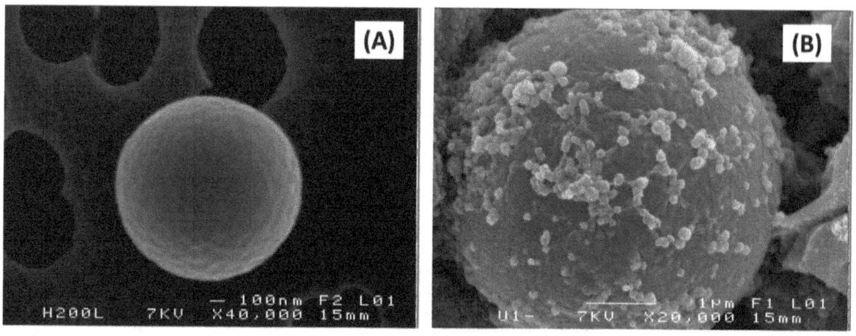

Fig. 13 Scanning electron microscopy images of (**a**) unprocessed milk fat globules, and (**b**) heat-treated milk fat globule covered by the colloid protein assemblies called the casein micelles

Sample Preparation for SEM

Step 1: Primary fixation with aldehydes (proteins).

Step 2: Secondary fixation with osmium tetroxide (lipids).

Step 3: Dehydration series with solvent (ethanol or acetone) to 100%.

Step 4: Drying by critical point dryer.

Step 5: Mounting on a stub.

Step 6: Sputter coating with conductive material.

Step 7: Observation with scanning electron microscope

The images presented in Fig. 13 illustrate the potentialities of SEM to characterize the surface of unprocessed milk fat globules and heat-treated milk fat globules.

2.6 Atomic Force Microscopy

Atomic force microscopy (AFM) has gained considerable interest for examining, at the nanoscale level, the components involved in the formation and stabilization of oil–water interfaces in emulsions and nanoemulsions. AFM allows imaging surfaces and provides information on the microstructure and topography in dried systems or in aqueous environments. AFM force spectroscopy provides information on the mechanical properties of soft matter materials and on interactions between components thanks to the grafting of tips. Although most of the AFM studies are performed on 2D planar surfaces, the challenging investigation of 3D lipid droplets is developing.

A typical AFM equipment consists of a sharp tip (also known as a probe) attached to the end of a flexible cantilever that acts as a spring (Fig. 14). The cantilever, also known as spring system/force sensor, is designed with a very low spring constant, so that it is very sensitive to any forces. The cantilever bends in the presence of attractive and repulsive forces and allows to calculate force from

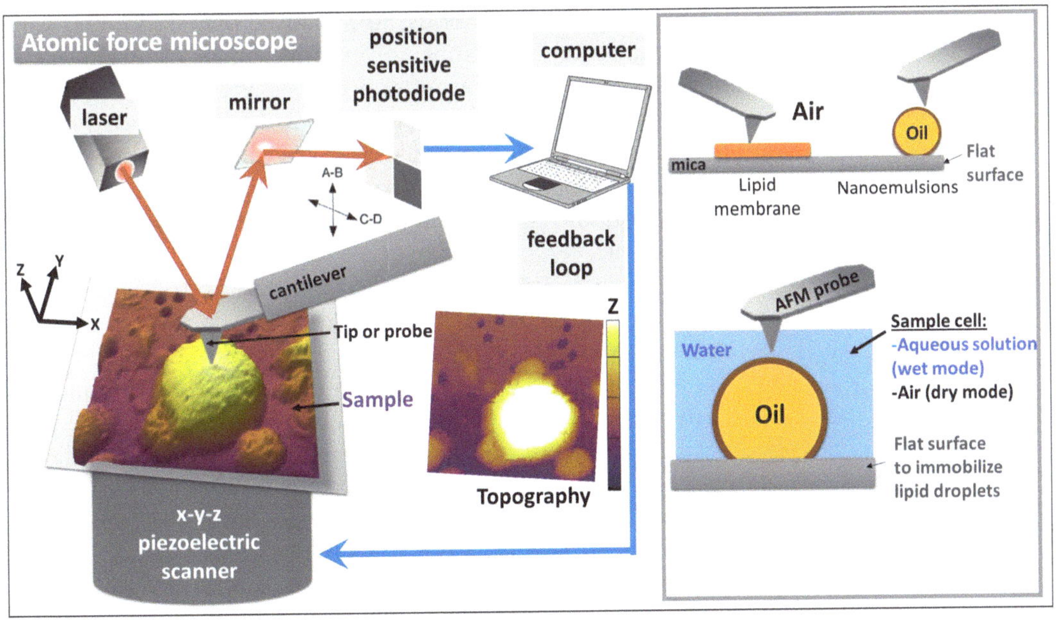

Fig. 14 Atomic force microscopy

its deflection following Hooke's Law: Force = (spring constant) × (deflection). The configurations of the AFM tips determine the resolution and quality of the images taken. The AFM tips can be classified according to their shape (e.g., square-base pyramid, rectangular-base pyramid, circular symmetric, spike), materials (e.g., silicine, silicone nitride, silicone oxide, high-density carbon), and coating materials (e.g., uncoated, gold-coated, platinum-coated, diamondlike-carbon coated). The AFM tips can have different functionalized/modified groups (e.g., -CO_2H, -NH_2, -OH, -CF_3, -CH_3, -NHS, etc.), and attached colloidal spheres. A photodiode detects the deflections of the cantilever during scanning over the sample surface via a laser beam focused on and reflected from the rear of the cantilever. A computer, which controls the AFM system, acquires the electrical signals from the photodiode to generate feedback signals to a piezoelectric scanner and the cantilever to maintain the tip at either a constant force or constant height above the sample and to display surface topographic images as well as force–distance curve measuring the interaction forces between the atoms on the tip and sample surface. The feedback loop controls the force between the sample and tip, and z-sample position. In AFM, the sample surface topography is constructed from the deflection of the cantilever, which is determined by the sample surface features, as the cantilever tip scans over the region of interest on the sample surface.

Depending on applications, AFM can be operated under various environmental conditions, for example, in air—« dry mode », in aqueous environment or under water—« wet mode » and under vacuum. The operation of AFM can be classified into three different modes, namely (1) the contact mode, (2) the non-contact mode, and (3) the tapping mode (from TappingMode^{TM}, Bruker), depending on the manner in which the AFM tip interacts with the sample surface.

In addition to imaging of the sample surface topography, the determination of the force–distance curve, which can be employed to investigate many surface material properties including surface forces, is another important function of the AFM technique. To perform the force measurement in most commercial AFM equipments, the cantilever deflection signals from the photodiode are monitored as the piezoelectric scanner moves the sample surface up and down (z-direction), by which the cantilever tip approaches, touches on, and retracts from the sample surface to complete a circular movement. A plot of the tip–sample interaction forces *versus* the tip–sample distance, known as a force–distance curve, provides quantitative information about the interfacial forces acting between the tip and the sample surface. The deflection of the cantilever from its original position is dependent on the distance between the tip and the sample surface. From the force–distance curve, the force interactions between the tip and sample can be determined when the spring constants of the cantilever are known. By functionalization and modification of the tip with substrates having different origins and natures, and by attaching colloidal particles/ droplets to the tip, the applicability of the AFM for surface force measurement has been enormously expanded from two rigid solid surfaces to two deformable soft droplets, and even between two surfaces that are premodified with the functional groups desired. In addition to surface forces, the surface elasticity of the materials analyzed can be extracted from the force–distance curve.

Concerning dietary lipid exploration, AFM has been recently applied (1) to surface heterogeneities in the organization of polar lipids in the milk fat globule membrane, investigated on dried monolayers and hydrated bilayers [11, 12], (2) to the surface topography of unprocessed vs. processed milk fat globules, (3) to the mechanical properties of lipid vesicles [2, 3], (4) to the mechanical properties of plant oil bodies [14].

Figure 15 gives examples of information that can be provided by AFM: topography of dried monolayers of polar lipids, hydrated bilayers of polar lipids, topography of lipid vesicles.

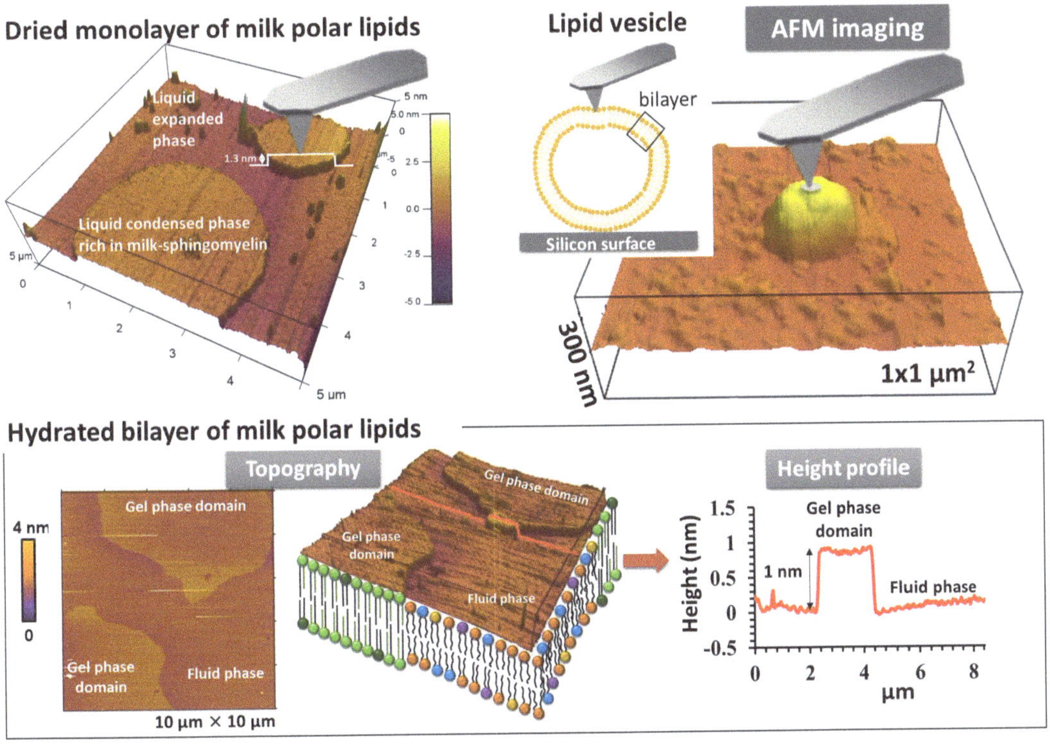

Fig. 15 Atomic force microscopy used for the study of air-dried monolayers of polar lipids, hydrated supported lipid bilayers composed of milk polar lipids, and lipid vesicles

References

1. Chevallier M, Riaublanc A, Lopez C, Hamon P, Rousseau F, Thevenot J, Croguennec T (2018) Increasing the heat stability of whey protein-rich emulsions by combining the functional role of WPM and caseins. Food Hydrocoll, 13th International Hydrocolloids Conference 76:164–172. https://doi.org/10.1016/j.foodhyd.2016.12.014
2. Et-Thakafy O, Delorme N, Gaillard C, Mériadec C, Artzner F, Lopez C, Guyomarc'h F (2017) Mechanical properties of membranes composed of gel-phase or fluid-phase phospholipids probed on liposomes by atomic force spectroscopy. Langmuir 33:5117–5126. https://doi.org/10.1021/acs.langmuir.7b00363
3. Et-Thakafy O, Delorme N, Guyomarc'h F, Lopez C (2018) Mechanical properties of milk sphingomyelin bilayer membranes in the gel phase: effects of naturally complex heterogeneity, saturation and acyl chain length investigated on liposomes using AFM. Chem Phys Lipids 210:47–59. https://doi.org/10.1016/j.chemphyslip.2017.11.014
4. Lopez C (2011) Milk fat globules enveloped by their biological membrane: unique colloidal assemblies with a specific composition and structure. Curr Opin Colloid Interface Sci 16:391–404. https://doi.org/10.1016/j.cocis.2011.05.007
5. Lopez C, Camier B, Gassi J-Y (2007) Development of the milk fat microstructure during the manufacture and ripening of Emmental cheese observed by confocal laser scanning microscopy. Int Dairy J 17:235–247. https://doi.org/10.1016/j.idairyj.2005.12.015
6. Lopez C, Cauty C, Guyomarc'h F (2015) Organization of lipids in milks, infant milk formulas and various dairy products: role of technological processes and potential impacts. Dairy Sci Technol 95:863–893. https://doi.org/10.1007/s13594-015-0263-0
7. Lopez C, Madec M-N, Jiménez-Flores R (2010) Lipid rafts in the bovine milk fat globule membrane revealed by the lateral segregation of phospholipids and heterogeneous distribution of glycoproteins. Food Chem

120:22–33. https://doi.org/10.1016/j.foodchem.2009.09.065

8. Lopez C, Maillard M-B, Briard-Bion V, Camier B, Hannon JA (2006) Lipolysis during ripening of emmental cheese considering organization of fat and preferential localization of bacteria. J Agric Food Chem 54:5855–5867. https://doi.org/10.1021/jf0602141
9. Lopez C, Novales B, Rabesona H, Weber M, Chardot T, Anton M (2021) Deciphering the properties of hemp seed oil bodies for food applications: lipid composition, microstructure, surface properties and physical stability. Food Res Int 150:110759. https://doi.org/10.1016/j.foodres.2021.110759
10. Lopez C, Schuck P (2021) Homogenization: a key mechanical process in interaction with product to modulate the organization of fat in spray-dried powders. In: Le Floch-Fouéré C, Schuck P, Tanguy G, Lanotte L, Jeantet R (eds) Drying in the dairy industry: from established technologies to advanced innovations. Routledge & CRC Press, pp 74–89
11. Murthy AVR, Guyomarc'h F, Lopez C (2016) The temperature-dependent physical state of polar lipids and their miscibility impact the topography and mechanical properties of bilayer models of the milk fat globule membrane. Biochim Biophys Acta Biomembr 1858:2181–2190. https://doi.org/10.1016/j.bbamem.2016.06.020
12. Murthy AVR, Guyomarc'h F, Paboeuf G, Vié V, Lopez C (2015) Cholesterol strongly affects the organization of lipid monolayers studied as models of the milk fat globule membrane: condensing effect and change in the lipid domain morphology. Biochim Biophys Acta (BBA) Biomembr 1848:2308–2316. https://doi.org/10.1016/j.bbamem.2015.06.014
13. Vignolles M-L, Jeantet R, Lopez C, Schuck P (2007) Free fat, surface fat and dairy powders: interactions between process and product. A review. Le Lait 87:187–236. https://doi.org/10.1051/lait:2007010
14. Yang N, Su C, Zhang Y, Jia J, Leheny RL, Nishinari K, Fang Y, Phillips GO (2020) In situ nanomechanical properties of natural oil bodies studied using atomic force microscopy. J Colloid Interface Sci 570:362–374. https://doi.org/10.1016/j.jcis.2020.03.011
15. Zouari A, Schuck P, Gaucheron F, Triki M, Delaplace G, Gauzelin-Gaiani C, Lopez C, Attia H, Ayadi MA (2020) Microstructure and chemical composition of camel and cow milk powders' surface. LWT 117:108693. https://doi.org/10.1016/j.lwt.2019.108693

Index

Christelle Lopez et al. (eds.), *Multidimensional Characterization of Dietary Lipids*, Methods and Protocols in Food Science, https://doi.org/10.1007/978-1-0716-3758-6, © INRAE 2024

Index

I

L

M

N

O

P

Q

R

S

SPRINGER NATURE

GPSR Compliance

The European Union's (EU) General Product Safety Regulation (GPSR) is a set of rules that requires consumer products to be safe and our obligations to ensure this.

If you have any concerns about our products, you can contact us on ProductSafety@springernature.com

In case Publisher is established outside the EU, the EU authorized representative is:

Springer Nature Customer Service Center GmbH
Europaplatz 3
69115 Heidelberg, Germany

FSC
www.fsc.org
MIX
Papier | Fördert
gute Waldnutzung
FSC® C083411

Kolibri 360 GmbH
Ferdinand-Jühlke-Straße 7
99095 Erfurt, Deutschland
produktsicherheit@kolibri360.de